A Handbook of Gene and Cell Therapy

Clévio Nóbrega • Liliana Mendonça
• Carlos A. Matos

A Handbook of Gene and Cell Therapy

Clévio Nóbrega
Department of Biomedical Sciences and Medicine & Centre for Biomedical Research (CBMR)
Universidade do Algarve
Faro, Portugal

Liliana Mendonça
Center for Neuroscience and Cell Biology
Universidade de Coimbra
Coimbra, Portugal

Carlos A. Matos
Centre for Biomedical Research (CBMR)
Universidade do Algarve
Faro, Algarve, Portugal

ISBN 978-3-030-41332-3 ISBN 978-3-030-41333-0 (eBook)
https://doi.org/10.1007/978-3-030-41333-0

Illustrations were made by Carlos A. Matos.

This Springer imprint is published by the registered company Springer Nature Switzerland AG
The registered company address is: Gewerbestrasse 11, 6330 Cham, Switzerland

Acknowledgments

Clévio Nóbrega is funded by the *Fundos Europeus Estruturais e de Investimento (FEEI), através do Portugal 2020 – Programa Operacional Regional do Algarve (CRESC 2020)* – and national funds through Fundação para a Ciência e Tecnologia (FCT), Portugal, Project ALG-01-0145-FEDER-29480 "SeGrPolyQ"; the French Muscular Dystrophy Association (AFM-Téléthon), Project #22424; and Ataxia UK. Liliana Mendonça is funded by the ERDF through the Operational Programme Competitiveness and Internationalization (COMPETE 2020), by national funds through FCT (Foundation for Science and Technology) under Project PTDC/BTM-ORG/30737/2017 (POCI-01-0145-FEDER-030737), by FCT national funds under Project CEECIND/04242/2017, and by the French Muscular Dystrophy Association (AFM-Téléthon, Trampoline Grant#20126). Carlos Matos is funded by the US National Ataxia Foundation.

Contents

1 Gene and Cell Therapy

The development of therapeutics to treat human disease has always been a major goal for biomedical research and innovation. Nevertheless, the complexity of the majority of human diseases poses an important and difficult obstacle to overcome. Moreover, the genetic contribution to these conditions complicates the targeting of endogenous and normal processes of cellular functioning, such as transcription and translation. In the past century, the understanding of DNA structure and the development of techniques to manipulate and recombine this molecule allowed the conception of new strategies that could use DNA as a therapeutic agent. In the 1970s, it was proposed for the first time that some human genetic conditions could be treated by the administration of exogenous DNA. The enormous technological advance in the field of biomedicine, with the human genome sequencing or the development of high-throughput techniques, for example, contributed to an effective application of gene therapy in the human context. Recently, the approval of several gene therapy medicines in Europe and the USA definitively established a new paradigm in human disease treatment and opened a new era for gene therapy.

1.1 The Concepts of Gene and Cell Therapy

As the name clearly implies, gene therapy refers to the use of genes as "drugs" to treat human diseases. In a simple way, **gene therapy** could be defined as a set of strategies modifying gene expression or correcting mutant/defective genes, which involves the administration nucleic acids - DNA or RNA - to cells. However, more elaborate and complete definitions for gene therapy can be found, especially the ones produced by regulatory agencies. According to the European Medicines Agency (EMA) and to the European Union (EU) directive 2001/83/EC, a gene therapy product consists on a biological medicinal product, which has the following characteristics [1]: (a) it contains an active substance, which includes or consists of a recombinant nucleic acid used in or administered to human beings with a view to regulating, repairing, replacing, adding, or deleting a genetic sequence; (b) its therapeutic, prophylactic, or diagnostic effect relates directly to the recombinant nucleic acid sequence it contains or to the product of the genetic expression of this sequence. Gene therapy medicinal products do not include vaccines against infectious diseases. For the US Food and Drug Administration (FDA), gene therapy is the administration of genetic material to modify or manipulate the expression of a gene product or to alter the biological properties of living cells for therapeutic use [2].

If the idea seems like science fiction, the truth is that in 1972 Friedmann and Roblin discussed this possibility in a*Science* paper entitled: "Gene therapy for Human Genetic Disease?" [3]. In this very interesting and advanced paper for their time, the authors postulated that gene therapy

C. Nóbrega et al., *A Handbook of Gene and Cell Therapy*,
https://doi.org/10.1007/978-3-030-41333-0_1

could be used in the future to ameliorate genetic diseases. In a very simple, but very accurate image, they describe how a mammalian cell could be modified by an exogenous source of DNA, following a similar path to that used by a virus (Fig. 1.1). Despite predicting important advantages with this therapy, authors clearly opposed any attempt to perform gene therapy in humans in a foreseeable future. The authors provided three important reasons for that: (i) the understanding on gene regulation and genetic recombination was still inadequate; (ii) the relation between gene and disease phenotype was not clear for many genetic disorders; and (iii) there was no information on gene therapy side effects. Moreover, authors also raised some important ethical concerns about gene therapy, including eugenics, which are still important questions today in gene therapy applications.

Less than 20 years later in 1990, the first therapeutic clinical trial using gene therapy started in the USA at the National Institutes of Health (NIH) in Bethesda, Maryland. Michael Blease and French Anderson led the clinical trial in two patients with adenosine deaminase (ADA) deficiency, a monogenic condition causing severe immunodeficiency. In this trial, the two patients, Ashanthi DeSilva and Cindy Kisik, had their autologous T-cells treated *ex vivo* with the correct ADA gene (using retroviral vectors), which were then reinfused [4]. The success of this trial remains controversial as (i) the response of the patients to the treatment was modest and (ii) patients simultaneously received enzyme replacement therapy [5]. Nevertheless, it became the first clinical trial of gene therapy in history and provided an important boost and enthusiasm for the field.

However, in 1999 a major drawback led to the partial suspension of gene therapy clinical trials and to the reevaluation of others in the USA. In that year, Jesse Gelsinger died after a severe adverse immune reaction to a adenoviral vector used in gene therapy treatment, only 4 days after the procedure [6]. Gelsinger had ornithine transcarbamylase (OTC) deficiency, which is fatal for most of the carriers, although, in his case, it was partially controlled through drugs and a low-protein diet. Later, it was identified that there were several non-authorized alterations in the clinical protocol and not enough information in the informed consent [7].

Despite this important drawback, several clinical trials using gene therapy continued. One of those trials conducted in Europe treated 10 boys with X-linked severe combined immunodeficiency (X-SCID) [8]. However, after the gene therapy treatment, four boys developed leukemia and one died 60 months after the intervention. Posterior studies showed that the therapeutic transgene was inserted near an oncogene, leading to severe complications starting around

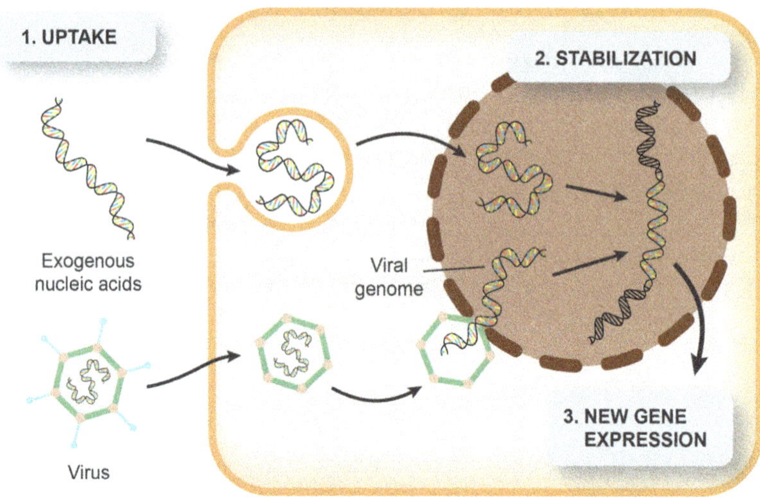

Fig. 1.1 Strategies for the genetic modification of a cell, as proposed by Friedmann and Roblin in 1972, in their *Science* paper entitled: "Gene therapy for Human Genetic Disease?". They posited that delivery of exogenous nucleic acids could be performed using different methods, by which exogenous genes could then be expressed in the modified cells.

30 months after the gene therapy. Nevertheless, in a 10-year follow-up of the intervention, the gene therapy was shown to have corrected the disease of the surviving participants [9]. In 2003, due to the adverse events in this study and several other concerns, the FDA suspended gene therapy clinical trials, arguing that not all the safety issues were addressed and more research was needed in the field [10]. Curiously, in the same year, China approved the first gene therapy product, Gendicine®, aiming to treat patients with tumors with p53 gene mutations [11]. However, the therapy was never approved in Europe, the USA, or Japan.

The first gene therapy medicine approved in these countries came in 2012, when Glybera® received marketing authorization in Europe for the treatment of lipoprotein lipase (LPL) deficiency, with the cost of 1.1 million euros, being, however, withdrawn from the market in the end of 2017 due to efficacy and low demand issues. Currently, several gene and cell therapy products are approved in Europe and the USA, including Strimvelis® to treat ADA-SCID, and more are in the pipeline for approval in the next years (see Sect. 1.11).

The concept and application of gene therapy is closely related to the idea of **cell therapy**, which can be roughly defined as an approach where cells are used as therapy or vehicle for therapy. Of course, cell therapy has been used for many years, considering, for example, blood transfusions and bone marrow transplants. Currently cell therapy per se has an enormous potential in regenerative medicine, even without genetic modifications of the cells. Nevertheless, the immunologic issues associated with cell transplants opened an opportunity for the combination of cell and gene therapy. In fact, several clinical trials have involved both gene and cell therapies, where defective (or not) cells are isolated from patients, treated with the therapeutic gene (using the appropriate vector), and then reinfused into the patient. Therefore, there is a clear overlap between both strategies Combined, they can be defined as a therapeutic intervention based on the administration of genetic material in order to modify or manipulate the expression of a gene product, altering the biological properties of living cells.

Despite all the drawbacks pointed before, research and the improvement in gene therapy knowledge and techniques continued. Important discoveries and advances like the RNA interference (RNAi) pathway, the Human Genome Project, or the production of induced pluripotent stem cells (iPSC) contributed to the continued interest in, and development of, gene therapy. More recently, the advance in gene editing techniques like TALENs or CRISPR provided a new boost in gene therapy, promising better, more accurate, and more effective forms to introduce or modifying genes.

For some, gene therapy was a promise that was never fulfilled, while others argue that the better is still to come. What offers no doubt is that gene therapy presents both advantages, like the possibility to effectively eradicate disease, and disadvantages, such as important ethical and safety issues that need to be addressed (Table 1.1). Currently, gene therapy is again in the spotlight of clinical and basic research, utilizing new techniques and taking advantage from the accumulated knowledge, the promising results of preclinical and clinical studies, and the interest of pharmaceutical companies due to the recent approval of several gene therapy products. Of

Table 1.1 Advantages and disadvantages of gene and cell therapy

Advantages	Disadvantages
Contributes to disease prevention	Modification of human abilities
Contributes to eradicate diseases	Changes the genetic pool
Helps to reduce the disease risk in future generations	Potential increase in diseases
Extends life expectancy	Safety problems
Avoids constant medication	High costs
Could replace defective cells	Ethical concerns
Unexplored potential	Short-time effect of some strategies
Allows a better understanding of how genes work	Might not be effective against complex diseases

course, this renovated interest makes gene therapy more prone to unethical procedures or poorly designed studies. Thus, important regulation procedures and careful attention to the studies are needed from the regulatory authorities, but also from scientists all over the world.

Designing a gene therapy study (in a preclinical or clinical context) is a complex process, where several variables must be considered ensuring the success and safety of the proposed therapy. Questions regarding the therapeutic target, the delivery system, or the immune response, for example, must be carefully studied and addressed before the application of the gene therapy. In the next sections, we will discuss some of the important issues that should be considered when designing a gene therapy study.

1.2 Types of Gene Therapy

The presence of genetic material in almost every cell in the human body makes these cells potential targets for gene therapy, including the germline cells. The major division between somatic and germline cells provides a categorization of gene therapy into two types, depending on the target cells. **Somatic gene therapy** refers to the interventions targeting the vast majority of human cells (somatic cells). On the other hand, we can speak of **germline gene therapy** if the targets of the intervention are the reproductive cells (Table 1.2). This very simple but clear classification advises that gene therapy directed to humans should be carried out exclusively in somatic cells. The germline gene therapy raises important ethical questions, being at least for now prohibited in Western countries [12]. Nevertheless, the advent of gene editing reopened the debate, and recently a panel of the US National Academy of Sciences considered the possibility of allowing embryo gene editing to prevent a disease, but only in rare circumstances and after further research [13].

Table 1.2 Main features of somatic gene therapy compared with germline gene therapy

Somatic gene therapy	Germline gene therapy
For the majority of the human cells	Changes will be transmitted to next generations
Alterations restricted to the patients	Unknown effect on future generations
Not passed on to future generations	Important bioethical issues
Less bioethical concerns	Technical difficulties in inserting genes in germ cells

Besides the important ethical and moral questions behind the gene editing of the germline, other technical questions also make it difficult: (i) currently the preimplantation diagnosis is able to identify and prevent several disease mutations, thus limiting the need for genome editing; (ii) the current procedures for zygote editing are not infallible, even in rodents; and (iii) offsite adverse and severe modifications can occur from the gene editing procedure. Despite the ethical and technical arguments, gene editing therapy was recently in the world spotlight, as in 2018 Chinese scientist He Jiankui claims that he performed gene editing in two human embryos, which were implanted and had already been born [14]. The scientific community and the world in general were astonished by this bold but highly questionable move, and the veracity of the claim is not completely assured.

For sure, the next few years will bring more debate and controversy on this matter, raising the need for rules and the maintenance of high ethical standards. Scientists and regulatory agencies are in the field, and the organization of world forums and summits on human gene editing will bring new regulation proposals. However, some controversies will certainly arise.

1.3 Gene Therapy Strategies

The application of gene therapy seems very straightforward if one thinks of genetic recessive disorders caused by a dysfunctional gene, where one normal copy of the gene could revert the disease phenotype and thus the only material to transfer is the correct gene (Fig. 1.2). This strategy, also called **gene augmentation therapy**, would be ideal to treat diseases caused by a gene mutation that leads to a malfunctioning or

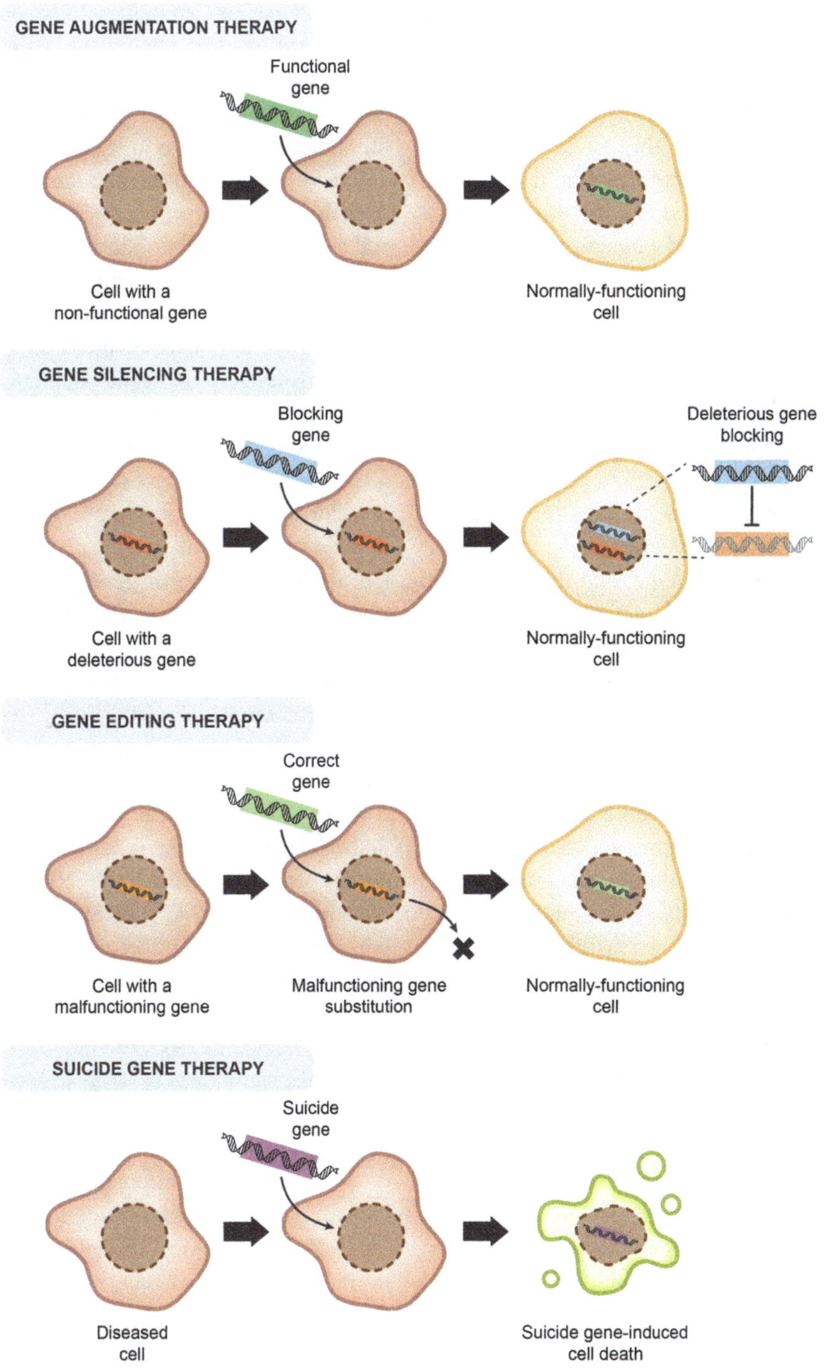

Fig. 1.2 Gene therapy strategies. In its simplest form, gene therapy could be performed by adding a new functional gene aiming to compensate a mutation or to improve cellular homeostasis – *gene augmentation therapy*. However, in some cases restoring a certain protein normal function may not be enough to mitigate the disease phenotype. For example, in genetic dominant diseases, mutant gene expression should instead be silenced, thus preventing the formation of the defective protein that causes the disease – *gene silencing therapy*. Recently, with the advent of gene editing tools, another strategy became available, in which a cell genome could be directly edited by removing a mutation or an entire gene and/or introducing a correct gene – *gene editing therapy*. All these three gene therapy strategies aim to revert cellular defects caused by malfunctioning genes. However, in certain diseases such as cancer, the aim is to cause cell death. In these cases, gene therapy could also be used, by introducing, for example, a toxic gene that will cause cellular apoptosis – *suicide gene therapy*.

deficiency of the resulting protein. In the treatment, adding a functional/normal version of the defective gene would theoretically guarantee the success of gene therapy. However, from a more practical point of view, this success is conditioned by at least two factors: (i) the levels of the normal protein produced by the inserted gene have to be sufficient and physiological, and (ii) the effects of the disease are still in a reversible state. This type of gene therapy was used in the first gene therapy clinical trial that was already mentioned for ADA-SCID, but it could also be effective for types of severe combined immunodeficiency or for cystic fibrosis (CF), among many others.

However, for many diseases, restoring the normal protein function would not be enough to revert the disease phenotype, and the expression of the mutant gene should actually be inhibited. This strategy, also named **gene silencing therapy** (or gene inhibition therapy) would be suitable, for example, for some genetic dominant diseases, some types of cancer or certain infectious diseases (Fig. 1.2). In the case of dominant diseases, the theoretical setup of this strategy would be to introduce a gene which could inhibit the expression of the mutant gene or that would interfere with the activity of the mutant protein. This approach became very feasible with the discovery of the RNAi pathway in 1998, by Andrew Fire and Craig Mello [15]. RNAi is an endogenous and conserved cellular pathway able to regulate gene expression through small RNA molecules that are complementary to mRNA (for more details, see Chap. 7). For gene therapy, the RNAi pathway offered an opportunity to use endogenous cellular machinery to control the expression of abnormal/defective genes. The gene silencing strategy has already been tested with success for several diseases in preclinical studies and currently is also being tested in clinical trials [16].

With the advent of **gene editing** techniques like TALENs or CRISPR, another strategy for gene therapy became available, aiming to edit the genome by removing a mutant gene and/or precisely correcting a gene (Fig. 1.2).

Of course, all strategies have problems and particularities that should be considered. For example, one of the main safety concerns with gene augmentation therapy is the possibility of the random insertion of the transgene, which could occur in problematic genome locations, such as the vicinity of oncogenes, tumor suppressor genes, or unstable genomic regions. On the other hand, the gene silencing or inhibition strategies, despite important successes, fail to completely shut down the expression of the target gene. Moreover, for gene silencing using the RNAi pathway, safety questions like off-target effects, long-term toxicity of the small RNA molecules or RNAi pathway saturation should also to be addressed and considered. Thus, the guided insertion of the transgene or the replacement of the abnormal/defective gene by a normal functioning gene appears as the ideal form of gene therapy, surpassing some of the problems presented by augmentation and silencing strategies of gene therapy. However, only recently have gene editing tools became easier to manipulate, allowing their use in the human gene therapy context. The enthusiasm in this field is so high that in 2016 a Chinese research group injected a person with cells edited by CRISPR-Cas9 [17]. Also in 2016, the first clinical trial using this system received a favorable opinion from an advisory committee at the US National Institutes of Health (NIH), aiming to be used in cancer therapy [18]. Nevertheless, potential off-target effects with these techniques or undesirable editing phenomena should also be considered and studied carefully.

The strategies mentioned above aim to restore the cellular homeostasis trying to revert the pathological abnormalities. However, in certain types of diseases such as cancer, the objective is to kill the defective cells. Gene therapy can also be used in this context, by using a transgene that (i) codifies for a highly toxic protein that kills the diseased cells or (ii) expresses a protein that marks the cell as a target for the immune system (Fig. 1.2). This type of gene therapy is sometimes called **suicide gene therapy** and is discussed in more detail in Chap. 9.

1.4 Choice of the Therapeutic Target

Another important issue in designing a gene therapy study is the choice of the target gene or cell, which entails a proper understanding of the genetic and molecular causes of a particular condition or disease. In the case of cell therapy, it is easy to conceive that the target will be the sick or defective cells. Nevertheless, it is crucial to consider important questions: (i) Do the cells used as therapeutics need to be treated with gene therapy? (ii) If stem cells are used, what would be the differentiation stage? (iii) What is the source of the cells?

In the case of gene therapy, the choice of the target is not so linear, as several options are available and their suitability depends on condition/disease pathogenesis. As mentioned before, it is easy to understand that, for a monogenic recessive disease, the gene therapy will consist in the addition of a "healthy" copy of the defective gene. However, in more complex pathologies, for example genetic dominant diseases, this strategy is not enough. In these diseases, one possible strategy for gene therapy would be the use of RNA and small RNA molecules to silence the expression of the abnormal causative gene. The different RNAi molecules, like siRNAs, shRNAs, and miRNAs, could be specifically designed to target the mRNA of the causative gene leading to

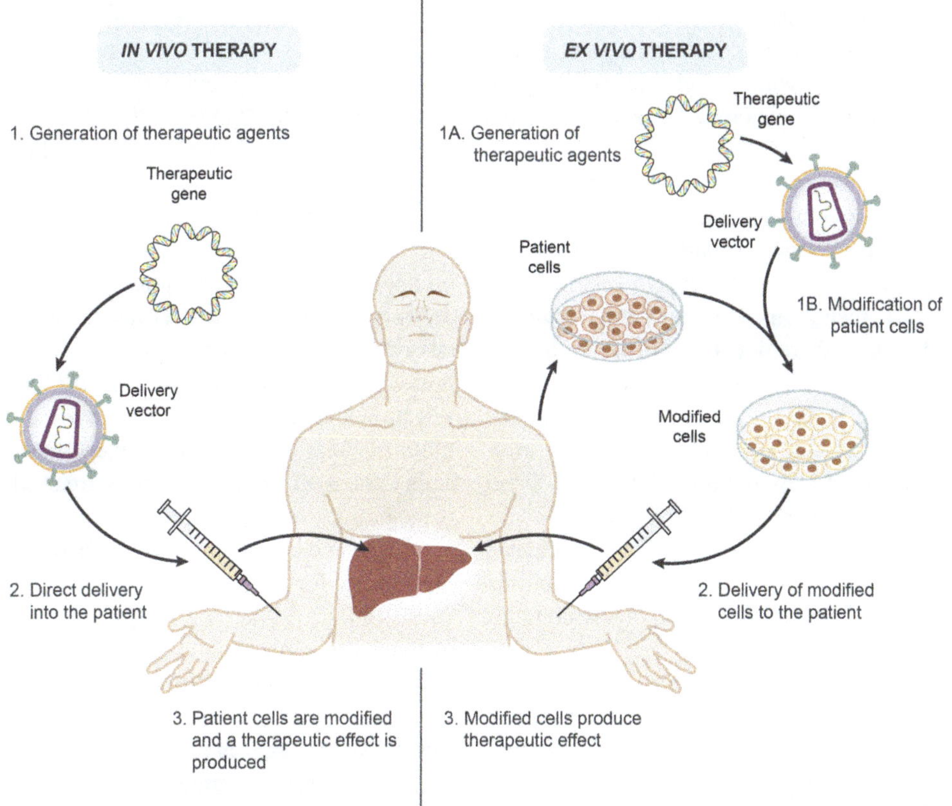

Fig. 1.3 Administration routes in gene therapy. An important consideration when designing a gene therapy application is to define the administration route to efficiently deliver the therapeutic gene to the target cells/organs. If the genes are directly delivered to the organism, the gene therapy is called *in vivo*. On the other hand, if the gene is delivered to cells outside the organism and then these manipulated cells are administrated to a subject, then the gene therapy is named *ex vivo*.

its cleavage or preventing its translation. Alternatively, a gene could also be used to treat dominant diseases aiming to improve the cellular function (like a gene to activate autophagy) or leading to cellular death (e.g., "suicide" gene therapy). Recently, the addition of a healthy copy of a gene also became an alternative for dominant diseases, if the abnormal copy is removed using gene editing tools.

Importantly, the disease pathophysiology should be carefully weighted when choosing a therapeutic target, as for many of the diseases affecting human health the use of cells would not be suitable.

Table 1.3 Comparison of *ex vivo* and *in vivo* administration routes used in gene therapy

In vivo (direct delivery)	*Ex vivo* (cell-based delivery)
Less invasive	More invasive
Technically more simple	Technically more complex
Vectors introduced directly	No vectors introduced directly
Safety check more difficult	Safety check easier
Reduced control of treated cells	More control of treated cells
Could be applied to a high number of diseases	Applied only to a small number of diseases
More definitive (depending on the delivery system)	Could be transient (cell lifetime)
Difficult to reach some cells/tissues	Possibility of accumulation of mutations
More off-target effects	Specificity for the treated cells

1.5 Administration Routes

The localization of the target cells/organs is probably the main factor in deciding the administration route, along with the choice of the gene delivery vehicle, which is normally named **vector**.

Broadly, we can consider two options of administration routes for gene therapy: the direct delivery of the genes to organisms, also named *in vivo* therapy, and the delivery of genes to cells, which are then transplanted to the organism, named *ex vivo* therapy (Fig. 1.3). In the *in vivo* administration, the therapeutic sequence is delivered directly to the target cells, organs, or the whole body, which could be a less invasive method but is more prone to have off-target effects. On the other hand, in *ex vivo* therapy cells are treated outside the body and then transplanted to the patients, allowing more control of the treated cells, but being technically more complex (Table 1.3).

Nevertheless, this rather simple categorization of the administration routes is in fact far more complex. For example, in the direct *in vivo* administration, important questions should be considered: (i) Are the target cells/organs accessible to a direct application? (ii) In a whole organism administration, which fraction of the therapy reaches the target cells/organs? (iii) Could the remaining fraction be toxic? These and other questions need to be considered when designing a gene therapy study and before its application. For example, when targeting the central nervous system, the direct delivery route should consider the blood-brain barrier (BBB) and its selectivity. One way to circumvent the BBB would be the intraparenchymal injection into the brain or the infusion into the cerebrospinal fluid (the gene delivery to the central nervous system is discussed in detail in Chap. 4). However, these routes are highly invasive and greatly limit their selection in human patients. The *ex vivo* administration is also complicated by the source of the cells to be used. If allogenic cells are used, there is the problem of immune compatibility, whereas autologous cells sometimes are defective and are not suitable for the therapy.

1.6 Delivery Systems

The delivery of exogenous genetic material into a cell or tissue is not a straightforward or easy process, as organisms developed several strategies and barriers to prevent it (see Chap. 4 for more details). Thus, one of the main issues to consider in a gene therapy strategy is the way to deliver the therapeutic sequence, that is, which delivery system is more suitable to ensure the success of the therapy. In a broad way, two main groups of delivery systems are currently considered: the viral and the non-viral systems (Fig. 1.4). The **viral** systems take advantage from the broad diversity of viruses and their innate ability to infect/transduce cells. The key advantage of these systems is their high efficiency, whereas the main drawback is the safety concerns on using modified viruses. On the other hand, **non-viral** systems include several chemical or physical methods, which have as their principal advantage their safety profile, whereas the main disadvantage is their relatively low efficiency (Table 1.4).

The choice of the correct/ideal delivery system for a given gene therapy is dependent on several variables, including the size of the gene, the expected effect, and the toxicity profile, among others. The different delivery systems used in gene and cell therapy are described in more detail in Chaps. 2 and 3.

1.7 Expression and Persistence of the Therapy

Another important concern on gene and cell therapy applications is the expression levels of the inserted transgene/sequence, as it is virtually impossible to introduce a single copy of the transgene into the target cells. Importantly, the number of copies introduced is often different among the target cells. Both factors lead to (i) expression differences between target cells and (ii) increased expression levels relative to basal conditions. Moreover, if the transgene is integrated (using, e.g., retroviral vectors), expression will be continuous, producing expression levels that might be different from physiological basal levels (probably much higher), which could lead to toxicity effects. Thus, the implementation of a gene therapy in a clinical setting must ensure very tight and consistent regulation of transgene expression, which could be achieved using regulatable promoters. A proper gene regulation system should display several features, including [19]: (i) a low basal expression of the transgene, (ii) the expression should be triggered by the administration of a molecule and be responsive to a wide range of doses, (iii) be specific to the target cells/organs, (iv) do not interfere with endogenous gene expression, and (v) allow a rapid and effective induction or repression of the transgene expression.

Gene regulation systems can be categorized into two main groups: (i) **exogenously**-regulated systems, which use exogenous compounds to regulate gene expression and which are the most widely used in gene therapy applications, and (ii) **endogenously-**controlled systems, which rely on internal stimuli to control the transgene expression. Within the first group, the tetracycline (Tet) regulation systems are the most exploited and used tool for controlling gene expression, although others have been developed, like the rapamycin-regulated or the RU486-regulated systems. In the second group of systems, the promoter is sensitive to physiological parameters and conditions, such as glucose levels or hypoxia. However, this endogenous regulation is difficult, and thus most of the systems used are based on the administration of exogenous molecules.

Tetracyclines and their derivates like doxycycline (dox) have been widely used in the clinical setting as antibiotics, binding to the bacterial 30S ribosomal subunit and thus inhibiting protein translation. The **Tet systems** have two variants, the Tet-off system, which was the first one developed and that is based on the negative control by tetracycline [20], and the Tet-on system, that is currently more used and which is based on the positive control of expression by tetracycline [21] (Fig. 1.5). Both systems are based on the bacterial Tet operon, namely, in the Tet repressor protein (TetR) and the tet operator (tetO) DNA elements. In the eukaryotic **Tet-off system**, the TetR was modified with a transcription activation domain (AD) from the VP16 protein of the herpes simplex virus, creating a tetracycline-controlled transcriptional activator (tTA). Moreover, the tetO sequences were fused with a TATA box-containing eukaryotic promoter to construct the tetracycline-responsive promoter (P_{tet}). In the absence of tetracycline (or its derivates), the tTA will bind to the tetO sites in the P_{tet}, thus activating the expression of the downstream transgene. On the other hand, the pres-

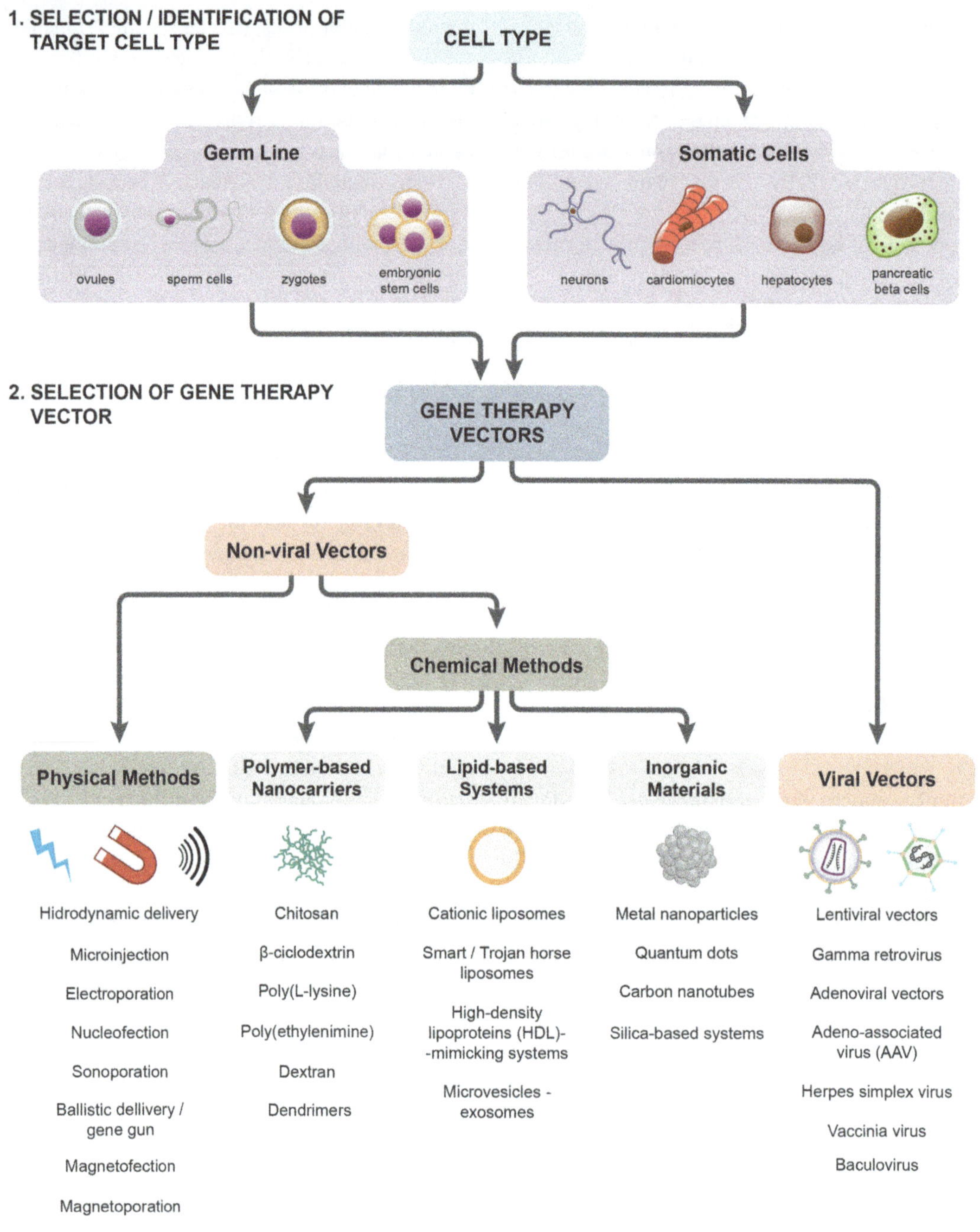

Fig. 1.4 Overview of the delivery systems used in gene therapy. Organisms and cells have developed several barriers to prevent the entry of exogenous genetic material. Therefore, overcoming these barriers to deliver the therapeutic gene is crucial to the success of gene therapy. In a broad manner, delivery systems for gene therapy can be classified into two groups: *non-viral vectors* and *viral vectors*. The first group refers to *physical and chemical methods*, such as microinjection or cationic liposomes. On the other hand, the second group is based on *engineered recombinant viruses* that are used to deliver the therapeutic transgene.

Table 1.4 Main advantages and disadvantages of non-viral and viral systems to used deliver therapeutic genes

Advantages	Disadvantages
Non-viral systems	
Easy production	Low efficiency/low expression
Low toxicity	No transgene integration
Unlimited cloning capacity	Low tropism
Viral systems	
High efficiency in gene transfer both *in vivo* and *ex vivo*	Possible immune/inflammatory response
Persistence of therapeutic strategy (in some cases)	Safety/toxicity concerns
High variety of cells able to be transduced	Limited cloning capacity
High variety of viruses to be engineered	Complex production
Natural tropism to infect/transduce cells	Limited tropism (in some cases)
Natural DNA transport mechanism into nucleus (in some cases)	Possibility of mutagenesis
	Limited knowledge on molecular mechanisms of infection

ence of tetracycline alters the conformation of the TetR domain of tTA, which prevents the binding to tetO sites and blocks the transgene expression (Fig. 1.5, upper panel). From a gene therapy clinical view, the Tet-off system is not very suitable, as long-term administration of tetracycline might be needed in order to repress transgene expression. To overcome this problem, the **Tet-on system** was developed, allowing the activation of transgene expression in response to the presence of tetracycline. This was achieved by a mutation in the TetR, which allows it to function in a reverse way (rTetR), binding to tetO in the presence of the effector. The system is completed by the fusion of AD to reverse-tTA (rtTA) that binds Ptet and activates the transcription of the downstream transgene in the presence of tetracycline (Fig. 1.5, lower panel). Several improvements on these and other systems were made and are already used in gene therapy clinical trials, namely in cancer [22].

1.8 Cell Targeting

In most human diseases, different cells and organs are not affected, and thus a gene or cell therapy must ensure preferential treatment of the affected cells and organs. Targeting specificity will increase the therapy efficacy, raising the concentration of the therapeutic molecule in the most affected cells/organs and avoiding its sequestration, dilution, or inactivation in non-target cells, at the same contributing for the increase of the safety profile of the therapy.

Taking into account the existing experience with conventional drugs, as well as with cellular transplants, several strategies could be employed in gene therapy aiming at specific targeting [23]: (i) **physical** strategies, where the molecule/cell is delivered locally into the target area; (ii) **physiological** strategies, based on natural physiological mechanisms of distribution; and (iii) **biological** strategies, based on biological alterations of the vehicles to achieve a specific localization (Fig. 1.6). Local delivery is probably the most straightforward way of administering a gene therapy; nevertheless, the procedure is extremely invasive, and particular cells/organs may be difficult to access. Taking advantage of physiological mechanisms such as blood circulation is another strategy for gene delivery. Despite the fact that it could be used in some contexts, systemic delivery has to deal with several physiological barriers, for example the BBB, when accessing the central nervous system.

The biological strategy implies the modification of the vector/cell, altering its entry proprieties or modulating post-entry features, for example by using a promoter specific to the target cells (Table 1.5) [24]. Depending on the type of vector (viral or non-viral), different modifications could be made in the surface of the vector (mode details on these modifications are detailed in Chaps. 2 and 3). As an example, the lentiviral vector envelope can be modified using glycoproteins from other viruses, thus altering its tropism. In the case of adeno-associated virus (AAV), different serotypes have tropism to different cells, providing a wide

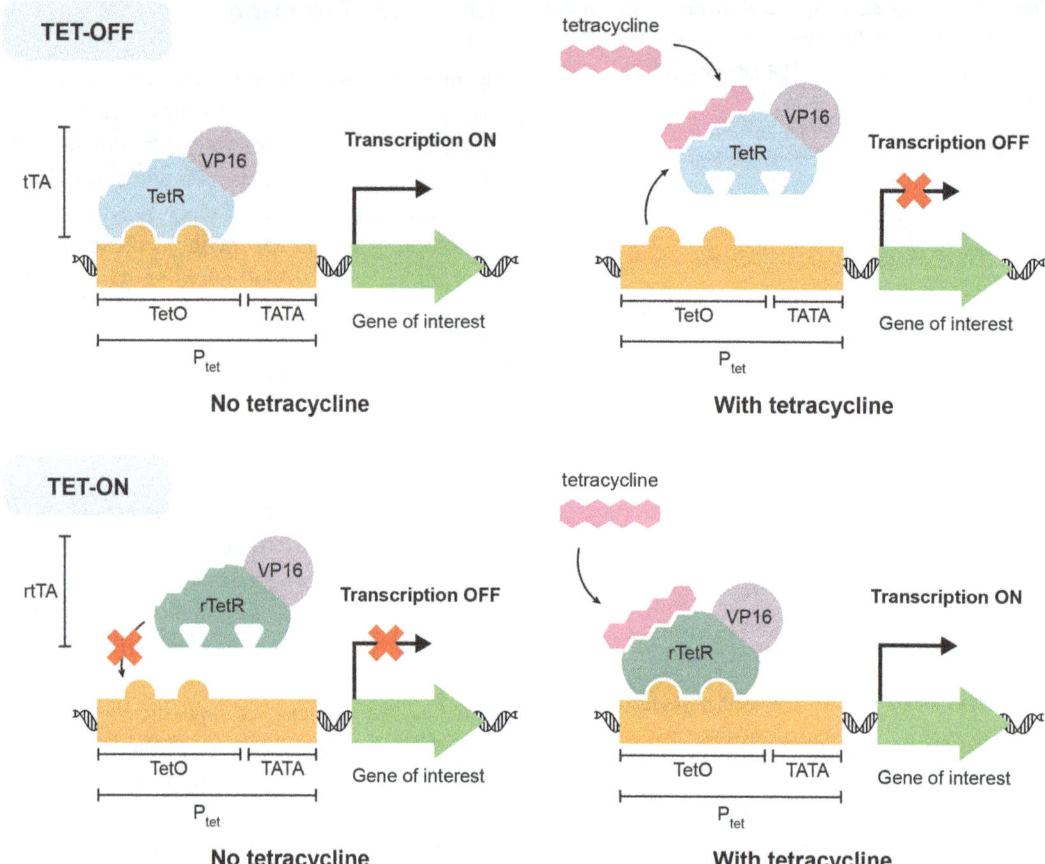

Fig. 1.5 Tetracycline (Tet) regulation system for the control of gene expression. This system is based on the *bacterial Tet operon* and the administration of *tetracyclines* or their derivates to control gene expression. The Tet system has two main variants: the *Tet-off* system (upper panel) and the *Tet-on* system (lower panel). In the Tet-off system, the TetR is modified with a transcription activation domain from the VP16 protein of the herpes simplex virus, creating a tetracycline-controlled transcriptional activator (tTA). Additionally, the tetO sequences are fused with a TATA box-containing eukaryotic promoter to construct the tetracycline-responsive promoter (P_{tet}). In the absence of tetracycline (or its derivates), the tTA will bind to the tetO sites in the P_{tet}, thus activating the expression of the downstream transgene. On the other hand, the presence of tetracycline alters the conformation of the TetR domain of tTA, which prevents the binding to tetO sites and blocks the transgene expression. On the contrary, the Tet-on version allows the activation of transgene expression in the presence of tetracycline. This is achieved by a mutation in the TetR, which allows it to function in a reverse way (rTetR), binding to tetO in the presence of the effector. The system is completed by the fusion of AD to reverse-tTA (rtTA) that binds Ptet and activates the transcription of the downstream transgene in the presence of tetracycline.

range of selectable choices. Other viruses like herpes simplex virus have a natural tropism to neurons, which makes them particularly suitable as delivery systems to the nervous system. In the case of non-viral vectors, especially the chemical-based vectors, the use of additional molecules helps to ensure a more directed targeting of particular cells. For example, the use of transferrin allows liposomes to enter the brain, surpassing the BBB.

1.9 Immune Response to the Therapy

Circumventing the immune response is a major issue in gene therapy (except if the goal is vaccination or tumor lysis), especially when using viral vectors (see Chap. 4 for more details). The immune system comprehends a complex array of mechanisms protecting the body against

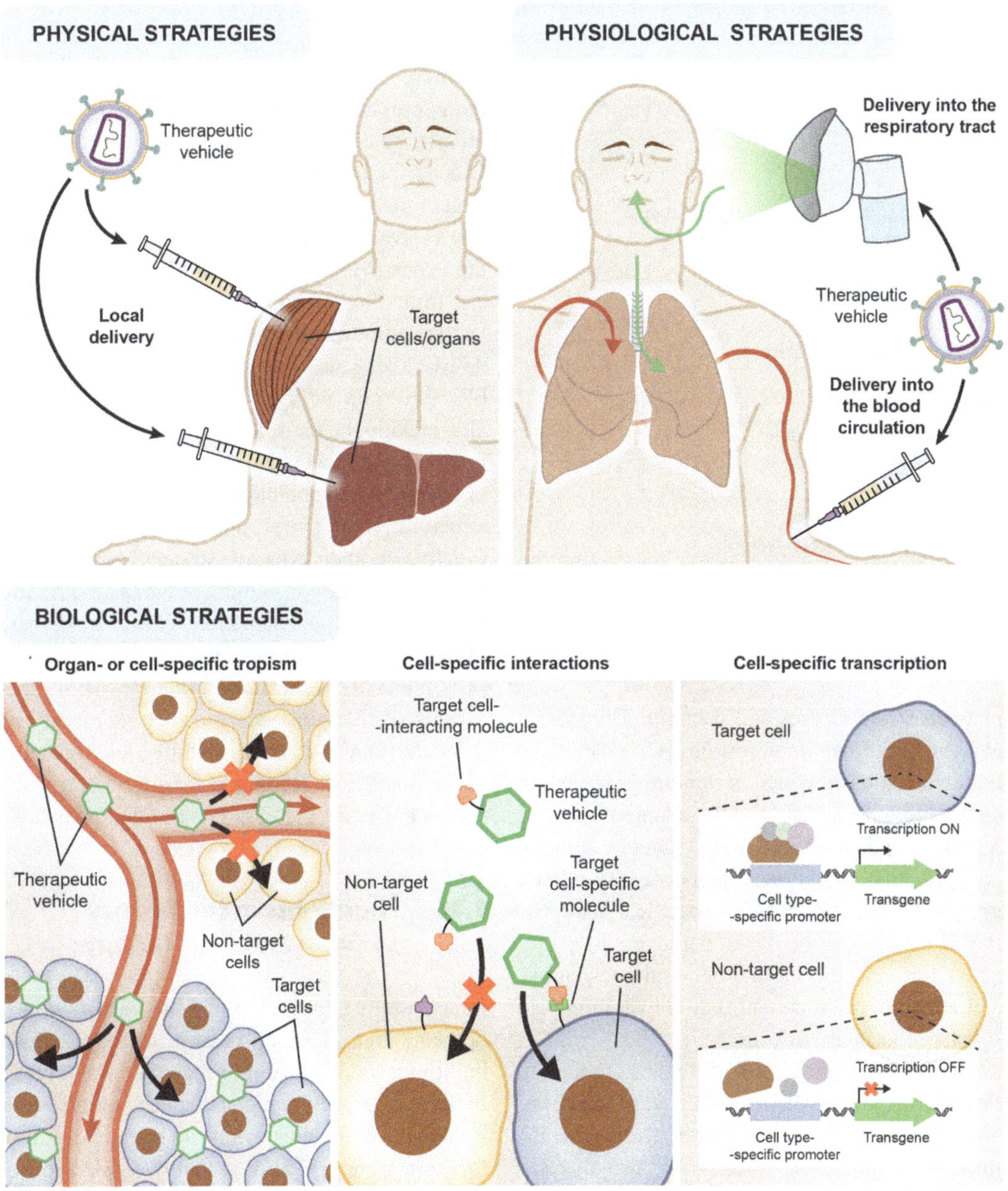

Fig. 1.6 Cell targeting strategies in gene therapy. The delivery of the therapeutic gene to the target cells/organs can be performed using *physical strategies*, in which there is a local and direct delivery of the gene using devices, such as catheters, or by using *physiological strategies*, in which endogenous physiological mechanisms, e.g., the blood circulation, are used to deliver the therapeutic gene. Finally, different *biological strategies* that take advantage of different biological mechanisms can also be used to target cells. For example, the therapeutic gene and vector can have a natural *tropism* to the target cell or organ, while not targeting other cells. Another strategy involves the use of *specific molecules* to be recognized by specific target cell receptors, thus specifically delivering the therapeutic gene to these cells. Finally, the use of *cell-specific promoters* could also limit the expression of the therapeutic gene to the target cells.

Table 1.5 Different gene promoters and their cell specificity

Promoter		Specificity
CMV	Cytomegalovirus	Ubiquitous
PGK	Phosphoglycerate kinase	
UbC	Ubiquitin	
hAAT	Human α-1-antitrypsin	Liver
TBG	Thyroxine-binding globulin	
Desmin	Desmin	Skeletal muscle
MCK	Muscle creatine kinase	
Synapsin	Synapsin	Neurons
CaMKII	Ca^{2+}/calmodulin-dependent protein kinase II	
mGluR2	Metabotropic glutamate receptor 2	
GFAP	Glial fibrillary acidic protein	Astrocytes
MBP	Myelin basic protein	Oligodendrocytes

pathogens like viruses and bacteria. The system is divided into two main responses: (i) the innate immune response, which is the initial, rapid and unspecific defense, and (ii) the adaptive immune response, which is stimulated later and more complex than the innate response. The latter response involves the specific, antigen-mediated, recognition of the pathogen, its elimination through humoral and/or cellular responses, and a memory component that allows improved resistance to future infections.

Both viral and non-viral vectors can induce an immune response, which could lead to the elimination of the vector and of the transduced cells, thus decreasing the efficacy of the therapy. Moreover, the production of proinflammatory cytokines and chemokines also has a harmful effect on the organism. Diverse factors influence the immune response to a vector, including [25] (i) the route of administration, (ii) the dose of the vector, (iii) patient-related factors (e.g., age, gender, immune status, drug intake), (iv) the type of promoters and/or enhancers used, and (v) the alterations made to the vector genome sequence and/or structure.

Several strategies were developed to ensure the success of gene therapy by overcoming the immune response [26]. One strategy is to avoid the expression of the delivered gene in antigen-presenting cells (APCs), such as dendritic cells, B-cells, or macrophages. The regulated expression of the transgene could also be used as a strategy to avoid the immune response, by delaying the expression of the gene until the tissue recovers from the inflammation associated with the vector administration. Another strategy is to deliver the genes into immune-privileged sites, like the brain or the eye. Additional strategies also include the modifications of the vector used, for example, performing genetic (viral vectors) or chemical and nonchemical modifications to its structure (both viral and non-viral vectors). Finally, another type of strategy is based on immunosuppression, similar to the procedure following organ transplantation. However, the use of immunosuppression strategies should be carefully planned, as they could interfere with other aspects of the gene therapy, such as modifying the vector internalization, stability, and transduction efficiency or leading to long-term complications such as the increased risk of malignancies.

1.10 Highlights in the History of Gene and Cell Therapy

Many events contributed to gene therapy development; thus, it is not easy to select the main highlights, taking also into account that for many of these important marks to happen, a lot of studies and developments were also made. Several of the important milestones that we selected for these highlights were already referred, including the first gene therapy clinical trial or the first approved gene therapy product in Europe. It is also important to note that several discoveries and studies were not related directly to gene therapy, but contributed decisively for its development, taking as an example the discovery of the RNAi mechanism. Furthermore, several advances related to cell therapy were also important for the gene therapy history, for example, the development of iPSC.

It is virtually impossible to date the beginning of gene therapy. However, two pioneer papers are probably in the genesis of the idea and possible implications of human gene therapy. In 1967, Marshall Nirenberg formulated the question: Will society be prepared?, recognizing that genetic messages could be synthesized chemically and then be used to program cells [27]. The author recognized the enormous potential of this approach but also the obstacles and ethical problems behind it. Later, in 1972 Friedmann and Roblin asked: Gene therapy for human genetic disease? [3]. If these two papers probably gave the starting shoot on the discussion about gene therapy and its implications, two important scientific advances contributed decisively for the concretization of gene therapy: the first virus-mediated gene transfer in 1968 [28] and the creation of the first recombinant DNA molecule in 1972 [29].

As pinpointed throughout this book, the delivery system of the gene is essential to the success of gene therapy and, since the first attempts, a lot of effort was directed to that development. Therefore, another important advance for gene therapy came with the construction of a retrovirus vector. In 1984, Cepko, Roberts, and Mulligan [30] reported the development of a murine retrovirus vector, which allowed the efficient introduction of DNA into mammalian cells. The first approved protocol to introduce an exogenous gene into humans was approved in 1988 and the study published in 1990 [31]. The study was not therapeutic, but rather described the introduction of a bacterial gene into tumor-infiltrating lymphocytes and the tracking of the persistence and localization of the cells after reinfusion into patients with advanced melanomas.

Also in 1990, the first gene therapy clinical trial for ADA-SCID took place [4] and launched gene therapy interventions in humans. The initial idea was to perform ex vivo gene therapy using autologous hematopoietic stem cells (HSCs) and retrovirus; however the preliminary studies in nonhuman primates were disappointing, with low levels of viral transduction and engraftment. Instead, the researchers used autologous T-cells treated with the functional ADA gene delivered by a gamma retrovirus vector, which were stimulated to divide *in vitro* and then the cells were reinfused into the patients. More than the success of the intervention, this clinical trial was a mark for gene therapy, as it proved that it could be applied in humans in a feasible and safe manner. Until now, dozens of ADA-SCID patients have been treated using gene therapy with enormous success, opening the way for the approval of Strimvelis® to treat ADA-SCID, which was the second gene therapy product approved in Europe.

In 1998, an important discovery that contributed decisively to the success of gene therapy was published. Scientists Andrew Fire and Craig Mello discovered RNAi [15], a mechanism of gene expression regulation, based on small RNA molecules (siRNAs and microRNAs) complementary to messenger RNAs (mRNAs), which activates a degradation pathway for those mRNAs. But what is the importance of RNAi to gene therapy? In the classical view of gene therapy, a functional or normal copy of a malfunctioning gene is introduced in a patient to treat a disease. Theoretically, in some recessive conditions, this intervention will lead to a complete cure, as one single functional copy may be enough to prevent the disease phenotype. However, for dominant conditions, the introduction of a normal gene is not enough to revert the disease. In this sense, RNAi offered the possibility to treat dominant diseases with gene therapy, by using these small RNA molecules. These molecules can be designed to be complementary to the target mRNA, leading to a reduction of target protein levels and in theory to the mitigation or even cure of the disease phenotype. The use of RNAi or antisense oligonucleotides greatly contributed to the development of gene therapy silencing strategies, aiming to abrogate or reduce the expression of a defective protein causing a disease.

In 1999, an important complication that negatively impacted the development of the gene therapy field occurred, with the death of Jesse Gelsinger in a clinical trial for ornithine transcarbamylase (OTC) deficiency. This is a metabolic disease that affects ammonia elimination, being fatal in the first days after birth. However, some patients have a partial OTC deficiency that may be controlled by a strict diet and pharmacological drugs. Jesse Gelsinger had this partial deficiency and was considered an ideal candidate for the gene therapy intervention. He received a dose of 3.8×10^{13} recombinant adenoviral vectors containing the normal OTC gene directly *in vivo* in the hepatic artery [7]. Jesse died 4 days after the intervention, due to a severe immune reaction to the vector, which induced shock syndrome, cytokine release, acute respiratory distress, and multiorgan failure. It is, however, important to refer that the other 17 subjects that participated in this study, including asymptomatic ones, presented transient mild adverse effects such as muscle aches and fevers. It is also very important to point out that by February 2000 (some months after Jesse death), more than 4000 subjects were already subjected to gene therapy in approximately 400 clinical trials, and Jesse was the only reported death [32]. Nevertheless, at that time, several clinical trials of gene therapy were halted, reviewed, or suspended, and, in the USA, FDA and the National Institutes of Health (NIH) promoted the development of new two programs, trying to enhance the monitoring of gene therapy clinical studies.

In 2000, a gene therapy clinical trial conducted in Europe was published in *Science*, reporting the treatment of 10 boys with a type of immunodeficiency (X-SCID) [8]. The treatment was successful, and 10 years after the trial, the disease was corrected [9]. However, the insertion site of the therapeutic gene led to the development of leukemia in four of the treated boys, and one died because of that. Following this event, FDA suspended gene therapy clinical trials in 2003.

Contrastingly, in 2003 China approved the first gene therapy product, Gendicine®, which started to be commercialized in 2004. The product consists of the p53 gene delivered in an adenoviral vector aiming to treat patients with head and neck squamous cell carcinoma [11]. The reported data showed very good therapeutic results and no major adverse side effects, and until 2013 more than 10,000 patients were treated with Gendicine® for different types of cancer [33]. Despite all of this, the product was never approved in Europe, the USA, or Japan.

The halt in the clinical trials did not stop the development and improvement of gene therapies and strategies. In 2004, EMA granted the first commercial Good Manufacturing Practice (GMP) certification in the EU for the production of commercial supplies of gene-based medicines. The product licensed was Cerepro®, which is an adenoviral vector with the gene thymidine kinase from herpes simplex virus aiming to treat malignant brain tumors. Despite several clinical trials, including a phase III trial, the product did not receive marketing authorization from EMA.

In 2004, an important advance for science in general and for gene therapy, in particular, was achieved. In a major international effort, the human genome was fully sequenced, thus completing the Human Genome Project [34] and providing the possibility of precisely locating all human genes. Among other important applications, human genome mapping provided a framework for the development of gene editing techniques for gene therapy.

In 2006, another major breakthrough was published by Shinya Yamanaka, which reported the development of induced pluripotent stem cells (iPSC) [35]. These pluripotent stem cells were reprogrammed from adult fibroblasts using four genes, now known as the Yamanaka reprogramming factors: *Oct3/4*, *Sox2*, *Klf4*, and *c-Myc*. The potential of this advance was soon perceived, namely for regenerative medicine. In the field of gene therapy, it expanded the possibilities of using autologous cells in *ex vivo* gene therapy applications. In 2014, the first application of iPSC-derived cells in humans started in Japan, for macular degeneration, which is the most prevalent retinal disease in aged people [36]. Despite the reported positive results, the clinical trial was halted 1 year later, due to safety concerns [37].

In the following years, constant research aiming at improving the safety of delivery vectors and at developing better assays for risk assessment led to the return of gene therapy clinical trials for different conditions, such as Leber's congentinal amaurosis, β-thalassemia, or hemophilia B [38]. The continuous push and effort by many researchers both in preclinical and clinical gene therapy studies led to the approval [39] in 2012 of the first gene therapy product in Europe, Glybera®. The product was based on the delivery of a functional copy of the lipoprotein lipase gene by an AAV vector. However, at the end of 2017, the company producing the product did not renew the marketing authorization, and it was removed from commercialization.

More recently, the development of techniques based on nucleases to precisely edit the genome brought a new advance to the gene therapy field, with the possibility of treating the genetic cause of a disease by directly editing or replacing the causative gene(s). These technologies utilize meganucleases, zinc finger nucleases (ZFN), transcription activator-like effector nucleases (TALENs), and the clustered regularly interspaced palindromic repeats (CRISPR) systems. ZFN was the first gene editing system reaching a phase I clinical trial [40]; however, currently most of the studies and research is directed to the CRISPR-Cas system. The very high efficiency and an easy and rapid construction provided an enormous boost in its use compared to the other gene editing plastforms. In 2016, the first clinical trial using the system was approved and launched in China [17].

Gene therapy has come a long way, following a path of successes and drawbacks since the first clinical trial back in 1990. A huge effort by scientists, clinicians, and biotech companies led to the development and continuous improvement of the techniques, systems, and safety of gene therapy applications. The future is of course unknown; nevertheless the recent approval of several gene and cell therapy products in Europe and the USA might indicate that gene therapy is here to stay.

1.11 Current Status of Gene Therapy

Until November 2017, more than 2600 approved gene therapy clinical trials took place around the world (Fig. 1.7) [41]. Most of these clinical trials were directed to treat cancer, followed by monogenic, infectious, and cardiovascular diseases (Fig. 1.8). The recent boost in gene therapy and its related safety and technical developments led to the approval of 13 gene therapy products in Europe and/or the USA (Table 1.6). Several others are in the final stages of development and could receive marketing authorization in the near future.

1.12 Ethical Questions and Concerns About Gene and Cell Therapy

Considering several issues already mentioned in this book, it is clear that gene therapy is very prone to raise important ethical questions, controversies, and debates. It is evident that genetic manipulation comes with many disadvantages, but also with enormous potential and opportunities (Table 1.7). Thus, gene therapy, as any other medical intervention involving human subjects, has to important requirements and considerations, such as (1) guaranteeing informed consent, considering the ethical principle of respect for persons; (2) having a favorable risk-benefit balance, considering the ethical principle of beneficence/non-maleficence; and (3) having a fair and rigorous selection of the research subjects, considering the ethical principle of justice.

However, the complex issues surrounding gene therapy, such as the possibility of altering the personal genetic information, raise specific ethical quandaries: Should germline gene therapy be allowed? How to distinguish between gene enhancement and gene therapy? How to regulate the application of gene therapy? Do we understand the full implications of gene alterations? Is gene therapy only available to people with higher monetary incomes? And so on.

A fundamental ethical question in gene therapy concerns the somatic versus germline gene therapy debate. From the accepted point of view, all gene therapies in human subjects should target somatic cells; however, even in this case, enhancement and safety problems should be contemplated. The prohibition on gene therapy targeting the germline cells ensures that its

Fig. 1.7 Gene therapy clinical trials approved by the end of 2017. More than 2500 clinical trials using gene therapy were performed worldwide until 2017.

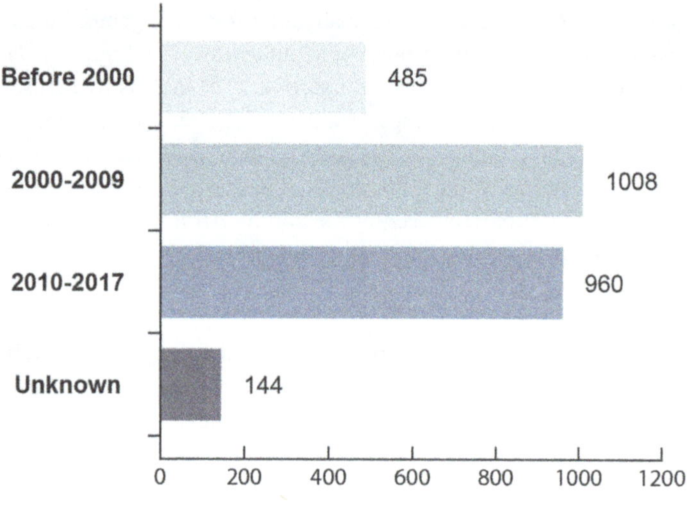

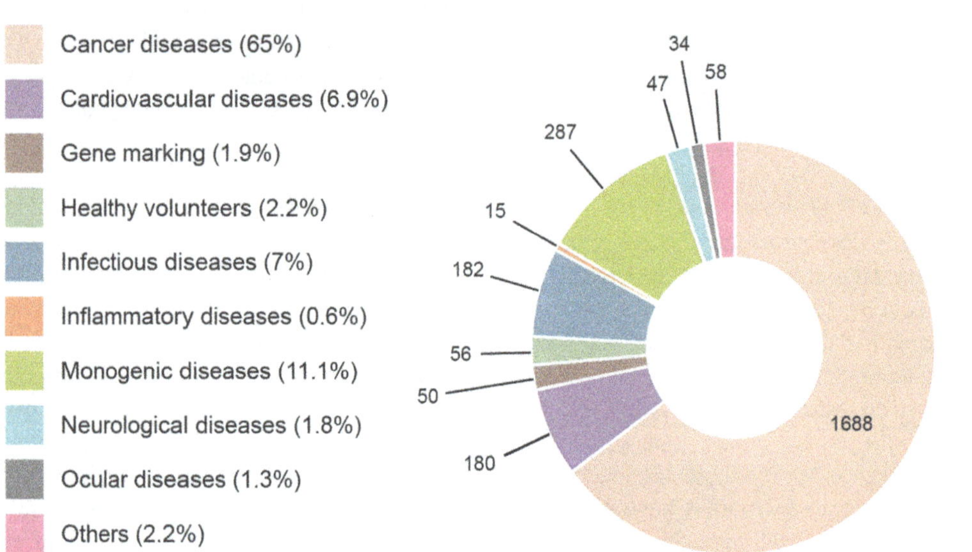

Fig. 1.8 Human conditions addressed in gene therapy clinical trials. Different gene therapy strategies have already been applied to different human conditions, especially to cancer (65% of all the gene therapy clinical studies performed), monogenic diseases (11.1%) and cardiovascular diseases (6.9%).

nonethical application is prevented altogether. This issue is now even more relevent, with the development of gene editing techniques allowing the direct and locally specific modification of genes in cells and patients, which fomented novel ethical questions and concerns around gene therapy. Even though the matter of not performing germline gene therapy in humans was a relatively consensual issue among the medical and scientific communities, the recent claim of human embryo editing sounded the alarm and reinforced the regulatory concerns. Although the gene editing procedure was not yet completely verified and certified, the very possibility that is true united the scientific and medical communities against that type of manipulation.

Something that undoubtedly emerges from these controversies and debates is the need for a clear regulation on gene therapy application in human subjects, which could be seconded by most of the developed and developing countries having the technology and the means to apply it.

Table 1.6 Current gene therapies approved for human use in Europe and the USA.

	Product	Administration route	Gene	Vector	Indication	Targeting	Approval in Europe	Approval in the USA	Price
1	Imlygic	*In vivo*	GM-CSF	HSV-1	Melanoma	Intralesional injection	Yes	Yes	$65,000
2	Kymriah	*Ex vivo*	CD19-specific CAR T	Lentiviral vector	Acute lymphoblastic leukemia (ALL)	Cell transplant	Yes	Yes	$475,000
3	Luxturna	*In vivo*	RPE65	Adeno-associated vector	Inherited retinal disease[a]	Subretinal injection	No	Yes	$425,000[b]
4	Yescarta	*Ex vivo*	CD19-specific CAR T	Y-Retroviral vector	Non-Hodgkin lymphoma (NHL)	Cell transplant	No	Yes	$373,000
5	Strimvelis	*Ex vivo*	ADA	Retroviral vector	ADA-SCID	Cell transplant	Yes	No	€594,000
6	Zalmoxis	*Ex vivo*	ΔLNGFR and HSV-TK Mut 2	Retroviral vector	Received a stem cell transplant	Cell transplant	Yes	No	€149,000[c]
7	Kynamro	*In vivo*	ASO	–	Homozygous familial hypercholesterolemia (HoFH)	Injection	No	Yes	$176,000
8	Spinraza	*In vivo*	ASO	–	Spinal muscular atrophy (SMA)	Intrathecal injection	Yes	Yes	$750,000[d]
9	Exondys 51	*In vivo*	ASO	–	Duchenne muscular dystrophy	Intravenous	No	Yes	$300,000
10	Onpattro (patisiran)	*In vivo*	RNAi	Lipid nanoparticles	Hereditary transthyretin-mediated amyloidosis	Intravenous infusion	Yes	Yes	$450,000[e]
11	Tegsedi (inotersen)	*In vivo*	ASO	–	Hereditary transthyretin-mediated amyloidosis	Subcutaneous	Yes	Yes	$450,000[e]
12	Zolgensma	*In vivo*	*SMN1*	AAV9	Spinal muscular atrophy (SMA)	Intravenous infusion	No[f]	Yes	$2.1 million
13	Zynteglo	*Ex vivo*	BA-T78-Q-globin	Lentiviral vector	β-thalassemia	Cell transplant	Yes	No	€1,600,000

CAR T chimeric antigen receptor T-cell
[a]Caused by mutations in both copies of the RPE65 gene
[b]Per eye
[c]Per infusion
[d]The first infusion and then 375,000 per year for life
[e]Per year
[f]Currently under review by EMA

Table 1.7 Main advantages and disadvantages of gene manipulation

Advantages	Disadvantages
Less risk of genetic diseases	Access could be limited to high income individuals
Possibility of preventing disease transmission to next generations	Reduction of genetic diversity
Increase in life expectancy	Unknown interaction between genes
Increase in the quality of life	Irreversiblte alteration of the human gene pool

However, the huge advance in technical aspects and their availability to almost everyone will certainly constitute an important and difficult challenge for authorities and for the scientific community.

This Chapter in a Nutshell

- Gene therapy encompasses a set of strategies for modifying gene expression or correcting mutant/defective genes, involving the administration of nucleic acids.
- Somatic gene therapy refers to the interventions targeting somatic cells.
- Germline gene therapy targets the reproductive cells, which could affect the progeny genetic information.
- There are several types of gene therapy: (a) gene augmentation, (b) gene silencing, (c) gene editing, and (d) suicide gene therapy.
- The direct administration of gene therapy is called *in vivo* therapy, whereas the treatment involving gene therapy in cells and then the transplant to the organism is called *ex vivo* therapy.
- Two main delivery systems exist, the viral and the non-viral vectors. The former systems have a high efficiency compared to the non-viral systems, although their safety profile is lower.
- There are several strategies to ensure the gene therapy expression, persistence, and targeting, for example using gene regulation systems.
- The immune response to gene therapy is also an important consideration in clinical trial design and implementation.
- The story of gene therapy is full of advances and setbacks; however, there are currently several gene therapy products approved in Europe and the USA.
- Several ethical problems arise with gene therapy, as it became evident in 2018, with the claim of the first human embryo gene editing procedure with the CRISPR-Cas system.

Review Questions

1. Which sentence better defines gene therapy?
   (a) A method to cure genetic disorders
   (b) A method to deliver a gene
   (c) A method to correct a defective gene
   (d) A method to silence the expression of a defective gene
   (e) All the previous options
2. Which of the following points is not essential in the development of a gene therapy clinical trial?
   (a) Delivery system
   (b) Target cell/organ
   (c) Immune response
   (d) Civil status
   (e) Therapy cost
3. To have a successful gene therapy, the healthy gene should be inserted in the target cell and:
   (a) Destroy the defective gene
   (b) Produce the correct amount of normal protein
   (c) Bind to mRNA molecules in the cell
   (d) Be inserted in the mitochondria
   (e) None of the previous
4. From the following, which is a disadvantage of gene therapy?
   (a) Eradicate diseases
   (b) Prevent diseases
   (c) High cost
   (d) Enormous potential
   (e) Correct genetic defects
5. The first clinical trial and the first death caused by gene therapy had as a disease target (choose the correct answers):
   (a) Adenosine deaminase deficiency

(b) Severe combined immunodeficiency linked to X chromosome
(c) Leber's congenital amaurosis
(d) Ornithine transcarbamylase deficiency
(e) Lipoprotein lipase deficiency

References and Further Reading

1. Directive 2001/83/EC of the European Parliament and of the Council of 6 November 2001 on the Community code relating to medicinal products for human use
2. What is gene therapy? U.S. Food & Drug Administration. Available at https://www.fda.gov/BiologicsBloodVaccines/CellularGeneTherapyProducts/ucm573960.htm
3. Friedmann T, Roblin R (1972) Gene therapy for human genetic disease? Science 175:949–955
4. Anderson WF, Blaese RM, Culver K (1990) The ADA human gene therapy clinical protocol: points to consider response with clinical protocol, July 6, 1990. Hum Gene Ther 1:331–362
5. Blaese RM, Culver KW, Miller AD, Carter CS, Fleisher T, Clerici M, Shearer G, Chang L, Chiang Y, Tolstoshev P et al (1995) T lymphocyte-directed gene therapy for ADA- SCID: initial trial results after 4 years. Science 270:475–480
6. Sibbald B (2001) Death but one unintended consequence of gene-therapy trial. CMAJ 164:1612
7. Steinbrook R (2008) The Gelsinger case. In: Emanuel EJ (ed) The Oxford textbook of clinical research ethics. Oxford University Press, Oxford
8. Cavazzana-Calvo M, Hacein-Bey S, de Saint Basile G, Gross F, Yvon E, Nusbaum P, Selz F, Hue C, Certain S, Casanova JL et al (2000) Gene therapy of human severe combined immunodeficiency (SCID)-X1 disease. Science 288:669–672
9. Hacein-Bey-Abina S, Hauer J, Lim A, Picard C, Wang GP, Berry CC, Martinache C, Rieux-Laucat F, Latour S, Belohradsky BH et al (2010) Efficacy of gene therapy for X-linked severe combined immunodeficiency. N Engl J Med 363:355–364
10. Marwick C (2003) FDA halts gene therapy trials after leukaemia case in France. BMJ 326:181
11. Zhang WW, Li L, Li D, Liu J, Li X, Li W, Xu X, Zhang MJ, Chandler LA, Lin H et al (2018) The first approved gene therapy product for cancer Ad-p53 (Gendicine): 12 years in the clinic. Hum Gene Ther 29:160–179
12. Hildt E (2016) Human germline interventions-think first. Front Genet 7:81
13. National Academic of Sciences and National Academy of Medicine (2017) Human genome editing, science, ethics, and governance. Report highlights
14. Cyranoski D (2018) CRISPR-baby scientist fails to satisfy critics. Nature 564:13–14
15. Fire A, Xu S, Montgomery MK, Kostas SA, Driver SE, Mello CC (1998) Potent and specific genetic interference by double-stranded RNA in Caenorhabditis elegans. Nature 391:806–811
16. Chakraborty C, Sharma AR, Sharma G, Doss CGP, Lee SS (2017) Therapeutic miRNA and siRNA: moving from bench to clinic as next generation medicine. Mol Ther Nucleic Acids 8:132–143
17. Cyranoski D (2016) CRISPR gene-editing tested in a person for the first time. Nature 539:479
18. Reardon S (2016) Welcome to the CRISPR zoo. Nature 531:160–163
19. Naidoo J, Young D (2012) Gene regulation systems for gene therapy applications in the central nervous system. Neurol Res Int 2012:595410
20. Gossen M, Bujard H (1992) Tight control of gene expression in mammalian cells by tetracycline-responsive promoters. Proc Natl Acad Sci U S A 89:5547–5551
21. Gossen M, Freundlieb S, Bender G, Muller G, Hillen W, Bujard H (1995) Transcriptional activation by tetracyclines in mammalian cells. Science 268:1766–1769
22. Goverdhana S, Puntel M, Xiong W, Zirger JM, Barcia C, Curtin JF, Soffer EB, Mondkar S, King GD, Hu J et al (2005) Regulatable gene expression systems for gene therapy applications: progress and future challenges. Mol Ther 12:189–211
23. Roth JC, Curiel DT, Pereboeva L (2008) Cell vehicle targeting strategies. Gene Ther 15:716–729
24. Powell SK, Rivera-Soto R, Gray SJ (2015) Viral expression cassette elements to enhance transgene target specificity and expression in gene therapy. Discov Med 19:49–57
25. Arruda VR, Favaro P, Finn JD (2009) Strategies to modulate immune responses: a new frontier for gene therapy. Mol Ther 17:1492–1503
26. Sack BK, Herzog RW (2009) Evading the immune response upon in vivo gene therapy with viral vectors. Curr Opin Mol Ther 11:493–503
27. Nirenberg MW (1967) Will society be prepared? Science 157:633
28. Rogers S, Pfuderer P (1968) Use of viruses as carriers of added genetic information. Nature 219:749–751
29. Jackson DA, Symons RH, Berg P (1972) Biochemical method for inserting new genetic information into DNA of Simian Virus 40: circular SV40 DNA molecules containing lambda phage genes and the galactose operon of Escherichia coli. Proc Natl Acad Sci U S A 69:2904–2909
30. Cepko CL, Roberts BE, Mulligan RC (1984) Construction and applications of a highly transmissible murine retrovirus shuttle vector. Cell 37:1053–1062
31. Rosenberg SA, Aebersold P, Cornetta K, Kasid A, Morgan RA, Moen R, Karson EM, Lotze MT, Yang JC, Topalian SL et al (1990) Gene transfer into humans--immunotherapy of patients with advanced melanoma, using tumor-infiltrating lymphocytes

modified by retroviral gene transduction. N Engl J Med 323:570–578
32. Verma IM (2000) A tumultuous year for gene therapy. Mol Ther 2:415–416
33. Kaufmann KB, Buning H, Galy A, Schambach A, Grez M (2013) Gene therapy on the move. EMBO Mol Med 5:1642–1661
34. International Human Genome Sequencing Consortium (2004) Finishing the euchromatic sequence of the human genome. Nature 431:931–945
35. Takahashi K, Yamanaka S (2006) Induction of pluripotent stem cells from mouse embryonic and adult fibroblast cultures by defined factors. Cell 126:663–676
36. Mandai M, Watanabe A, Kurimoto Y, Hirami Y, Morinaga C, Daimon T, Fujihara M, Akimaru H, Sakai N, Shibata Y et al (2017) Autologous induced stem-cell-derived retinal cells for macular degeneration. N Engl J Med 376:1038–1046
37. Scudellari M (2016) How iPS cells changed the world. Nature 534:310–312
38. Wirth T, Parker N, Yla-Herttuala S (2013) History of gene therapy. Gene 525:162–169
39. Yla-Herttuala S (2012) Endgame: glybera finally recommended for approval as the first gene therapy drug in the European union. Mol Ther 20:1831–1832
40. Tebas P, Stein D, Tang WW, Frank I, Wang SQ, Lee G, Spratt SK, Surosky RT, Giedlin MA, Nichol G et al (2014) Gene editing of CCR5 in autologous CD4 T cells of persons infected with HIV. N Engl J Med 370:901–910
41. The Journal of Gene Medicine, Gene Therapy Clinical Trials Worldwide. Available at http://www.abedia.com/wiley/

2 Non-viral Vectors for Gene Therapy

Presently, more than 3400 genes have been associated with diseases [1], some of these pathologies are debilitating, mortal, and without any effective therapeutic options, and this number is expected to increase in the next decade as genomic studies advance. Gene therapy is a therapeutic strategy that seeks to target the genes behind the disorders in order to cure patients. Thus, this approach holds the promise of establishing therapies that treat the causes, rather than the symptoms, by manipulating deficient genes, removing or silencing pathologic genes, or adding missing genes. However, these ambitious strategies have been hampered by several problems, such as the lack of safety and efficiency of some of the developed approaches, which can be explained by the very nature of most gene-manipulating tools. Naked DNA plasmids, for example, are rapidly degraded in biological fluids, they are unable to efficiently cross the cellular membranes and therefore reach the target cells and they activate the immune system, which is programmed to identify and eliminate vehicles containing foreign genetic information [2]. Thus, although a large number of preclinical and clinical studies have been performed using naked nucleic acids, the use of **delivery vectors** yields better results, namely because they protect the nucleic acids from nuclease degradation and increase intracellular delivery. Still, despite the big efforts made in the last decades with gene therapy strategies, very few gene therapy-based drugs or therapeutic products have reached the market, revealing that much still has to be done in the field of vectors for gene therapy.

In fact, one of the major hurdles to the efficacious clinical application of gene therapy agents is the lack of efficient and safe delivery vehicles, which have to meet four basic criteria: (i) to protect the nucleic acids against degradation by nucleases; (ii) to deliver the nucleic acids into the cytoplasm or nucleus of the target cells; (iii) to be safe (i.e., trigger low levels of genotoxicity and immunogenicity); and (iv) to be economically viable (i.e., acceptable costs on a per patient basis). Gene therapy vectors are classically organized into two major groups, the non-viral and the viral vectors. The present chapter is focused on the **non-viral vectors**, which are subdivided into two categories: the physical and chemical methods. **Physical methods** (Fig. 2.1, Table 2.1) comprehend hydrodynamic delivery, microinjection, electroporation, nucleofection, sonoporation, gene gun, magnetofection, magnetoporation, microneedles, etc. **Chemical vectors** (Fig. 2.2, Table 2.2) include the polymeric-based systems, the lipid-based systems, metal nanoparticles, quantum dots, and graphene-based systems such as carbon nanotubes, silica nanoparticles, etc.

C. Nóbrega et al., *A Handbook of Gene and Cell Therapy*,
https://doi.org/10.1007/978-3-030-41333-0_2

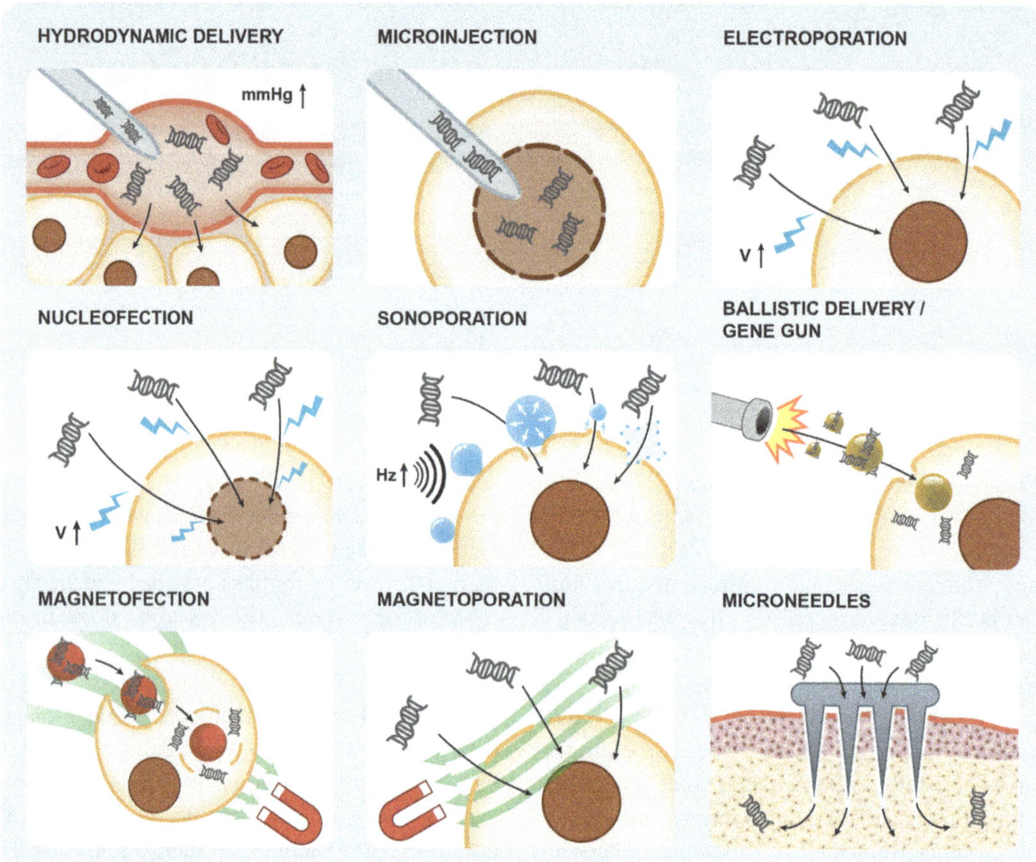

Fig. 2.1 Physical methods used for delivery of gene therapy tools. Different physical methods are available, namely, (i) *hydrodynamic delivery* that uses hydrostatic pressure enhancement to enlarge fenestrae and consequently tissue permeability; (ii) *microinjection*-mediated direct transfer of genetic material into the cell interior; (iii) *electroporation* and *nucleofection*, in which genetic material transfer is facilitated by electric pulses that cause transient pores in the cellular membrane and/or nuclear envelope; (iv) *sonoporation,* whereby delivery is triggered by transient cell membrane pores formed by ultrasounds; (v) *ballistic delivery* by accelerated nucleic acid-coated submicron metal particles; (vi) *magnetofection* and *magnetoporation*-mediated gene delivery, based on the application of an external magnetic field promoting the guiding of the magnetic agent-containing vector to the target cells or promoting transient cellular pore formation, respectively; and (vii) *microneedles* used to deliver therapeutic agents across the skin.

2.1 Physical Methods

2.1.1 Hydrodynamic Delivery

The hydrodynamic derlivery technique is based on the enhancement of the pressure implemented in a tissue in order to increase the tissue permeability. Generally, this is performed through the rapid delivery of the nucleic acids in a high solution volume into the bloodstream. This rapid enhancement of the blood volume leads to a rapid rise in venous or arterial pressure (depending on the injection site). This hydrostatic pressure increase will enlarge the fenestrae in the blood vessels (in the liver sinusoids, for example) and cause transient pores formation that result in the permeability enhancement of the tissue [3, 4].

This technique has been used with success in rodents; nevertheless, its application in humans is very limited. Regarding hydrodinamyc delivery to the liver in particular, the major concerns with are the transient elevation of hepatic

Table 2.1 Main advantages and disadvantages of physical methods of gene therapy delivery

	Advantages	Disadvantages
Hydrodynamic delivery	Simple and efficient transfection	Volumes required are too large for human application
Microinjection	Small amount of DNA required	Difficult to perform and low performance
Electroporation	Efficient transfection and fast method	Difficult to apply *in vivo*: difficulty to access target sites with electrodes and can cause tissue damage and inflammation
Nucleofection	Efficient in cells that are difficult to transfect and fast method	Can cause high cell death and has a limited *in vivo* application
Ultrasound	Noninvasive strategy; can reach deeper organs and can be directed to specific areas	Low efficiency of gene transfer and possible damaging of target cells
Ballistic gene delivery/gene gun	Gene delivery to cells that are difficult to transfect	Limited to a few-millimeter-deep penetration and limited to superficial tissues and organs (muscles and skin)
Magnetofection	Allows targeted cargo release and controlled release over time	Requires specialized equipment and agglomeration of magnetofection reagents may occur after the magnetic field is removed
Magnetoporation	Used in cells that are difficult to transfect and requires no direct contact of electrodes	Can cause cell death
Microneedles	Small amount of DNA required	Difficult to perform and with a low performance

enzymes levels in the bloodstream, temporary cardiac dysfunction, high rise of venous pressure, and liver congestion caused by the injection of the high solution volume [4].

2.1.2 Microinjection

Microinjection consists in the direct transfer of genetic material into the cell interior. Generally, a glass micropipette with a submicron-sized diameter tip (less than 0.5 μm) is prefilled with the genetic material solution and is introduced into the target cell, under observation in a microscope.

This technique may sound very straightforward; however it requires expensive equipment and extensive training of the operator, being technically very demanding. Additionally, (i) only a very small volume can be injected, otherwise additional stress can be triggered in the treated cells leading to intracellular environment and cell membrane disorganization; (ii) leakage of intracellular components at the injection site might occur, causing cell death; and (iii) the method is limited by cell size, being mostly applied to large cells, such as oocytes [5, 6].

2.1.3 Electroporation

Electroporation, also known as electropermeabilization and electrotransfer, is based on the application of a series of electrical pulses leading to a transient disruption of the cell membrane, allowing the passage of the genetic material into the cytosol of the cells. The exact mechanism underlying genetic material transfer into the cells is not known. However, it is known that, after employing an effective voltage (for *in vivo* application, the voltage is in the order of 200 V/cm), cell membrane becomes permeable, which is assumed to happen through the formation of hydrophilic membrane pores and consequent movement of genetic material through these pores as a local electrophoretic effect [3, 7]. The optimal duration and intensity of the electric pulses are different for different cell types and tissues to be transfected. For example, for skeletal muscle, better results come from protocols that first apply a pulse of high voltage, which will open membrane pores, followed by several low-voltage pulses, which will electrophorese the nucleic acids into the cells [3].

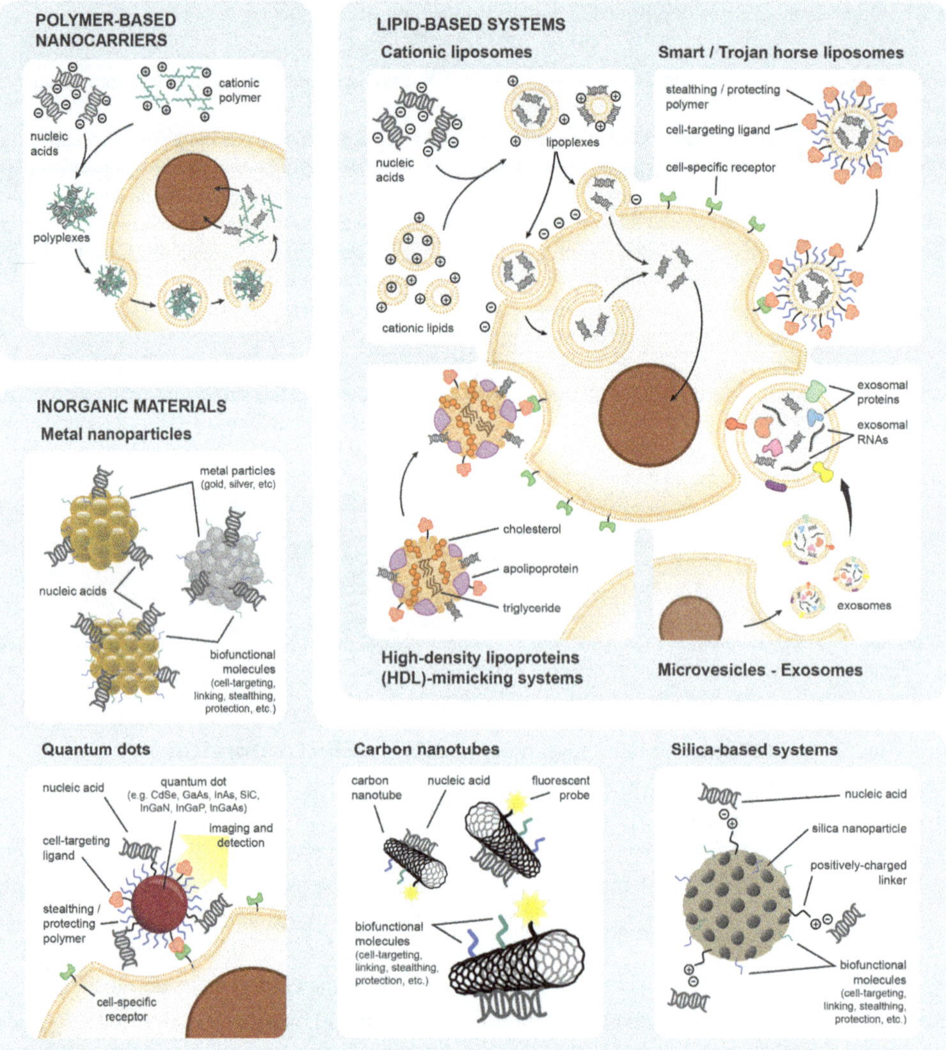

Fig. 2.2 Chemical methods used for delivery of gene therapy tools. Chemical vectors include the polymeric-based systems, the lipid-based systems, metal nanoparticles, quantum dots, carbon nanotubes and silica-based nanoparticles. *Polymer-based vectors* are composed by a large number of repeated units bonded together that interact with the negatively charged nucleic acids, originating compact polyplexes that protect nucleic acids. *Lipid-based systems* comprise different delivery agents such as *cationic lipoplexes* formed through electrostatic interaction between the cationic lipids and negatively charged nucleic acids, and *Smart/Trojan horse liposomes* that can evade the immune system through the inclusion of hydrophilic polymers and that contain cell-targeting ligands to target their cargo to a specific cell population. *High-density lipoprotein (HDL)-mimicking systems* are HDL that are artificially prepared in order to efficiently deliver nucleic acids. *Exosomes* are membrane-based vesicles secreted by cells that can also be engineered to deliver nucleic acids. *Inorganic materials* such as metal nanoparticles, quantum dots, carbon nanotubes, and silica-based systems also promote efficient delivery of nucleic acids. *Metal nanoparticles* made up of noble metals such as gold and silver can be biofunctionalized and applied under specific conditions such as events triggered by light. *Quantum dots* are semiconductor crystals used as fluorescence probes to which it is possible to conjugate nucleic acids and a targeting ligand allowing the combination of the fluorophore properties and delivery functions in one system. *Carbon nanotubes* are made up of one or more sheets of graphene with cylindrical shape that form stable complexes with nucleic acids and to which it is possible to attach biomolecules to target them to a specific cell population and/or fluorescent probes to track them. *Silica-based systems* are silicon dioxide (SiO_2, silica)-based nanoparticles that can be functionalized through the introduction of cationic components (polymers, lipids, etc.) that will complex the negatively charged nucleic acids.

Table 2.2 Main advantages and disadvantages of polymer- and lipid-based systems of gene therapy delivery.

	Advantages	Disadvantages
Lipoplexes/ polyplexes	High transfection levels; possibility of choice of ideal lipids/polymers depending on administration route and targeted cells; large nucleic acid cargo capacity; easy to manufacture and low cost	Interaction with plasmatic proteins can reduce half-life time and cause embolism
Smart liposomes	Functionalization leading to cell targeting and reduced interaction with plasmatic proteins	More complex and expensive production
High-density lipoprotein (HDL)-mimicking systems	Biodegradable, do not trigger immune reactions, and are less captured and removed from blood circulation by the reticuloendothelial system	More complex and expensive production
Microvesicles-exosomes	Biodegradable, do not trigger immune reactions	Heterogeneity between batches related to size and composition (namely with respect to lipid, protein, and noncoding RNA content); poor pharmacokinetic profile in a systemic admnistration

This technique has been used with success for *in vitro* delivery of nucleic acids, as well as in some *in vivo* applications, such as for solid tumors in the skin and liver. The major limitations of this technique are the substantial cellular damage and consequent cytotoxicity that is triggered, as well as the limited *in vivo* application to local regions [7].

2.1.4 Nucleofection

Nucleofection technology is based on the same physical principles as electroporation; an electrical pulse is applied to the cells in order to open membrane pores through which the genetic material is internalized. However, in this method the exogenous material is directly delivered to the nucleus, being, therefore, a method used for cells that are more difficult to transfect. Moreover, the combination of the electric pulse with specific buffers significantly improves the outcomes.

This methodology requires a low cell number, can be performed within minutes, and is very efficient in delivering the nucleic acids. On the other hand, in some cell types this method can trigger a high mortality [8–10].

2.1.5 Ultrasound and Sonoporation

Ultrasound is a mechanical vibration with a frequency above human audible range (20 kHz) that can be applied for nucleic acid delivery purposes. Ultrasound's main action mechanism is the cavitation phenomenon. Cavitation is the process of formation and subsequent collapse of low-pressure voids (bubbles) driven by an acoustic field. These bubbles oscillate, grow, and implode releasing energy. The bubbles collapse causes sonoporation and local temperature and pressure rise. Sonoporation is the transient permeabilization (caused by pores formation) of the plasma membrane mediated by ultrasounds, which allows the transfer of nucleic acids into the cells [3, 11].

Microbubbles and some nucleation agents, such as ultrasound contrast agents, have been used to enhance the cavitation process, consequently increasing membrane permeabilization and leading to an improved gene transfer. Microbubbles are small bubbles (<10 μm) made of water-insoluble gas encapsulated in a biocompatible material that when irradiated by ultrasounds oscillate, generating enough fluid flow (microflow) in the vicinity of the membranes to cause sonoporation [3, 11].

Ultrasound-mediated gene delivery is a noninvasive strategy that can reach deeper organs and can be directed to a specific area.

2.1.6 Ballistic Gene Delivery/Gene Gun

Ballistic gene delivery (or gene gun) is based on the bombardment of tissues by accelerated nucleic acid-coated particles. These submicron-sized particles, usually gold particles, are first covered with the nucleic acids to be delivered to the cells and then they are accelerated in order to penetrate through the cell membrane, delivering their cargo into the cytosol [3, 5]. Several systems for the particles acceleration are available; however, most of the devices use gas shock, produced either by a chemical explosion or a discharge of pressurized gas [5].

This strategy allows the delivery of genetic material to a large surface and is also very useful in the case of cells that are difficult to transfect, such as plant cells. In fact, the gene gun was originally developed to transform plants [12] that, given the rigidity of their cell wall, are more difficult to penetrate at the cellular level and to be genetically manipulated. Although this is an interesting dellivery strategy, the technology has the disadvantage of being limited to a few-millimeter-deep penetration and therefore is restricted to a more localized intervention [3].

2.1.7 Magnetofection and Magnetoporation

These delivery approaches are based on the application of an external magnetic field, which will promote the atomic dipole alignment of the material with the field, causing a magnetic moment within the material and its magnetization.

Magnetofection, also known as magnetic transfection, is based on the incorporation of a magnetic agent in the vector, in order to deliver the nucleic acids under the influence of a magnetic field. A magnetic compound can be transported across the bloodstream and be concentrated in a particular organ of the body. Additionally, the cargo release can be controlled over time with the guidance of the magnetic field. The major advantage of this strategy is the enhancement of the accumulation of nucleic acids at a specific site, which, for example, works well for solid tumor masses. However, in a situation that requires spreading of the therapeutic agent, such as metastatic, highly invasive, and infiltrative cancers, in which the cells are not accessible to a magnet, this technique has limited application [13, 14].

The method of magnetoporation is also based on the use of magnetic fields to deliver nucleic acids; in this case the field is used to promote the formation of transient pores in the cellular membranes, through which the nucleic acids cross to the cell interior. This is possible because magnetic fields change the transmembrane potential of cells and, above a certain threshold, an increased transmembrane potential can trigger electroporation. The advantages of magnetoporation, as compared to electroporation, are as follows: (i) it requires no direct contact of electrodes with the material to be permeabilized, being, therefore, noninvasive ("needle-less"); (ii) it is easier and faster to apply; (iii) it is less expensive because it requires no disposables; and (iv) it has a bigger tissue penetration capacity, allowing access to anatomical areas unreachable by electroporation [14, 15].

2.1.8 Microneedles

The use of microneedles to deliver therapeutic agents across the skin can be applied for DNA vaccine administration and to the delivery of nucleic acids in several skin pathologies, such as cutaneous cancers, wounds and hyperproliferative disorders. This delivery strategy provides means to overcome the skin barriers and to deliver the therapeutic agents directly into the dermal layers, allowing both localized and systemic delivery and avoiding the first-pass metabolism in the liver. The length of the micron-sized needles, placed in the surface of a solid support like silicon, is designed to penetrate the subcutaneous layer, creating micron-sized channels through the skin that allow the localized deposit of large molecular weight molecules, without causing bleeding or pain and without reaching nerve fibers and blood vessels in the dermis [16, 17].

2.2 Chemical Systems

2.2.1 Polymer-Based Nanocarriers

Polymers are, by definition, substances whose molecular structures are composed of a large number of repeated units bonded together. Moreover, polymeric systems can be subdivided into two groups: (a) **natural polymers**, such as the polysaccharide chitosan, proteins, and peptides, and (b) **synthetic polymers**, such as cyclodextrins and poly(ethylenimine) [18].

Synthetic **cationic polymers** constitute the polymeric vehicles that are most often used in gene therapy. This is explained by their interaction with the negatively charged nucleic acids, originating **polyplexes** (designation for the complex formed by a polymer and the nucleic acid) capable of protecting the cargo and enabling its intracellular delivery [18, 19]. There are several natural and synthetic polymers used in gene delivery, such as chitosan, β-cyclodextrin, poly(L-lysine), poly(ethylenimine), dextran, and dendrimers [20, 21]. Additionally, these polymers can be engineered in order to improve their delivery efficiency, namely by adding cell-targeting ligands, like transferrin or epidermal growth factor (EGF), that will increase cellular internalization, or by adding shielding reagents, such as polyethylene glycol (PEG), that will increase the bloodstream circulation times, enhancing the amount of circulating delivery vectors and consequently gene-modifying tools that reach the target cells [22]. Some of these polymers are commercially available in ready-to-use **polyfection** (gene therapy transfer using a polymer) reagents.

2.2.2 Lipid-Based Systems

Presently, there are three main lipid-based vehicles used for gene therapy: (a) liposomes, (b) high-density lipoprotein (HDL)-mimicking systems, and (c) microvesicular systems (which include exosomes). Therefore, the pharmacological effect of the nucleic acids carried in these systems is a function of the pharmacokinetic, biodistribution, and drug release characteristics of those carriers.

Liposomes

Liposomes are micro- or nanoparticles composed of one or more lipid bilayers, with an aqueous core. These particles can be formed through a self-assembly process after addition of an ethanolic solution of lipids to an aqueous solution of nucleic acids. These vehicles were introduced as drug delivery agents in the 1970s [23] and have been adapted for nucleic acid delivery; their efficiency as delivery tools is dependent on physicochemical characteristics, such as lipid composition, size, net charge, loading efficiency, and stability [18, 23, 24]. There are different types of liposomes, with different physicochemical characteristics and consequently more appropriate for different applications; some examples are (a) cationic liposomes and (b) "Smart"/Trojan horse liposomes.

Cationic Liposomes

Cationic liposomes have a net positive charge and thus spontaneously associate with negatively charged nucleic acids through electrostatic interactions. This results in the entrapment of the large nucleic acid molecules into smaller **lipoplex** (designation for the complexes made up of liposomes and nucleic acids) particles [18, 25]. Cationic liposomes successfully mediate nucleic acids delivery *in vitro* because they can be formulated so as to exhibit a net positive charge, which triggers their association with the negatively charged cell membranes. Additionally, they possess fusogenic properties, increasing the escape of the carried nucleic acid molecules trapped in the endosomes, upon internalization, into the cytosol (endosomal escape), thus significantly decreasing their lysosomal degradation [18, 25, 26].

Since 1987, when Felgner and collaborators used cationic liposomes for the first time for gene therapy applications [27], many different cationic lipids have been synthesized and used for nucleic acid delivery purposes. Cationic lipids are composed of a positively charged polar head group, a hydrophobic region, and a linker connecting the polar and the non-polar group. These domains play an important role on the transfection ability and on the toxicity of the resulting lipoplexes. For example, the cationic head group, which will complex the negatively charged nucleic acid, is composed of either single or multiple groups of primary, secondary, tertiary, or quaternary amines or guanidine or imidazole groups. The liposomes made with multivalent lipids condense and protect the nucleic acids more efficiently than monovalent lipids [28, 29]. However, increasing the net positive charge may result in a lipid-nucleic acid interaction so strong that it reduces complex dissociation, consequently decreasing the efficiency of the nucleic acid release to the cytosol. Additionally, multivalent cationic lipids are more prone to form micelles, which compromise the liposomal/lipoplexes stability and also lead to higher toxicity of the resulting complexes. Cytotoxicity is also dependent on the stability and biodegradability of the lipid used to form the liposomes; thus lipids with strong linkages are normally more toxic [29, 30] because they are more difficult to be metabolized. One example of such lipids is the dioleoyloxypropyl (trimethylammonium) chloride (DOTMA), which has ether linkages and is more toxic than lipids with more labile ester linkages such as 1,2-dioleoyl-3 (trimethylammonium) propane (DOTAP) [18].

The main advantages of positively charged lipoplexes as compared to more complex systems, such as Smart/Trojan horse liposomes, are (i) the high efficiency in delivering nucleic acids *in vitro*; (ii) having no restrictions on the size of the nucleic acid to be delivered; (iii) their simplicity, given that these lipoplexes are easier to manufacture; (iv) the lower costs associated with these systems; and (v) some of them being commercially available [18]. Concerning the drawbacks, one of their major limitations is that, generally, they cannot be considered for *in vivo* applications through intravenous administration. This is explained by their cytotoxicity, instability, tendency to aggregate (which can lead to microemboli and tissue ischemia occurrences), and poor biodistribution (poor distribution through the organism, which will not enable the therapeutic compounds to reach their molecular and cellular targets) [25, 29–31]. This *in vivo* limitation extends to an inefficient *in vivo* transfection, also caused by the interaction of the positively charged lipoplexes with blood components, such as serum proteins, which decrease the availability of lipoplexes that reach and transfect the target cells. At the same time, this triggers lipoplex destabilization through modification of the lipoplex surface size and charge. These charge and size changes enhance lipoplex accumulation in the lungs, liver, and spleen, which is also a cause of the poor biodistribution of these particles. Opsonization (activation of the complement system, which activates phagocytosis and an inflammatory response) of positively charged lipoplexes with plasma proteins and lipoproteins, such as albumin and high-density lipoprotein (HDL), causes their rapid plasmatic clearance by the mononuclear phagocyte system (MPS) players, the phagocytic cells [18, 25, 30, 31]. Thus, lipoplexes are efficient *in vitro* nucleic acid delivery tools; however, they have important *in vivo* limitations, and consequently new lipid-based delivery systems have been developed to overcome the *in vivo* biodistribution limitations of the lipoplexes.

A successful example of lipoplexes used as drug vehicles for clinical practice is the recently approved patisiran (Onpattro™), developed by Alnylam Pharmaceuticals [32]. This drug was approved for the treatment of hereditary transthyretin-mediated (hATTR) amyloidosis (hATTR amyloidosis) in adults with stage 1 or stage 2 polyneuropathy and is the first ever RNAi therapeutic approved in the EU. Onpattro acts by specifically downregulating the transthyretin (TTR) messenger RNA (via RNA interference with siRNA) and subsequently leading to a reduction in serum TTR protein levels and tissue TTR protein deposits, which are associated with the pathology. The siRNA molecules are complexed in lipoplexes and delivered intravenously to the liver, the primary place of TTR protein production. Patisiran significantly improves multiple clinical manifestations of these patients [32–34].

Smart/Trojan Horse Liposomes

The ideal lipid-based systems for nucleic acids and drug delivery would consist in (i) long-circulating liposomes with extended permanence in the bloodstream and (ii) with the capability to specifically target the cells/organs to be treated. This would (i) increase their ability to reach the target cells in therapeutic doses and (ii) decrease the secondary effects caused by the delivery of therapeutic agents to healthy cells.

Long-circulating liposomes are frequently called "Stealth" or "Trojan horse" liposomes, given their ability to evade the immune system and remain in the bloodstream for longer periods. The most frequent strategy used to increase the permanence of liposomes in the bloodstream is through the modification of the liposomes' surface by introducing hydrophilic polymers, such as PEG. The incorporation of PEG in the particle surface will sterically prevent the interaction and binding of blood components (like the complement system elements) to the liposome surface, preventing its opsonization and removal from the bloodstream by the reticuloendothelial system. Other polymers have also been used to prolong blood circulation times such as poly(acrylamide), poly(vinylpyrrolidone), poly(acryloylmorpholine), poly(2-ethyl-2-oxazoline), poly(2-methyloxazoline), phosphatidyl polyglycerols, polyvinyl alcohols, and others [18, 35, 36].

Smart liposomes are characterized by their ability to deliver their cargo to a specific cell population or under the influence of a specific stimulus. Targeting nanocarriers to a specific cell population is often carried out with the aid of specific ligands that are incorporated into the nanocarrier surface. These ligands specifically recognize and interact with certain cell surface components allowing the cargo to be selectively delivered to those cells. One example of such ligands is the transferrin (Trf) protein, given that the Trf receptor is overexpressed in cancer cells, allowing, therefore, the preferential delivery of the Trf-targeted nanocarriers to cancer cells [35–38]. Stimulus-sensitive liposomes deliver their cargo after the influence of a specific stimulus. There are several stimuli able to destabilize liposomes and promote the release of their content. These stimuli can be endogenous factors such as pH and redox conditions, or exogenous factors such as magnetic fields, ultrasound, and light. Temperature can either be an endogenous or exogenous stimulus. Thus, different controlled release liposomal-based systems can be formulated in order to specifically respond and deliver their content upon stimulation.

Some stimuli allow the distinction between normal and pathological tissues; one example of these endogenous factors is the intratumoral pH value that in solid tumors is significantly lower (approximately 6.5). Thus, pH-sensitive liposomes that are destabilized below pH 7.4 will allow specific cargo delivery to cancer cells. Additionally, tumors are also characterized by a local temperature rise. Thus, thermosensitive liposomes can be designed in order to maintain their cargo at the physiological temperature of 37 °C and release their cargo at temperatures higher than 40–45 °C [35, 39].

Exogenous stimuli can be applied from the outside in the local to be treated in order to trigger the release of the liposomal content. These carriers have an on-off drug release because the structure (lipid bilayer) is affected when stimulated and this allows achieving a temporal and spatial controlled release of its cargo [35, 39].

High-Density Lipoprotein (HDL)-Mimicking Systems

High-density lipoproteins (HDL) are a group of lipoproteins including chylomicrons, very low-density lipoproteins (VLDL), and the low-density lipoproteins (LDL). **Lipoproteins** are spherical particles composed of a hydrophobic core comprising triglycerides and cholesteryl esters covered by a shell of apolipoproteins, esterified cholesterol, and phospholipids. The different classes of lipoproteins differ in their size, lipid and apolipoprotein composition, and in their specific functions. Generically, lipoproteins are responsible for the transport of lipids in blood circulation, targeting different cell types through specific cellular receptors [40].

Endogenous HDL are nanoparticles with 6–13 nm in diameter, composed of apolipoproteins and lipids, with a density of 1.063–1.210 g/mL that function has a transport system [41, 42] for hydrophobic molecules such as cholesterol. The most abundant apolipoprotein in HDL is apolipoprotein A-1 (Apo A1), representing 70% of the protein content, and is essentially responsible for supporting the size and shape of these lipoproteins. The major role of HDL is to transport cholesterol from the peripheral tissues to the liver for catabolism in the process of reverse cholesterol transportation. However, HDL are associated with other activities, namely the transportation of miRNA, carotenoids, vitamins, and hormones; they are also associated with antioxidative, anti-inflammatory, anti-apoptotic, anti-atherogenic, and immunogenic activities. Additionally, HDL have been related to a positive influence on the regulation of glucose homeostasis [40–43]. As these particles are endogenous, they are biodegradable, do not trigger immune reactions, and are not captured and removed from circulation by the reticuloendothelial system [40], features that make them very promising delivery systems.

SR-B1 is the natural receptor for HDL; this receptor has the highest expression in the liver, which is the main target organ of HDL. Therefore, many of the HDL-based systems developed so far have been formulated to target the liver. However, SR-B1 can also be found in adrenal gland, ovarian, and placental tissues [42]; additionally, it has been reported that this receptor is also highly expressed in some cancers, such as prostate, breast, colorectal, ovarian, and nasopharyngeal cancer, which is explained by the requirement of high cholesterol levels on the part of the highly proliferating cancer cells. Thus, HDL can also be formulated to deliver anti-cancer drugs [42]; however, in this particular case, the use of HDL to deliver toxic anti-cancer drugs may trigger substantial hepatotoxicity caused by the accumulation of the drug in the liver, mediated by the natural delivery of HDL to this organ. One way to overcome this problem is to modify the HDL in order to add additional targeting ligands mediating cell-specific cargo delivery to its surface, such as folic acid and targeting antibodies, like the CD20 antibody that targets B-cell lymphoma cells [42, 44].

To deliver the negatively charged nucleic acids with HDL it is necessary to introduce in the formulation cationic lipids, such as N, N-dimethyl-N, N-dioctadecylammonium bromide (DDAB), which will form complexes with the genetic material, allowing its incorporation in these particles. Thus, HDL can be artificially prepared in order to efficiently deliver nucleic acids [44–46] to a specific cell population evading the reticuloendothelial system and promoting no immune reaction.

Microvesicles-Exosomes

Exosomes are membrane-based vesicles with 40–100 nm that belong to the group of extracellular vesicles, which also include microvesicles and apoptotic bodies. These vesicles are produced by almost all types of cells, being released to the extracellular space with the purpose of acting as intercellular communication vectors promoting the transport of different molecules between different cells at big distances. The cargo content and the membrane composition of exosomes are dependent on the cells of origin and their current state, namely on their differentiation stage and stress. Thus, in addition to their potential use as transport vehicles, exosomes also hold the potential of being implemented in clinical practice as disease's biomarkers. In fact, exosomes are enriched in particular proteins, RNAs, and lipids. Some of these are used as exosomal markers, enabling the confirmation of exosome preparation/isolation. These include the Alix, flotillin, TSG101, and CD63 proteins and raft lipids such as cholesterol, ceramide, and glycerophospholipids with long and saturated fatty acyl chains [47–49]. Other proteins have been used as biomarkers; for example, it is known that, in Parkinson's disease, exosomes act as spreading vehicles of the disease, performing the transport of α-synuclein from diseased cells to healthy cells [50]. Thus, the presence of this protein in exosomes is an indication of the disease.

Because exosomes are vesicles naturally produced by cells and their size allows a reasonable cargo transport between cells, exploring these vesicles as drug and nucleic acid delivery agents is logical. However, these vesicles still have several drawbacks, in particular the heterogeneity between batches and within the same sample, as different subspecies exist among exosomes isolated from the same batch of cells. The differences are related to size and composition, namely with respect to lipid, protein, and noncoding RNA content. As most of the identified noncoding RNAs in exosomes have unknown biological functions, and therefore the full impact of these molecules in the target cells is unknown, there is also a risk of triggering unpredictable secondary molecular effects. Additionally, in a systemic administration, exosomes have a poor pharmacokinetic profile, because they are captured in the liver, spleen, and lungs [47, 48].

2.2.3 Inorganic Materials

Many inorganic compounds (Table 2.3) have been used to make inorganic nanoparticles, which have different physical and chemical characteristics and capabilities depending on their composition. The most often used inorganic materials are magnetic compounds such as iron oxides, previously described in the magnetofection section, metal nanoparticles, quantum dots, graphene-based systems such as carbon nanotubes, and silica nanoparticles. The advantages of these inorganic-based delivery systems are their easy functionalization, low cytotoxicity, and high biocompatibility [51].

Metal Nanoparticles

Metal nanoparticles have been used as gene delivery vectors and noble metals such as gold and silver are some of the materials that are commonly used in their composition. These metals have characteristic surface plasmon resonance [52, 53] that allows specific sensoring; additionally, their surface is easy to be biofunctionalized and to be applied in specific applications such as light-based triggered events [51]. The major advantages of these systems are that they are easy to synthesize, they have a well-defined composition and good biocompatibility profile, and they are easy to track upon administration using techniques such as fluorescence resonance energy transfer (FRET) [54].

Quantum Dots

Quantum dots (QD) are semiconductor crystals with sizes ranging from 1 to 20 nm [51, 54]; these systems are clusters of hundreds to thousands of atoms arranged in binary (e.g., CdSe, GaAs, InAs, SiC) or ternary compounds (e.g., InGaN, InGaP, InGaAs) [55]. QDs are used as fluorescence probes and have several advantages as compared to organic fluorophores, such as high brightness, longer fluorescence lifetime, and better photostability [51, 54]. These systems can be engineered in order to become very efficient biosensing platforms with high specificity and

Table 2.3 Main advantages and disadvantages of inorganic materials used in gene therapy delivery

	Advantages	Disadvantages
Metal nanoparticles	Easy to synthesize; well-defined composition; good biocompatibility; easy to track upon administration	Require specialized equipment
Quantum dots	High specificity and sensitivity of tracking; can be functionalized to specific intracellular delivery of nucleic acids	Complex production
Carbon nanotubes	Small size; chemical inertness; high drug capacity; controlled drug release ability	Complex production; poor solubility in aqueous solutions, reducing the applications in biological systems
Silica-based systems	Good drug loading capacity; can be functionalized to have increase circulating times, targeting properties, and cellular uptake; good storage stability; low toxicity; unexpensive and easy preparation in large amounts	Could lead to hemolysis and metabolic deregulation

sensitivity used in bioimaging to detect specific targets such as nucleic acids [55]. Additionally, they can be designed for gene delivery purposes; the classical approach is the conjugation of the nucleic acids and a targeting ligand on the surface of the quantum dot core. This will combine in one system the fluorophore properties and the therapeutic functions (theranostics) [56], allowing, for example, the tracking of the intracellular delivery of the nucleic acids [54, 55] mediated by these systems.

Carbon Nanotubes

Carbon nanotubes are made up of one or more sheets of graphene with cylindrical shape. This system can be formulated as (i) multiwalled carbon nanotubes, exhibiting two or more cylindrical graphene sheets centrically arranged and displaying 4–30 nm of diameter, or as (ii) single-walled carbon nanotubes, composed of a single graphene sheet, having 0.4–3 nm of diameter [51, 54]. The small size, chemical inertness, high drug loading capacity and controlled drug release ability are some of the biggest advantages of this system. Additionally, it is also possible to attach biomolecules and fluorescent probes to the carbon nanotubes, improving their delivery properties and enabling the study of the cellular delivery process [54, 57]. Moreover, carbon nanotubes can form stable complexes with nucleic acids mediating their efficient delivery. However, the poor solubility in aqueous solutions reduces the applications in biological systems [51].

Silica-Based Systems

Silicon dioxide (SiO_2), or silica, nanoparticles have been widely used to deliver nucleic acids, drugs, and dyes. These particles can be modified in order to manipulate their size, shape, and porosity; it is also possible to functionalize their surface through the conjugation of several molecules such as targeting ligands and polymers that confer stealth properties. Thus, it is possible to modify them in order to obtain nanoparticles with long circulating times, favorable targeting properties, good drug loading capacity, adequate cellular uptake profiles and low toxicity. Additionally, these systems have good storage stability and are cheap and of easy preparation in large quantity [53, 54, 58]. Nevertheless, probably the most attractive property of these systems is the ability to store and release a big variety of drugs and to provide a big surface to store the drugs and nucleic acids, allowing the incorporation of hydrophilic and hydrophobic molecules; the latter are particularly difficult to be delivered by other systems [53].

A classical surface modification that is made to silica nanoparticles is the introduction of cationic components (such as PEI, dendrimers, and cationic lipids), which will complex the negatively charged nucleic acids; the genetic material will, therefore, be adsorbed at the nanoparticle surface and, consequently, will not be totally protected from nucleases. This problem was overcome by the design of silica nanoparticles with bigger pores (>15 nm) whose surface was functionalized in order to introduce positively charged primary amine groups, providing a large loading capacity and nuclease protection [59]. The loading capability is a function of the roughness, pore size, and the nature of the surface functionalization; accordingly, thiol-modified particles, followed by mixed thiol- and amino-functionalized silica nanoparticles, have the highest loading ability.

Besides the incorporation of targeting ligands, it is also possible to functionalize silica-based systems in order to obtain nanoparticles responding to pH, redox changes, and external stimuli, such as near-infrared light, which can be used in the development of less invasive diagnose methods [58].The major disadvantages of silica-based nanoparticles are (i) the observed hemolysis caused by its interaction with the surface of the phospholipids of the red blood cell membranes and (ii) the induction of metabolic changes resulting in melanoma [60–62].

This Chapter in a Nutshell

- Gene-manipulating tools such as siRNA or DNA have a very limited bioavailability, because they are degraded by nucleases, do not efficiently cross cell membranes, and are eliminated from the bloodstream.
- For the clinical implementation of gene therapy, the development of vectors that efficiently

and safely deliver the nucleic acids is required.

- Non-viral vectors are promising gene therapy vectors, which have been extensively engineered and investigated. They are simple, safe, easy to produce and economically viable systems.
- Nevertheless, these vectors have limited efficiency and due to their transient expression they are not suitable for some gene therapy applications.
- Non-viral vectors can be broadly divided in two main categories: physical methods and chemical systems.
- The physical methods rely on physical phenomena to counteract the membrane barrier of the cells, thus facilitating intracellular delivery of the genetic material.
- The chemical systems are based on the combination of nucleic acids to other chemical molecules to overcome extra- and intracellular barriers to gene delivery.

Review Questions

1. Alix, flotillin, TSG101, and CD63 proteins are:
   (a) Markers used to identify high-density lipoprotein-mimicking systems
   (b) Markers used to identify microvesicles such as exosomes
   (c) Markers used to identify high-density lipoprotein-mimicking systems and microvesicles such as exosomes
   (d) Used to confer stealth properties to "smart liposomes"
   (e) None of the above
2. Polyethylene glycol (PEG) is:
   (a) A hydrophilic polymer used as a shielding reagent in polymer-based and lipid-based systems
   (b) Added to the formulation with the purpose of promoting cellular targeting
   (c) Added to the formulation with the purpose of decreasing the bloodstream circulation times
   (d) A steric promoter of the interaction and binding of blood components (like the complement system elements) to the vector surface
   (e) None of the above
3. Examples of "smart liposomes" are:
   (a) Liposomes able to deliver their cargo to a specific cell population
   (b) Liposomes able to deliver their cargo under the influence of a specific stimulus
   (c) Liposomes incorporating ligands that specifically recognize and interact with certain cell surface components allowing the cargo to be selectively delivered
   (d) Liposomes taking advantage of specific differentiating conditions such as endogenous pH levels and the redox environment or exogenous factors such as magnetic fields, ultrasound, and light to deliver their cargo
   (e) All of the above
   (f) None of the above
4. There are several physical methods used to deliver nucleic acids, such as:
   (a) Hydrodynamic delivery, in which hydrostatic pressure rise will enlarge the fenestrae enabling the delivery
   (b) Microinjection, which is a very straightforward method applied to normal-sized cells
   (c) Electroporation and nucleofection, that are based on the use of an electric pulse to open transient pores in the cell membrane and nuclear membrane, respectively
   (d) All of the above
   (e) Answers (a) and (c) are correct
   (f) None of the above
5. Many inorganic compounds have been used to make inorganic nanoparticles, including:
   (a) Quantum dots, that commonly include noble metals such as gold and silver in their composition
   (b) Single-walled carbon nanotubes, composed of a single graphene sheet, having 0.4–3 nm of diameter
   (c) Silica-based systems, that are made up of silica (SiO_3) and that can be used to deliver nucleic acids, drugs, and dyes
   (d) All of the above
   (e) None of the above

References and Further Reading

1. Yin H, Kauffman KJ, Anderson DG (2017) Delivery technologies for genome editing. Nat Rev Drug Discov 16(6):387–399
2. Cullis PR, Hope MJ (2017) Lipid nanoparticle systems for enabling gene therapies. Mol Ther 25(7):1467–1475
3. Wells DJ (2004) Gene therapy progress and prospects: electroporation and other physical methods. Gene Ther 11(18):1363–1369
4. Huang M, Sun R, Huang Q, Tian Z (2017) Technical improvement and application of hydrodynamic gene delivery in study of liver diseases. Front Pharmacol 8:591
5. Hapala I (1997) Breaking the barrier: methods for reversible permeabilization of cellular membranes. Crit Rev Biotechnol 17(2):105–122
6. Navarro J, Risco R, Toschi M, Schattman G (2008) Gene therapy and intracytoplasmatic sperm injection (ICSI) – a review. Placenta 29(Suppl B):193–199
7. Lambricht L, Lopes A, Kos S, Sersa G, Preat V, Vandermeulen G (2016) Clinical potential of electroporation for gene therapy and DNA vaccine delivery. Expert Opin Drug Deliv 13(2):295–310
8. Kobayashi N, Rivas-Carrillo JD, Soto-Gutierrez A et al (2005) Gene delivery to embryonic stem cells. Birth Defects Res C Embryo Today 75(1):10–18
9. Hamm A, Krott N, Breibach I, Blindt R, Bosserhoff AK (2002) Efficient transfection method for primary cells. Tissue Eng 8(2):235–245
10. Stroh T, Erben U, Kuhl AA, Zeitz M, Siegmund B (2010) Combined pulse electroporation--a novel strategy for highly efficient transfection of human and mouse cells. PLoS One 5(3):e9488
11. Delalande A, Postema M, Mignet N, Midoux P, Pichon C (2012) Ultrasound and microbubble-assisted gene delivery: recent advances and ongoing challenges. Ther Deliv 3(10):1199–1215
12. Klein TM, Wolf ED, Wu R, Sanford JC (1987) High-velocity microprojectiles for delivering nucleic acids into living cells. Nature 327:70–73
13. Plank C, Zelphati O, Mykhaylyk O (2011) Magnetically enhanced nucleic acid delivery. Ten years of magnetofection-progress and prospects. Adv Drug Deliv Rev 63(14-15):1300–1331
14. Estelrich J, Escribano E, Queralt J, Busquets MA (2015) Iron oxide nanoparticles for magnetically-guided and magnetically-responsive drug delivery. Int J Mol Sci 16(4):8070–8101
15. Kardos TJ, Rabussay DP (2012) Contactless magneto-permeabilization for intracellular plasmid DNA delivery in-vivo. Hum Vaccin Immunother 8(11):1707–1713
16. Coulman SA, Barrow D, Anstey A et al (2006) Minimally invasive cutaneous delivery of macromolecules and plasmid DNA via microneedles. Curr Drug Deliv 3(1):65–75
17. McCaffrey J, Donnelly RF, McCarthy HO (2015) Microneedles: an innovative platform for gene delivery. Drug Deliv Transl Res 5(4):424–437
18. Mendonca LS, de Lima MCP, Simoes S (2012) Targeted lipid-based systems for siRNA delivery. J Drug Deliv Sci Technol 22(1):65–73
19. Thomas M, Klibanov AM (2003) Non-viral gene therapy: polycation-mediated DNA delivery. Appl Microbiol Biotechnol 62(1):27–34
20. El-Say KM, El-Sawy HS (2017) Polymeric nanoparticles: promising platform for drug delivery. Int J Pharm 528(1-2):675–691
21. Luten J, van Nostrum CF, De Smedt SC, Hennink WE (2008) Biodegradable polymers as non-viral carriers for plasmid DNA delivery. J Control Release 126(2):97–110
22. Park TG, Jeong JH, Kim SW (2006) Current status of polymeric gene delivery systems. Adv Drug Deliv Rev 58(4):467–486
23. Lasic DD (1998) Novel applications of liposomes. Trends Biotechnol 16(7):307–321
24. Drummond DC, Noble CO, Hayes ME, Park JW, Kirpotin DB (2008) Pharmacokinetics and in vivo drug release rates in liposomal nanocarrier development. J Pharm Sci 97(11):4696–4740
25. Morille M, Passirani C, Vonarbourg A, Clavreul A, Benoit JP (2008) Progress in developing cationic vectors for non-viral systemic gene therapy against cancer. Biomaterials 29(24-25):3477–3496
26. Safinya CR, Ewert KK, Majzoub RN, Leal C (2014) Cationic liposome-nucleic acid complexes for gene delivery and gene silencing. New J Chem 38(11):5164–5172
27. Felgner PL, Gadek TR, Holm M et al (1987) Lipofection: a highly efficient, lipid-mediated DNA-transfection procedure. Proc Natl Acad Sci U S A 84(21):7413–7417
28. Zhang S, Xu Y, Wang B, Qiao W, Liu D, Li Z (2004) Cationic compounds used in lipoplexes and polyplexes for gene delivery. J Control Release 100(2):165–180
29. Lv H, Zhang S, Wang B, Cui S, Yan J (2006) Toxicity of cationic lipids and cationic polymers in gene delivery. J Control Release 114(1):100–109
30. Karmali PP, Chaudhuri A (2007) Cationic liposomes as non-viral carriers of gene medicines: resolved issues, open questions, and future promises. Med Res Rev 27(5):696–722
31. Dass CR, Choong PF (2006) Selective gene delivery for cancer therapy using cationic liposomes: in vivo proof of applicability. J Control Release 113(2):155–163
32. Hoy SM (2018) Patisiran: first global approval. Drugs 78(15):1625–1631
33. Adams D, Gonzalez-Duarte A, O'Riordan WD et al (2018) Patisiran, an RNAi therapeutic, for hereditary transthyretin amyloidosis. N Engl J Med 379(1):11–21
34. Rizk M, Tuzmen S (2017) Update on the clinical utility of an RNA interference-based treatment: focus on Patisiran. Pharmgenomics Pers Med 10:267–278

35. Torchilin V (2009) Multifunctional and stimuli-sensitive pharmaceutical nanocarriers. Eur J Pharm Biopharm 71(3):431–444
36. Siafaka PI, Ustundag Okur N, Karavas E, Bikiaris DN (2016) Surface modified multifunctional and stimuli responsive nanoparticles for drug targeting: current status and uses. Int J Mol Sci 17(9):1440
37. Mendonca LS, Firmino F, Moreira JN, Pedroso de Lima MC, Simoes S (2010) Transferrin receptor-targeted liposomes encapsulating anti-BCR-ABL siRNA or asODN for chronic myeloid leukemia treatment. Bioconjug Chem 21(1):157–168
38. Mendonca LS, Moreira JN, de Lima MC, Simoes S (2010) Co-encapsulation of anti-BCR-ABL siRNA and imatinib mesylate in transferrin receptor-targeted sterically stabilized liposomes for chronic myeloid leukemia treatment. Biotechnol Bioeng 107(5):884–893
39. Liu D, Yang F, Xiong F, Gu N (2016) The smart drug delivery system and its clinical potential. Theranostics 6(9):1306–1323
40. Rensen PC, de Vrueh RL, Kuiper J, Bijsterbosch MK, Biessen EA, van Berkel TJ (2001) Recombinant lipoproteins: lipoprotein-like lipid particles for drug targeting. Adv Drug Deliv Rev 47(2-3):251–276
41. Luthi AJ, Patel PC, Ko CH, Mutharasan RK, Mirkin CA, Thaxton CS (2010) Nanotechnology for synthetic high-density lipoproteins. Trends Mol Med 16(12):553–560
42. Simonsen JB (2016) Evaluation of reconstituted high-density lipoprotein (rHDL) as a drug delivery platform – a detailed survey of rHDL particles ranging from biophysical properties to clinical implications. Nanomed Nanotechnol Biol Med 12(7):2161–2179
43. Mo ZC, Ren K, Liu X, Tang ZL, Yi GH (2016) A high-density lipoprotein-mediated drug delivery system. Adv Drug Deliv Rev 106(Pt A):132–147
44. Damiano MG, Mutharasan RK, Tripathy S, McMahon KM, Thaxton CS (2013) Templated high density lipoprotein nanoparticles as potential therapies and for molecular delivery. Adv Drug Deliv Rev 65(5):649–662
45. Rui M, Qu Y, Gao T, Ge Y, Feng C, Xu X (2017) Simultaneous delivery of anti-miR21 with doxorubicin prodrug by mimetic lipoprotein nanoparticles for synergistic effect against drug resistance in cancer cells. Int J Nanomedicine 12:217–237
46. Uno Y, Piao W, Miyata K, Nishina K, Mizusawa H, Yokota T (2011) High-density lipoprotein facilitates in vivo delivery of alpha-tocopherol-conjugated short-interfering RNA to the brain. Hum Gene Ther 22(6):711–719
47. Ferguson SW, Nguyen J (2016) Exosomes as therapeutics: the implications of molecular composition and exosomal heterogeneity. J Control Release 228:179–190
48. Morishita M, Takahashi Y, Nishikawa M, Takakura Y (2017) Pharmacokinetics of exosomes-an important factor for elucidating the biological roles of exosomes and for the development of exosome-based therapeutics. J Pharm Sci 106(9):2265–2269
49. Simons M, Raposo G (2009) Exosomes--vesicular carriers for intercellular communication. Curr Opin Cell Biol 21(4):575–581
50. Russo I, Bubacco L, Greggio E (2012) Exosomes-associated neurodegeneration and progression of Parkinson's disease. Am J Neurodegener Dis 1(3):217–225
51. Riley MK, Vermerris W (2017) Recent advances in nanomaterials for gene delivery-a review. Nanomaterials 7(5):94
52. Lopatynskyi A, Guiver M, Chegel V (2014) Surface plasmon resonance biomolecular recognition nanosystem: influence of the interfacial electrical potential. J Nanosci Nanotechnol 14(9):6559–6564
53. Anselmo AC, Mitragotri S (2015) A review of clinical translation of inorganic nanoparticles. AAPS J 17(5):1041–1054
54. Loh XJ, Lee TC (2012) Gene delivery by functional inorganic nanocarriers. Recent Pat DNA Gene Seq 6(2):108–114
55. Zhang Y, Wang TH (2012) Quantum dot enabled molecular sensing and diagnostics. Theranostics 2(7):631–654
56. Xie J, Lee S, Chen X (2010) Nanoparticle-based theranostic agents. Adv Drug Deliv Rev 62(11):1064–1079
57. Loh XJ, Lee TC, Dou Q, Deen GR (2016) Utilising inorganic nanocarriers for gene delivery. Biomater Sci 4(1):70–86
58. Mebert AM, Baglole CJ, Desimone MF, Maysinger D (2017) Nanoengineered silica: properties, applications and toxicity. Food Chem Toxicol 109(Pt 1):753–770
59. Kim MH, Na HK, Kim YK et al (2011) Facile synthesis of monodispersed mesoporous silica nanoparticles with ultralarge pores and their application in gene delivery. ACS Nano 5(5):3568–3576
60. Bharti C, Nagaich U, Pal AK, Gulati N (2015) Mesoporous silica nanoparticles in target drug delivery system: a review. Int J Pharm Investig 5(3):124–133
61. Kettiger H, Sen Karaman D, Schiesser L, Rosenholm JM, Huwyler J (2015) Comparative safety evaluation of silica-based particles. Toxicol in Vitro 30(1 Pt B):355–363
62. Huang X, Zhuang J, Teng X et al (2010) The promotion of human malignant melanoma growth by mesoporous silica nanoparticles through decreased reactive oxygen species. Biomaterials 31(24):6142–6153

Viral Vectors for Gene Therapy 3

The selection of the delivery system is one of the most critical points for the success (and safety) of gene therapy. First, the chosen vector must ensure that the gene is delivered to the correct target cells, or at least predominantly to those cells. Second, the introduced transgene has to be activated, that is, it must go to the nucleus, be transcribed, and then translated. In a one-time cure setting, ideally the transgene should be integrated into the host genome or endure episomally, ensuring a long-term persistent expression. The ideal delivery system should meet several criteria, including (i) a good safety profile; (ii) easy production; (iii) good stability in target cells, and (iv) a high transgene capacity. As mentioned, the delivery systems can be roughly divided into two main categories: non-viral and viral systems. The non-viral methods were described in the previous chapter, whereas in this chapter the viral vectors used in gene therapy will be explained.

Viruses have evolved for millions of years, aiming to efficiently introduce and replicate their genetic material in host cells. Taking advantage of this feature, gene therapy developed tools and procedures to engineer viruses and their genomes, in order to artificially deliver nucleic acids. In fact, the main advantage of viral-based systems for gene therapy is their high efficiency in delivering transgenes, surpassing anatomical and cellular barriers. For this reasons, viral vectors have been broadly used since the first gene therapy clinical trial in 1990. Data from *The Journal of Gene Medicine* highlights that until November 2017 around 68% of gene therapy clinical trials worldwide used viral vectors as a delivery system (Fig. 3.1) [1]. Recently, viral vectors became more than a promise in gene therapy with the approval of several gene therapy products using these delivery vectors (see Chap. 1 for a list of currently approved gene therapy products). Moreover, several other products using viral systems are in the final stages of clinical trial development [2], highlighting that new products could be approved in the next years. This broad use also emphasizes that viral vectors are now considered safe for human use, despite their natural infectious profile.

Nevertheless, the safety issues were and still are major concern in the development and use of viral vectors. For that, several engineering solutions were developed trying to enhance viral vectors' safety, without compromising their efficiency, such as (i) avoiding their replication; (ii) promoting their inactivation; and (iii) attenuating their natural toxicity. Besides their efficiency, another important advantage of viral vectors is the wide range of existing viruses, which have different features, like the type of genetic material, natural tropism, and size, which provide an enormous offer of systems for gene delivery. On the other hand, besides their safety concerns in gene therapy, viral vectors also have other disadvantages, such as their limited cloning capacity (Table 3.1).Several criteria are used to classify

C. Nóbrega et al., *A Handbook of Gene and Cell Therapy*,
https://doi.org/10.1007/978-3-030-41333-0_3

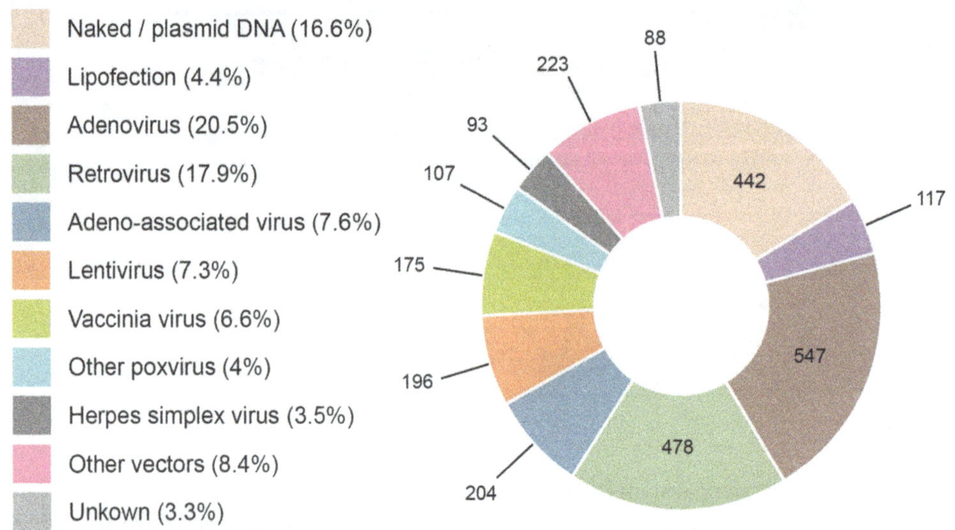

Fig. 3.1 **Vectors used in gene therapy clinical trials** until 2017, at which time viral vectors accounted for 68% of all vectors used in these studies.

Table 3.1 Main advantages and disadvantages of viral vectors used in gene therapy.

Advantages	Disadvantages
High efficiency in gene transfer both *in vitro* and *in vivo*	Potential for immune and/or inflammatory response triggering
Long-term persistence (in some cases)	Limited cloning capacity
Broad cell targets	Complex production
Broad range of viruses to use	Limited tropism to certain types of cells (in some cases)
Natural tropism towards infection	Possibility of insertional mutagenesis
Evolved mechanisms of endosomal escape	Molecular infection mechanisms not fully understood

virus [3], including (i) the nature and form of their nucleic acid (DNA or RNA), (ii) symmetry of the capsids, (iii) presence of an envelope, (iv) the size of the virion and (v) the site of viral replication (cytoplasm or nucleus). Thus, choosing and engineering a virus for gene therapy must consider these and other features in order to ensure a successful and safe delivery of the transgene. In this section, we detail several aspects of the viral vectors most commonly used in gene therapy: lentivirus, gamma retrovirus, adenovirus, adeno-associated virus, herpes simplex virus, vaccinia, and baculovirus.

3.1 Lentiviral Vectors

Lentivirus belongs to the *Retroviridae* family (retroviruses), which is constituted by RNA-type viruses characterized by the use of viral reverse transcriptase (RT) and the ability to insert their viral genetic information into the host genome through the action of viral integrase (IN) [4]. The name lentivirus derives from the long period between infection and the disease onset. Contrary to other retroviruses, lentivirus can also replicate in nondividing cells. They are enveloped virus, with a size ranging from 80 to 120 nm in diameter (Fig. 3.2). Lentiviruses have different host species, including primates, differing in their genome structure and entry receptors, among other features. As the majority of the lentiviral vectors used in gene therapy are based on **human immunodeficiency virus 1** (HIV-1), the following descriptions are centered on this virus.

HIV-1 has a single-stranded positive RNA genome with around 9 kb size, with three main open reading frames encoding for nine viral proteins (Fig. 3.3, Table 3.2) [5]. Each viral particle has two identical, positive-sense copies of the genomic RNA, each one encoding the full information needed for viral replication, although

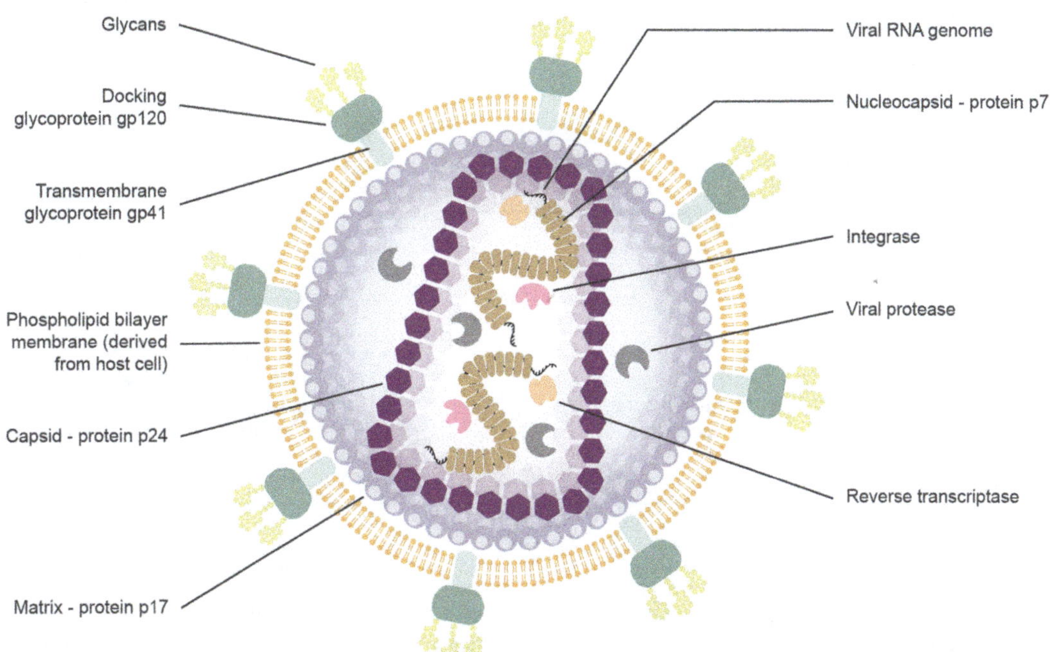

Fig. 3.2 Lentivirus morphology highlighting its main structural components. HIV-1 is a retrovirus with two identical copies of a single-stranded positive RNA molecule. Virions are coated by a phospholipid envelope derived from the host cell, with viral glycoproteins (gp41 and gp120) that are important for the viral entry in the cell. The viral genome is further protected by a capsid formed by p24 protein, and the virus has two additional structural proteins, p17 and p7. All these structural proteins are important for essential functions in the viral replicative cycle. Other components found in the virus include the reverse transcriptase, which mediates the converstion of the viral RNA genome into viral DNA; the integrase, which mediates the genomic integration of the proviral DNA into the host cell genome; and the viral protease that is responsible for processing of the gag and gag-pol polyproteins during viral maturation.

recent studies suggest that both copies are required for is successful completion [6]. The viral genome is edged by long terminal repeats (LTRs), comprising each one by three regions, U3, R and U5, which are essential for transcription, reverse transcription and integration.

3.1.1 Replicative Cycle

The HIV-1 replicative cycle (Fig. 3.4) starts with the binding of the envelope glycoprotein gp120 to the CD4 receptor (or co-receptor CXCR4 or CCR5) of the host cell [7]. After this attachment, the viral envelope fuses with the cellular membrane, resulting in the entry of the viral core into the cell. There, the viral capsid proteins are uncoated, releasing the RNA genome and viral enzymes (RT, IN, and PR). Still in the cytoplasm, the RNA molecule is reverse transcribed into a double-stranded proviral DNA molecule by the RT enzyme. This newly synthesized DNA molecule forms a complex with proteins (both viral and cellular) forming the pre-integration complex (PIC), which shuttles to the nucleus. The proviral DNA integrates into the cell genome (mediated by the IN enzyme) and starts to be transcribed into the viral mRNAs (by cellular RNA polymerase II) [8]. After some splicing events (Fig. 3.5) [9], the viral mRNAs are exported to the cell cytoplasm, where the viral proteins are produced. The viral proteins and two full-length RNA transcripts are assembled near the cellular membrane, and the viral particles (virions) are released by budding. Maturation of the viral particles occurs outside the cells.

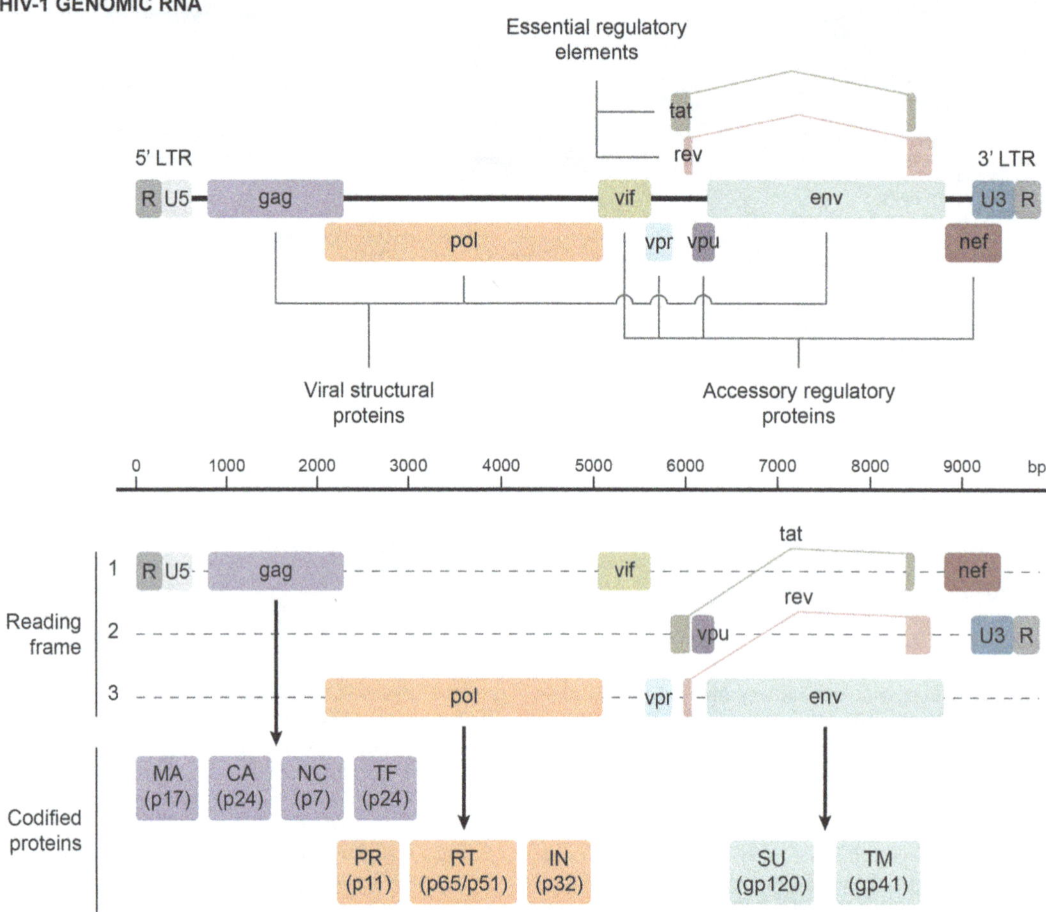

Fig. 3.3 Organization and main elements of a lentivirus genome, based on HIV-1. The RNA genome is about 9 kb long with three main coding segments, *gag* (group antigen), *pol* (polymerase), and *env* (envelope), flanked by two long terminal repeats (LTRs). Each LTR is divided into three regions, U3, R and U5, although the complete LTR sequence is only generated during reverse transcription and present in the proviral DNA. The LTRs are functional elements of the viral genome, being essential for viral transcription and therefore for the replicative cycle. The *gag* gene codes for structural proteins: matrix (MA), capsid (CA), nucleocapsid (NC) and transframe polypeptide (TF). The *pol* gene codes for the protease (PR), the reverse transcriptase (RT) and the integrase (IN). The third essential gene, *env*, codes for envelope proteins, the surface protein (SU) and the transmembrane protein (TM). These elements are common for all retroviruses; however, lentivirus genome is more complex and has additional elements. *Tat* and *rev* are essential regulators of the viral gene expression, coding for RNA-binding proteins that enhance transcription and act to induce late-phase gene expression, respectively. Additionally, the lentivirus genome includes four accessory regulatory elements, *vif*, *vpr*, *vpu* and *nef*, which are important for the replication and pathogenesis *in vivo*.

3.1.2 From Lentivirus to Lentiviral Vectors

Despite being based on HIV-1, lentiviral vectors have been used for the past years as an effective and safe system to deliver genes both in gene therapy-based preclinical studies and in clinical trials. Accounting for this use are several advantages, which are important features for gene therapy vectors, such as their mediated long-term expression or the possibility of transducing non-dividing cells. Further advantages and disadvantages of their use in gene therapy applications are summarized in Table 3.3.

Table 3.2 Coding segments of retroviruses and their functionality.

	Gene	Functionality	Protein	Function
Common to all retrovirus	*gag*	*Structural*	Capsid protein (CA)	Formation of the capsid
			Matrix protein (MA)	Formation of the inner membrane layer
			Nucleoprotein (NC)	Formation of the nucleoprotein/RNA complex
	pol	*Replication*	Protease (PR)	Proteolytic cleavage of gag precursor protein; release of structural proteins and viral enzymes
			Reverse transcriptase (RT)	Transcription of viral RNA into proviral DNA
			RNase H	Degradation of viral RNA in the viral RNA/DNA replication complex
			Integrase (IN)	Integration of proviral DNA into the host genome
	env	*Envelope*	Surface glycoprotein (SU)	Attachment of virus to the target cell
			Transmembrane protein (TM)	Anchorage of gp120, fusion of viral and cell membrane
Specific of complex lentivirus	*tat*	*Regulatory*	Transactivator protein	Activator of transcription of viral genes
	rev		RNA splicing regulator	Regulation of the export of non-spliced and partially spliced viral mRNA
	nef	*Accessory*	Negative regulating factor	Enhancement of infectivity of viral particles
	vif		Viral infectivity protein	Critical for infectious virus production *in vivo*
	vpr		Virus protein r	Component of virus particles, interaction with p6, facilitates virus infectivity
	vpu		Virus protein unique	Efficient virus particle release, control of CD4 degradation, intracellular trafficking modulation

Since the first use of lentiviral vectors in gene therapy, several modifications have been introduced in order to increase safety, without compromising efficiency [10]. Because lentiviruses have high rates of mutation and recombination, even the vector production could be dangerous, especially with the possibility of generating **replication-competent lentiviruses** (RCLs). One of the developed strategies to increase lentiviral vector safety production, trying to overcome the RCL issue, was to separate the essential lentivirus genes in different plasmids and delete some accessory genes. This strategy is the main difference between the four generations of lentiviral vectors that were developed in the last years (Fig. 3.6) [11], although many other features and alterations can be found inthe different generations (Table 3.4). Another important advance in lentiviral vectors' development for gene therapy applications was the pseudotyping engineering, in which non-lentiviral envelope proteins are used in the process of producing lentiviral vectors. For example, one of the most popular alterations is the replacement of the native Env glycoprotein with the viral VSV-G protein, from the vesicular stomatitis virus envelope glycoprotein G. This modification increased viral stability, which permitted improvements in production and an increase in the viral titer yielded. Moreover, this protein has a ubiquitous receptor, which allowed the targeting of a much broader set of cells.

First Generation

Replication-deficient lentiviral vectors of the first generation are produced using three different plasmids: (i) the packaging construct, containing all the essential genes (except the *env* gene) and all the regulatory and accessory proteins, (ii) an *env* plasmid encoding a viral envelope glycoprotein, and (iii) a transfer vector containing the desired transgene and all the essential *cis*-acting

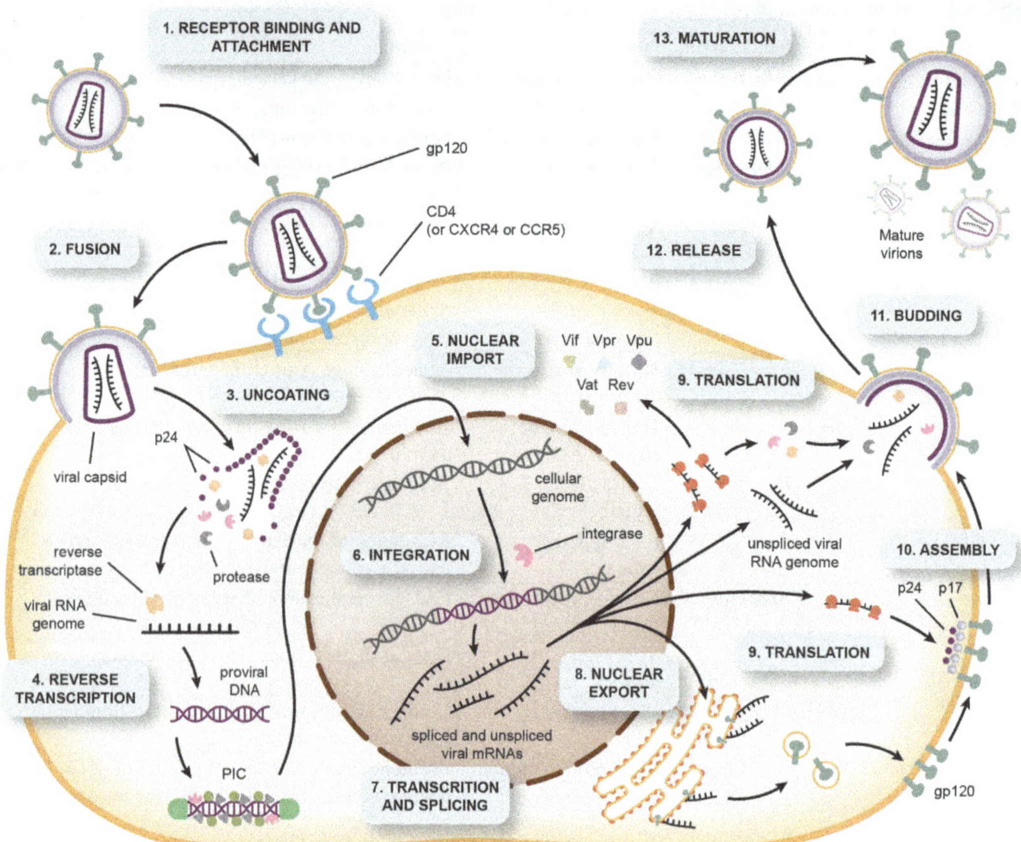

Fig. 3.4 Main phases of the replicative cycle of a lentivirus, based on HIV-1 cycle. Viral particle entry into the host cell is mediated by the gp120 glycoprotein of the envelope that binds to cell receptor CD4 (1). This leads to a conformational alteration in TM (gp41) that facilitates the fusion of the virus with the cellular membrane (2). The viral core enters the cell cytoplasm, where uncoating and release of the viral RNA takes place (3). The next step is the reverse transcription of the viral RNA into proviral DNA, mediated by the RT reverse transcriptase (4). After this step, the pre-integration complex (PIC) is assembled, being composed of the RT, IN, Vpr, MA and the proviral DNA. This complex is transported through the actin microfilaments and enters into the nucleus through the nuclear pore complex (5). Inside the nucleus, the proviral DNA is integrated into the cell genome, through the action of the IN integrase (6). After integration, the proviral DNA is normally transcribed by RNA polymerase II, and the transcription of viral genes is mediated by the U3 region of the 5'LTR (7). A single viral transcript is formed, some copies will eventually be packed into newly formed viral particles, but others will serve as template for the translation of viral proteins. Therefore, in the latter case single transcripts undergo splicing to generate the different viral mRNA segments, which are then exported from the nucleus (8). In the cytoplasm, the viral mRNAs are translated into different viral proteins, which are formed from precursor peptides that undergo cleavage by viral and cellular proteases (9). The assembly of the viral particles occurs in the cellular membrane and initiates with the interaction between precursor structural proteins and the packaging signal present in the unspliced viral RNA (10). The next stages are viral budding and release, wherein viral particles emerge from the cell surface, coated by portions of the cellular membrane containing the viral glycoproteins that were produced in the endoplasmic reticulum (11, 12). Finally, outside the cells the viral particles mature through cleavage of the viral polyproteins into individual proteins and reorganization of the structural proteins into the final viral arrangement (13).

elements (LTRs, Ψ, RRE). The first two plasmids do not have the packaging signal (Ψ) or any LTRs to prevent their inclusion in the produced vector particles.

Second Generation

Despite the important advance provided by the three-system plasmid of the first generation, some safety issues were still important as RCLs

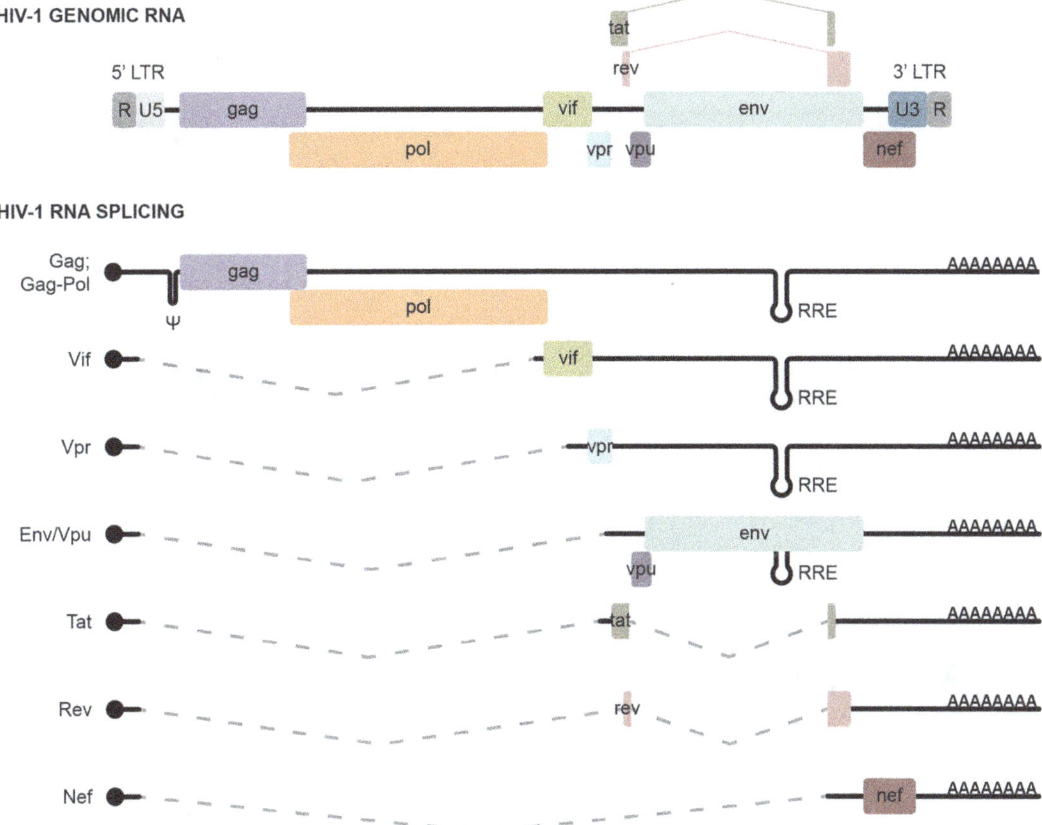

Fig. 3.5 **Lentivirus mRNA splicing transcripts**, based on the HIV-1 genome. The integrated proviral DNA is transcribed into a single mRNA transcript of approximately 9 kb size. This single transcript is then spliced into different transcripts, which can be grouped into three main clusters: (i) the primary mRNA that codes for the Gag-Pol polyprotein; (ii) intermediate mRNAs (~4 kb long) resulting from the splicing of the primary transcript, which code for polyprotein Env or accessory proteins Vif, Vpr and Vpu; and (iii) short transcripts (~2 kb long) resulting from additional splicing events and yielding mRNAs that code for the Tat, Rev and Nef proteins.

Table 3.3 Main advantages and disadvantages of lentiviral vectors for gene delivery.

Advantages	Disadvantages
Transduction of dividing and nondividing cells	Need for pseudotyping
Integration into the host genome	Possible generation of replication-competent lentivirus (RCLs)
Long-term expression	Insertional mutagenesis
Possibility for no integration	Safety problems due to the presence of viral regulatory proteins
Cloning capacity of around 8 kb	
Low immunogenicity	

could still be generated. To further increase safety, in the second generation of lentiviral vectors, the accessory genes (*vif*, *vpu*, *vpr*, *nef*) were deleted while maintaining the number of plasmids. This modification did not negatively affect vector titer or infectivity, but drastically reduced the chance of generating RCLs.

Third Generation

The next advance in lentiviral vectors brought the elimination of an essential gene (*tat*) and the introduction of the *rev* gene in a fourth plasmid. The need for Tat protein was overcome by replacing the U3 region in the 5'LTR by a strong heterologous promoter, such as the CMV (cyto-

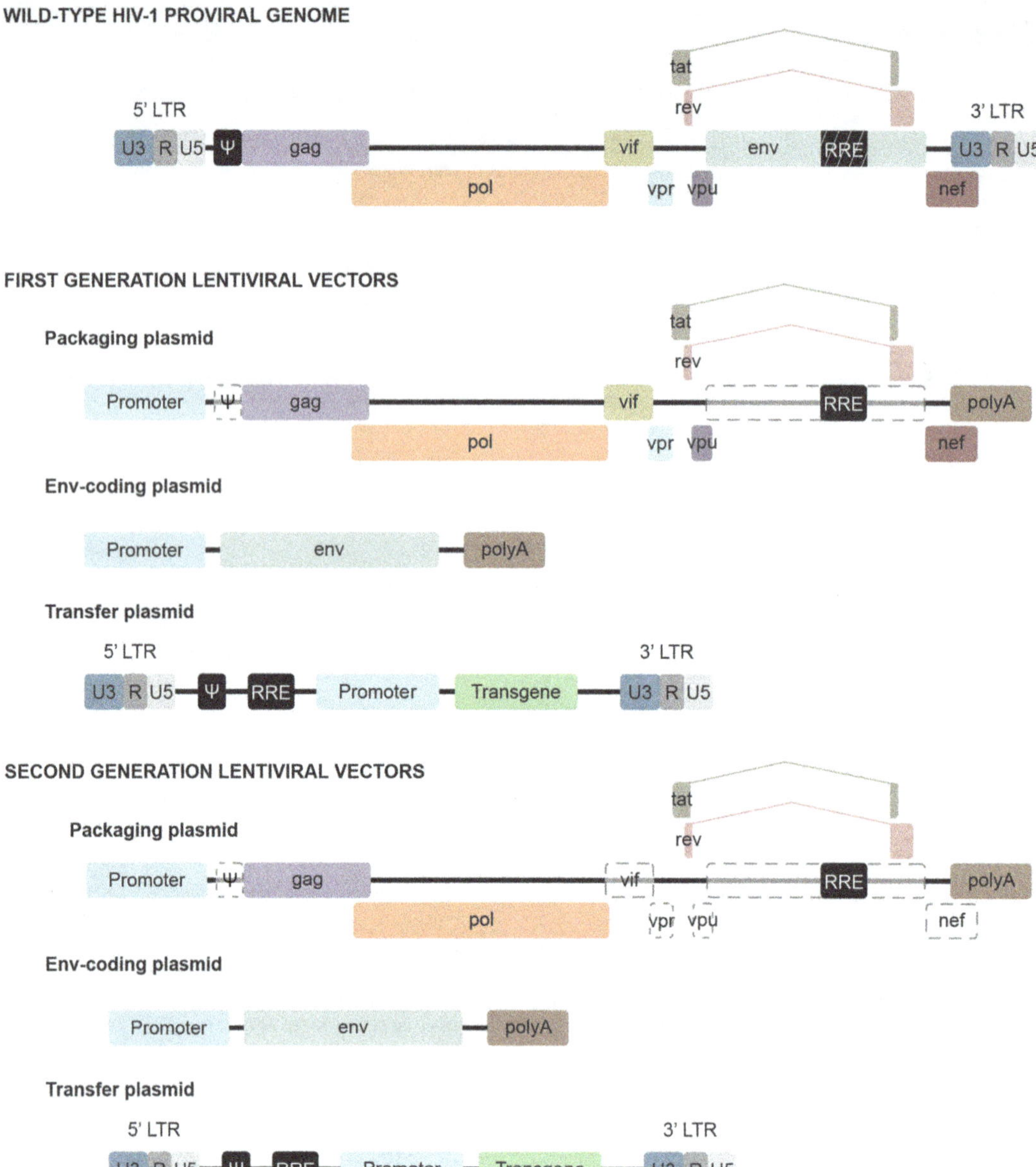

Fig. 3.6 Different lentiviral vector generations used in gene therapy and their modifications engineered from the wild-type HIV-1. The first generation of lentiviral vectors was produced using three plasmids encoding the viral proteins. In these vectors, only the transgene contained the essential *cis*-acting elements (LTRs, Ψ, RRE) that allowed their packaging in the produced lentiviral particles. In the second generation of lentiviral vectors, the number of plasmids was maintained, but in order to improve their safety the accessory genes were deleted. In the third generation, the *tat* gene was eliminated and the viral genome divided into an additional plasmid containing the *rev* gene. This also included the deletion of the U3 region of the 3'LTR, generating self-inactivating virus. Finally, a fourth generation of lentiviral vectors was developed to further reduce the risk of recombination between the plasmids used. For that, the segments *gag* and *pol* were codon optimized. In this fourth generation, all the modifications of the previous generations were maintained.

Table 3.4 Main features of the different generations of lentiviral vectors

	First generation	Second generation	Third generation	Fourth generation
Constructs	3	3	4	4
Packaging plasmids	1	1	2	2
Accessory genes	Yes	No	No	No
tat and *rev* genes	Same plasmid	Same plasmid	*tat* absent, *rev* in a separate plasmid	*tat* absent, *rev* in a separate plasmid
gag and *pol* genes	Same plasmid	Same plasmid	Same plasmid	Same plasmid
3'LTR deletion	No	No	Yes	Yes
Codon optimization	No	No	No	*gag* and *pol* genes

Table 3.5 Main advantages and disadvantages of third generation lentiviral vectors.

Advantages	Disadvantages
Able to transduce slowly dividing and nondividing cells	Production of high titers is more difficult
Delivered transgenes are more resistant to transcriptional silencing	Still some possibility of generating replication-competent lentiviruses
Suitable for several ubiquitous or tissue-specific promoters	
Safety increased by self-inactivation	

megalovirus) promoter. The addition of a new plasmid, separating the *rev* gene, further increased the safety of the viral vectors production by reducing the possibility of RCL generation. An additional improvement was also added to these lentiviral vectors with the deletion of part of the U3 region in the 3'LTR, generating SIN (self-inactivating) vectors, thus decreasing the risk of activating nearby genes in the integration process. These lentiviral vectors have only three of the nine wild-type HIV-1 genes, which significantly increases their safety profile. The four constructs used for their production are (i) a packaging plasmid with the *gag* and *pol* genes; (ii) a plasmid with *rev* gene, (iii) a plasmid with the *env* (or VSV-G or other pseudotyped protein), and (iv) a plasmid with the transgene (and a strong heterologous promoter).

Fourth Generation

The potential formation of RCLs using the third generation lentiviral vectors is very reduced, however, theoretically, it is still possible that homologous recombination between the transfer and the packaging plasmids takes place. To solve this problem, codon optimization was implemented in the *gag* and *pol* genes, thus eliminating the homology between plasmids. This fourth generation of lentiviral vectors displays improved biosafety; however, viral titers were negatively affected. Maybe because of this, it has not been extensively used, and the previous lentiviral vector generations are still more commonly applied, especially the third generation due to its important advantages (Table 3.5).

3.1.3 Additional Improvements to Lentiviral Vectors

Several other improvements in lentiviral vector design were introduced in the different generations to improve efficiency, to facilitate their production or, as already mentioned, to improve their biosafety (Fig. 3.7) [12]. One of those improvements was the introduction of a cis-acting polypurine tract (cPPT), to increase the viral vector transduction efficiency both *in vitro* and *in vivo*. Another strategy, improving lentiviral vector expression, was the introduction of the woodchuck hepatitis virus post-

	WILD-TYPE LENTIVIRUS	LENTIVIRAL VECTORS MODIFICATIONS
Viral genes	Nine genes, encoding structural proteins and regulatory elements	Elimination of non-essential elements and splitting of genes into several vectors
Tropism	Restricted	Altered by inclusion or modification of glicoproteins and addition of antibodies
Infection	Dividing and non-dividing cells	Introduction of cPPT, increasing transduction efficiency
Integration	Stable integration of proviral DNA into target cells	Development of non-integrative vectors, by mutation of the LTRs and the viral integrase
Gene expression	Ubiquitous	Improved by introduction of WPRE Directed to particular cells through specific promoters
Expression repression	Epigenetic modifications may cause repression of integrated DNA	Introduction of chromatin insulators
Viral replication	Virus replicates by hacking host cell machinery	Splitting of viral genes into several vectors and mutation of LTRs

↑Efficiency ↑Production ↑Biosafety

Fig. 3.7 **Modifications introduced to lentiviral vectors** relative to the wild-type HIV-1 virus, in order to improve their safety profile while maintaining their efficiency and production in high titers.

transcriptional regulatory element (WPRE), which increases the number of unspliced RNA molecules, leading to an increase of the transgene expression in target cells. One important issue resulting from lentiviral vector-mediated integration is the possibility of transgene expression repression, due, for example, to epigenetic events. To address this problem, chromatin-insulator sequences were introduced in the lentiviral vectors. Insulators have the ability to block enhancer-promoter interactions and/or serve as barriers against silencing effects. The further engineering of lentiviral vectors also included the fusion of proteins or antibodies to envelope glycoproteins to alter their tropism and retarget the vectors to specific cells. Another commonly used strategy to direct and specify transgene expression is to include a tissue-specific promoter in the transgene construct. This strategy will not select the cells to transduce, but rather limit the expression of the transgene to a certain cell type due to the presence of the specific promoter. Limiting expression of the transgene to particular cells is particularly useful in the context of the central nervous system, where the presence of different cells types makes specific transduction difficult.

SIN Vector Design

Another feature included in most of the lentiviral vectors is the inclusion of a self-inactivating long terminal repeat (SIN-LTR), in which the U3

WILD-TYPE HIV-1

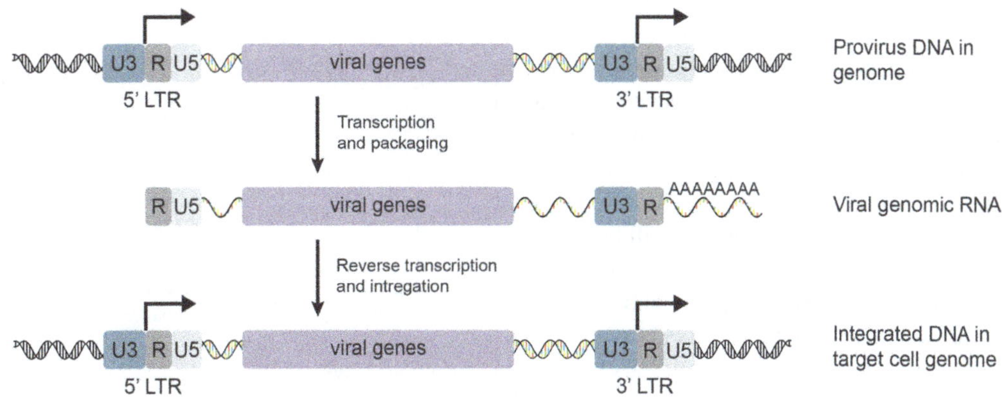

SIN LENTIVIRAL VECTOR

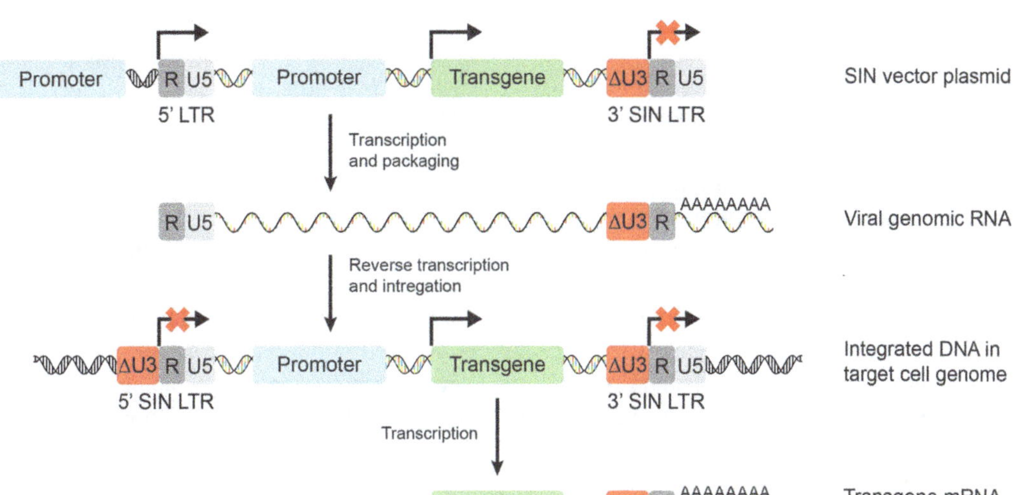

Fig. 3.8 Self-inactivating lentiviral vectors. In wild-type HIV-1 virus, the viral RNA is transcribed under the control of an enhancer-promoter sequence that is located in the U3 region of the LTRs, being required for the production of new viral particles and the continuation of the replicative cycle. During the process of reverse transcription the U3 region of the 3'LTR is transposed to the 5'LTR. In self-inactivating vectors, the U3 region of the 3'LTR is partially deleted in order to block its enhancer-promoter function. Upon integration, the proviral DNA will harbor the partially deleted U3 region in both LTRs, thus preventing the further continuation of the replicative cycle.

region in the 3'LTR has been partially deleted (Fig. 3.8) [13]. This feature reduces the risk of lentiviral vector recombination with wild-type viruses, thus increasing its safety. In the wild-type HIV-1, the U3 region acts as a viral enhancer/promoter, being essential for viral replication, especially for the formation of both the 5' and 3'LTR of new viral particles. However, in the recombinant lentiviral vectors (replicative-deficient), this region is dispensable, as the packaged RNA expression is driven by the 5'LTR region and the transgene expression by the respective promoter. The partially deleted U3 region will be transfered to both the 5' and 3'LTR regions during integration, so that any viral particle progeny that is formed will harbor two inactivated LTRs and transgene expression will be only driven by the internal promoter.

Non-integrative Lentiviral Vectors

In the context of gene therapy, one of the main advantages of using lentiviral vectors is their ability to stably integrate the desired transgene in the target cells genome. However, integration can lead to the development of undesired consequences, such as the activation of oncogenes. In fact, for some applications the development of **non-integrative vectors** would still be effective, having as an additional benefit a better safety profile [14]. To attain this goal, lentiviral vectors were designed containing mutations in regions that are essential for the proviral DNA integration process, namely in, (i) the U3 region of the 5'LTR, (ii) the U5 region of the 3'LTR, and (iii) the integrase protein itself.

Lentiviral Vector Pseudotyping

As mentioned, the inclusion of different glycoproteins into the envelope of lentiviral vectors alters and increases the variety of target cells that can be transduced, thus improving their tropism. Like it was referred, **pseudotyping** is the process of producing viral vectors (or virus) containing envelope proteins from a different virus [15]. Several glycoproteins from different virus have been used and studied to pseudotype lentiviral vectors providing different advantages, such as increasing their efficiency or reducing their toxicity (Table 3.6).

Table 3.6 Common viral glycoproteins used for lentiviral vector *pseudotyping*.

Species/envelope	Advantages
Vesicular stomatitis virus (VSV-G)	Wide tropism
	Facilitates production using ultracentrifugation
Feline endogenous retrovirus (RD114)	More efficient and less toxic than VSV-G in hematopoietic cells
Ebola	Efficiently transduces airway epithelium
Rabies	Retrograde transport in neuronal axons
Lymphocytic choriomeningitis virus (LCMV)	Low toxicity
Ross River virus	Transduces hepatocytes, glial cells, and neurons
Influenza virus hemagglutinin	Transduces airway epithelium
Moloney murine leukemia virus 4070 envelope	Able to transduce most cells

3.1.4 Lentiviral Vector Production

Any delivery system used for gene transfer, either non-viral or viral, has to be produced/built preferentially with a safe and simple method of low technical complexity that, of course, yields the system in high amounts. In the case of lentiviral vectors, the safety issue was a major driving force in the development of production methods and the main reason why the viral genome was separated into different constructs. However, this important advance brought a new challenge, as all the constructs must join together in the factory cells to produce effective lentiviral vector particles. To accomplish this, two transfection methods are used: **transient** transfection [16] and **stable** transfection [11]. In the first approach, all the plasmids (three or four, depending on the generation used) are simultaneously transiently transfected into a packaging cell line (Fig. 3.9). Several transfection agents cab be used, like calcium phosphate, polyethylenimine (PEI) or cationic lipids. The second approach involves the use of stable packaging cell lines already containing one or more viral genes, thus reducing the number of plasmids to be transfected [4]. Both approaches have advantages and drawbacks, and their choice is mainly dependent on the laboratory conditions, on the safety regulations and on the viral titer needed.

Other important issues to be considered in the production of lentiviral vectors include the method of viral particle purification and concentration, the procedure for titer assessment, the choice of producer cell line and the production scale (Fig. 3.10). For research purposes, the production of lentiviral vectors is well-established in human embryonic cells kidney 293 (HEK293) or HEK293T adherent cells using transient transfection methods. However, for clinical trials, the quantity and quality of the vectors needed are

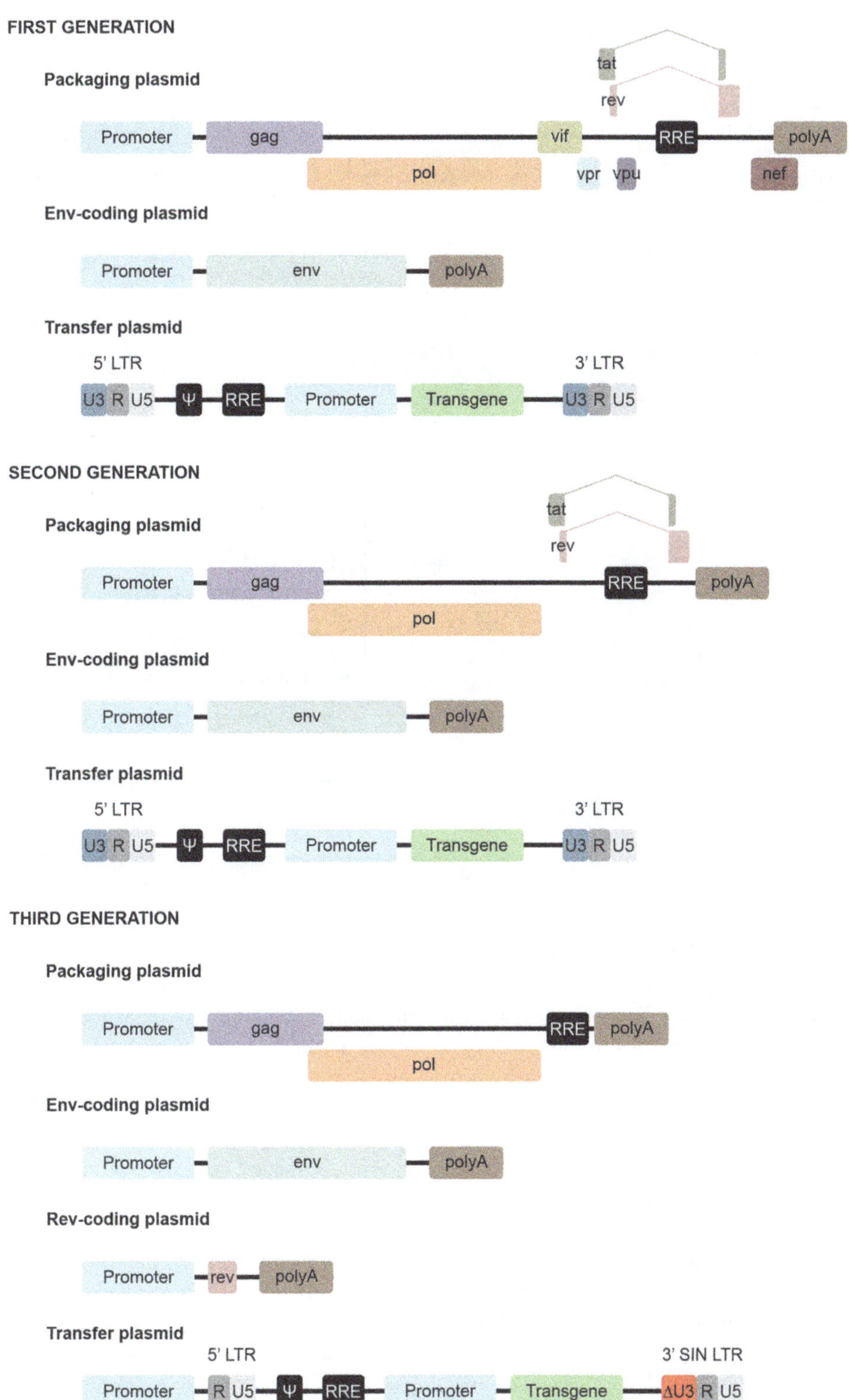

Fig. 3.9 Plasmids used to produce lentiviral vectors in different vector generations. Depending on the generation, lentiviral vectors can be produced by transfecting three or four plasmids into producing cells. Alternatively, a stable cell line expressing one or two viral genes could be used in the production, thus reducing the number of plasmids to be transfected.

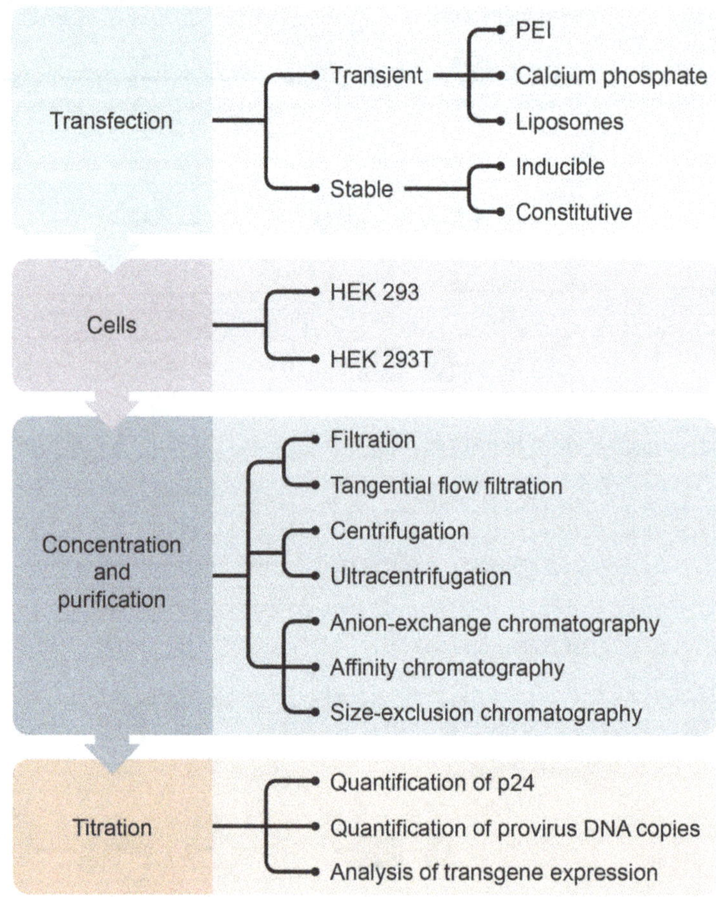

Fig. 3.10 Overview of the main steps of lentiviral vector production. Several methods can be used to transfect viral plasmids, including delivery mediated by liposomes. The produced viral particles can be concentrated and purified from the producing cells using different techniques, among which ultracentrifugation is the most common. Finally, several methods to quantify the lentiviral vectors titer can be used, although the quantification of p24 is the most common.

much higher. Thus, large-scale methods were developed using both adherent and suspension cell cultures. The process is quite expensive and complex, and only a few sites in Europe and the USA are able to produce GMP lentiviral vector particles in large scale [11].

Concentration and purification methods are both very important in the production of lentiviral vectors and are closely related to the production scale needed (research or clinical trial). Even though, probably the method most commonly used for concentration is the centrifugation of culture medium after an initial filtration. This method can also be combined with anion-exchange chromatography for an additional purification step. However, these methods are not suitable for large-scale productions and *in vivo* application, as often they are not free from culture medium/process contaminants. Therefore, in large-scale productions, several methods are combined to achieve three main goals: (i) an initial purification of the vectors; (ii) an intermediate purification to remove specific impurities, and (iii) final refining to remove trace contaminants.

Finally, the last step in the production of lentiviral vectors, which is the assessment of the viral production titer, is also very important, in order to adjust vector doses and to evaluate the transduction efficiency. There is a wide range of methods used for this purpose, for example (i) measuring the quantity of one vector component (commonly the viral p24 protein is used); (ii) measuring the number of provirus DNA copies in infected cells; or (iii) measuring transgene or reporter gene expression in infected cells.

Several modifications and improvements are continuously being tested aiming to increase lentiviral vector quality and quantity and also to increase safety in the production process.

3.1.5 Lentiviral Vectors in Clinical Trials

There is already one gene therapy product approved in Europe and the USA based on the delivery by lentiviral vectors. Kymriah® is an *ex vivo* gene therapy delivering CD19-specific CAR T-cells, indicated for the treatment of acute lymphoblastic leukemia. Due to their important advantages, lentiviral vectors were and are used with success in several clinical trials, particularly for rare diseases. More recently, they were also applied in gene therapy studies for more frequent genetic and acquired diseases, including β-thalassemia, Parkinson's disease and cancer. Tables 3.7 and 3.8 detail two examples of gene therapy clinical trials using lentiviral vectors [17, 18].

3.2 Gamma Retrovirus

Gamma retrovirus also belongs to the *Retroviridae* family (retroviruses), like the lentivirus. However, their genome organization is simpler [19], which makes its genetic engineering easier than for lentivirus (Fig. 3.11). As vectors for gene therapy, they share more or less the same advantages and disadvantages of the lentivirus. However, contrary to those, the gamma retrovirus only infects dividing cells, and maybe for that, currently, they are less used in gene therapy than lentivirus [20]. Moreover, lentiviral vectors seem to be safer than gamma retrovirus in relation to the risk of insertional mutagenesis. The gamma retrovirus was mainly used in the earlier gene therapy studies, until the previously mentioned incident with a X-linked severe combined immunodeficiency (SCID) trial (Table 3.9) [21] where insertional oncogenesis was observed in some of the trial subjects.

Table 3.7 Example 1 of a gene therapy clinical trial using lentiviral vectors as the delivery system of the therapeutic gene.

Study	Palfi, S., *et al.* (2014) Long-term safety and tolerability of ProSavin, a lentiviral vector-based gene therapy for Parkinson's disease: a dose escalation, open-label, phase 1/2 trial, *Lancet* 383, 1138–1146
Disease	Parkinson's disease, which is a common neurodegenerative disease mainly characterized by motor impairments, resulting from the progressive degeneration of dopaminergic neurons in the substantia nigra that project axons to the striatum, where dopamine is released.
Therapeutic gene	ProSavin (genes encoding key enzymes in the dopamine metabolism, tyrosine hydroxylase, AADC, and cyclohydrolase 1) in three doses (low 1.9×10^7 TU; mid 4.0×10^7; and high 1×10^8)
Delivery vector	Lentiviral vector (based on equine infectious anemia virus) produced by triple transient transfection methods in HEK293T cells, purified and concentrated by anion-exchange chromatography and hollow fiber ultrafiltration
Clinical trial	Phase I/II (long-term safety and tolerability), open-label and 12-month follow-up study, in France and the UK
Inclusion criteria	48–65 years; disease manifestation for 5 or more years and 50% or higher motor response to oral dopamine treatment
Type of administration	Local delivery of ProSavin bilaterally into the striatum
Clinical outcome	Despite some adverse events, the study found that ProSavin was safe and well tolerated. No detectable antibody response against any of the ProSavin transgene products was detected. There was an improvement in motor behavior in all the treated patients, and 11 of the patients had a reduction in the daily levodopa administration at 6 and 12 months after the therapy.

Table 3.8 Example 2 of a gene therapy clinical trial using lentiviral vectors as the delivery system of the therapeutic gene.

Study	Cartier, N., *et al.* (2009) Hematopoietic stem cell gene therapy with a lentiviral vector in X-linked adrenoleukodystrophy, *Science* 326, 818–823
Disease	X-linked adrenoleukodystrophy (ALD), which is a fatal demyelinating disease of the CNS caused by mutations in the ABCD1 gene that codifies for ALD protein, an adenosine triphosphate-binding cassette transporter localized in the peroxisomes membrane. Most of the affected boys die before reaching adolescence.
Therapeutic gene	Wild-type ABCD1 gene under the control of the MND (myeloproliferative sarcoma virus enhancer, negative control region deleted, dl587rev primer binding site replaced) promoter. Re-infusion of 4.6×10^6 and 7.2×10^6 cells per kilogram, for patient 1 and patient 2, respectively
Delivery vector	Replication-defective HIV-1-derived self-inactivating lentiviral vector (CG1711 hALD), in a concentration of 941×10^7 infectious particles/ml
Clinical trial	Interventional study in France
Inclusion criteria	Two ALD patients (aged 7.5 and 7 years) with progressive demyelination and no matched donors
Type of administration	*Ex vivo* lentiviral-mediated transfer of the ABCD1 gene into CD34+ cells from ALD patients
Clinical outcome	No adverse events were reported. Hematopoietic recovery occurred at days 13–15 after transplant and was sustained thereafter. Nearly complete immunological recovery occurred between 9 and 12 months. The progression in cerebral demyelination was stopped at 12–14 months after gene therapy for patient 1 and at 16 months for patient 2. A follow-up at 36 months revealed that no changes in the extent of cerebral demyelinating lesions were observed.

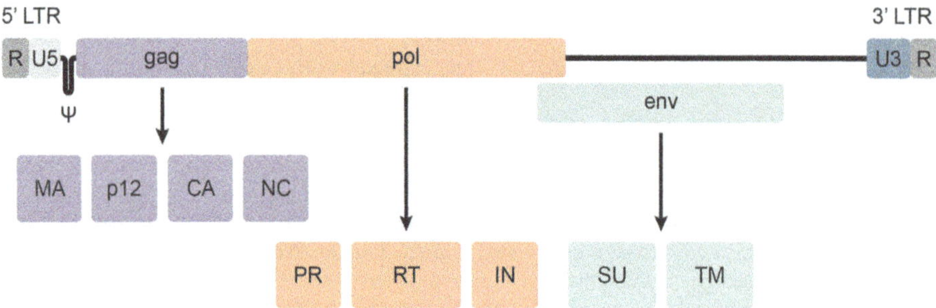

Fig. 3.11 **Organization of a gamma retrovirus genome**, based on the murine leukemia virus (MLV). The RNA molecule is about 8.3 kb long with the three retroviral main coding segments *gag* (group antigen), *pol* (polymerase), and *env* (envelope), flanked by two long terminal repeats (LTRs). Contrary to lentiviruses, the gamma retrovirus genome does not have additional coding segments. The *gag* gene codes for structural proteins: matrix (MA), capsid (CA), nucleocapsid (NC), and p12. The *pol* gene codes for the protease (PR), the reverse transcriptase (RT) and the integrase (IN). The third essential gene, *env*, codes for envelope proteins, the surface protein (SU) and the transmembrane protein (TM).

3.3 Adenoviral Vectors

Adenovirus belongs to the *Adenoviridae* family, comprising more than 100 viruses able to infect humans and several animal species. They are non-enveloped viruses, with a double-stranded linear DNA molecule as genetic material. The viral capsid has icosahedral symmetry (20 facets) a diameter of 70–100 nm and each icosahedron is composed of 240 proteins, named hexons (Fig. 3.12) [22].

Table 3.9 Example of a gene therapy clinical trial using gamma retroviral vectors as the delivery system of the therapeutic gene.

Study	Cavazzana-Calvo, M., *et al.* (2000) Gene therapy of human severe combined immunodeficiency (SCID)-X1 disease, *Science* 288, 669–672
Disease	Severe combined immunodeficiency-X1 (SCID-X1) is an autosomal recessive disease caused by mutations in the IL2RG (interleukin 2 receptor subunit gamma) gene, which encodes for a critical protein of the immune system.
Therapeutic gene	Wild-type IL2RG cDNA introduced in CD34+ autologous cells
Delivery vector	Moloney retrovirus-derived defective vector
Clinical trial	Interventional study, with a follow-up of 10 months in France
Inclusion criteria	Two patients with 8 and 11 months
Type of administration	*Ex vivo* gene therapy mediated by gamma retrovirus
Clinical outcome	After 10 months, transgene expression was detected in T- and NK-cells in both patients. This was accompanied by clinical improvement in both patients, leading to normal development without any further treatment.

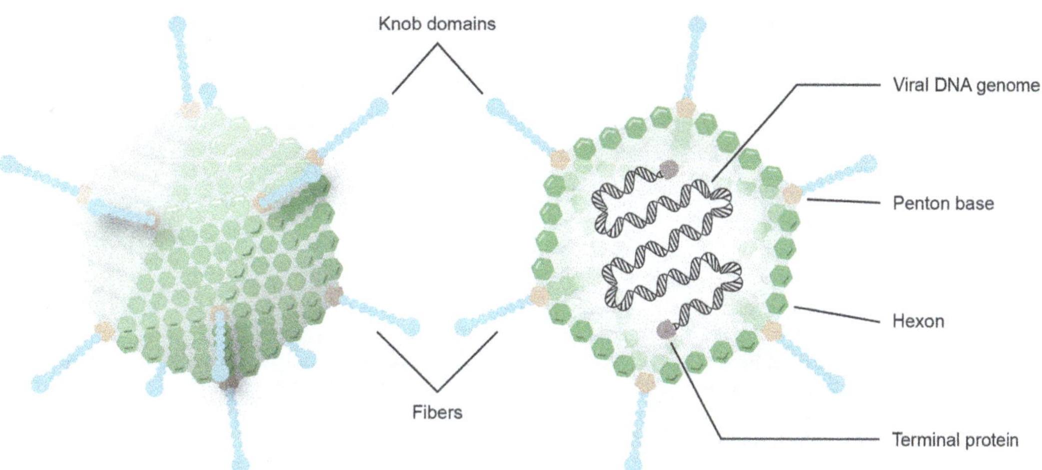

Fig. 3.12 Adenovirus morphology highlighting the icosahedral symmetry of the capsid. The genome of adenovirus consists in a double-stranded linear DNA molecule, with a terminal protein in each side of the helix, which act as primers for viral replication. They are non-enveloped virus, with a diameter of 70–100 nm, and each icosahedral capsid is composed of 240 proteins, named hexons. Each of the 12 vertexes is formed by another protein, named penton, constituting a base for a projecting fiber, which is important for the viral attachment to the host cell. The extremity of the fiber terminates with a globular domain, named knob.

The adenovirus genome size is around 34–43 kb, flanked by two inverted terminal repeats (ITRs), which act as the DNA replication origin. The genome encodes for approximately 35 proteins, generally expressed in 2 phases from 7 transcription units, which are mostly transcribed by RNA polymerase II (Fig. 3.13, Table 3.10): (i) four early transcriptional units (E1, E2, E3, and E4), which are important for the early infection stage ensuring the control of the host cell and DNA replication, (ii) two delayed early transcriptional units (IX and IVa2) that are important after replication, and (iii) one late transcriptional unit (MLTU, subdivided into L1, L2, L3, L4, and L5), mainly responsible for viral assembly, and release of the viral particles from the host cells [23].

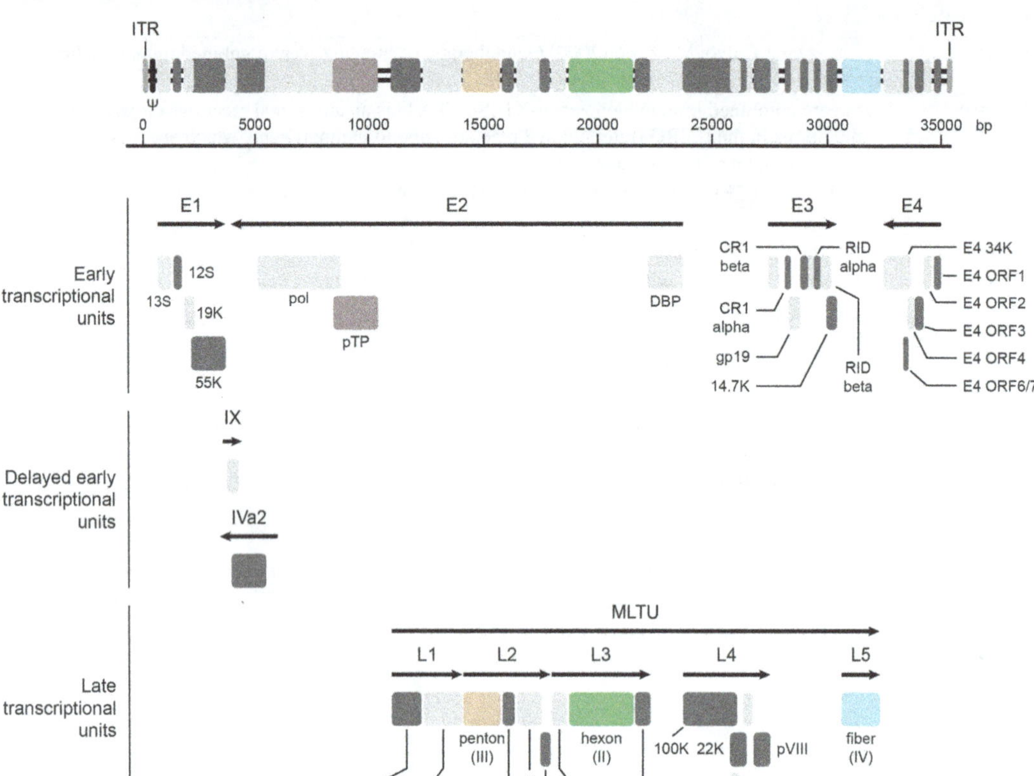

Fig. 3.13 Organization of the adenovirus genome, based on serotype 5. The double-stranded linear DNA molecule has 36 kb long, flanked by an inverted terminal repeat (ITR) in each extremity, and coding for approximately 35 proteins. The different coding segments are grouped, according to their transcription, in early transcriptional units (E1, E2, E3, and E4), which are activated upon infection; delayed early transcriptional units (IX and IVa2), encoding for capsid proteins; and a late transcriptional unit (MLTU), which encodes for structural proteins and is also important for virus assembly and release from the host cell.

3.3.1 Replicative Cycle

The adenovirus replicative cycle comprises different stages (Fig. 3.14):

1. Adsorption, which initiates viral entry into the host cell and is mediated by the surface receptor CAR (coxsackievirus and adenovirus receptor).
2. Internalization of the virus, via clathrin-dependent endocytosis.
3. Partial uncoating and escape from the endosome; the remaining capsid structures carry the genome to the nucleus.
4. Targeting of the nucleocapsid to the nucleus of the host cell.
5. Viral gene expression, whereby viral mRNAs are transcribed and viral proteins are synthesized.
6. Viral genome replication.
7. Virion assembly/maturation, through which the viral genome and proteins are assembled and packaged in the cytoplasm.
8. Release of the new infectious virus; the newly formed virions accumulate in the cell and are released upon cell lysis [22].

Table 3.10 Adenovirus transcription units and their main functions.

Transcription unit	Function
E1	Activates transcription of cellular and viral genes; induces S phase of the cell; inhibits cell apoptosis
E2	Codes for proteins essential to viral DNA replication
E3	Codes for proteins blocking the natural cell response to infection
E4	Codes for proteins involved in DNA replication, mRNA transport, and splicing
IX	Activates late gene expression
IVa2	Important for the activation of late promoter
MLTU	Codes for structural capsid proteins; promotes viral particles assembly and genome packaging

3.3.2 From Adenovirus to Adenoviral Vectors

So far, adenoviral vectors have been the most commonly used viral delivery system in gene therapy and were actually the vector chosen in the first gene therapy clinical trial. Some of their important advantages have contributed to this situation, including an easy production and high expression levels of the transgene (Table 3.11). On the other hand, one important disadvantage of their application is the pre-existing immunity (which could limit the therapeutic efficacy), as 80% of the human adult population has been naturally exposed to adenovirus serotypes 2 and 5, which are the most common serotypes in which adenoviral vectors production is based.

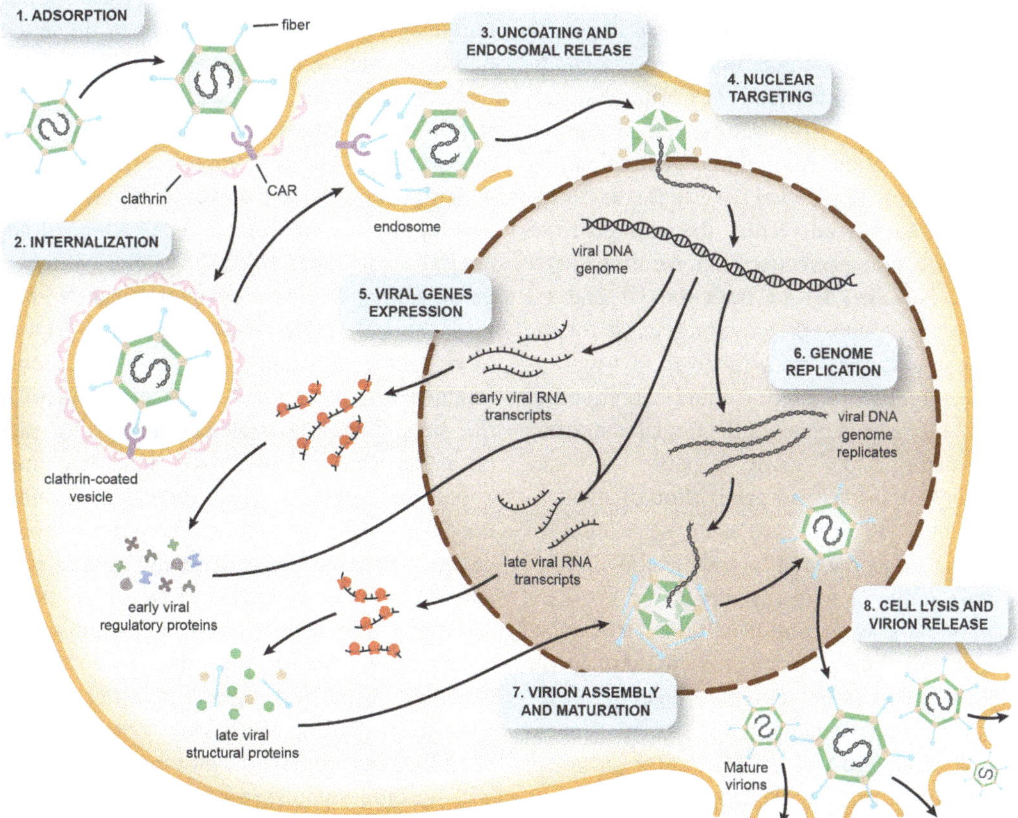

Fig. 3.14 Adenovirus replicative cycle. The cycle begins with the attachment of the virus to the host cell receptors, mediated by the fiber (1), and after that the virus is internalized in clathrin-coated vesicles (2). The next steps are the escape from the endosome, the partial uncoating of the viral capsid (3), and the targeting of the nucleocapsid to the nucleus (4). Next, the viral genes are transcribed (5) and the viral DNA is replicated (6). Viral assembly starts in the cell nucleus (7) and culminates with cell lysis and the release of the viral particles (8).

Table 3.11 Main advantages and disadvantages of adenoviral vectors for gene delivery.

Advantages	Disadvantages
Transduction of dividing and nondividing cells	Transient expression of the transgene
Cloning capacity of up to 8 kb	High levels of pre-existing immunity
No integration	Stimulation of a strong immune and inflammatory response
High expression levels of the transgene	
Production at high titers	
Suitable as oncolytic vector	
Broad host range	
Efficient transduction	

As was the case for lentiviral vectors, the development of adenoviral vectors was improved over time, with the introduction of several modifications leading to the existence of three different classes of adenoviral vectors: first generation, second generation and third generation, or *gutless*, vectors (Fig. 3.15) [24]. In the **first generation** of adenoviral vectors, the main goal was to produce non-replicative and therefore safer vectors. For that, E1 or both the E1 and E3 regions were removed, providing space for the transgene cassette. This resulted in non-replicative vectors, but the presence and expression of the remaining viral genes still stimulated a strong inflammatory and immune response. To avoid this problem, the **second generation** of adenoviral vectors was envisioned with the deletion of additional genome regions and allowing an increase in the cloning capacity, of up to 14 kb. However, these second-generation adenoviral vectors still induced some toxicity due to the immunogenic and inflammatory potential of the remaining viral genes. Finally, a **third generation** of adenoviral vectors, also called *gutless* or *helper-dependent*, was developed, being characterized by the complete deletion of the viral genome, with exception of the two ITRs and the packaging signal. *Gutless* adenoviral vectors have some advantages, even when compared to first-generation adenoviruses, which make them even more suitable as a gene therapy vector: (1) they are devoid of viral genes, which increases their safety profile; (2) they have an increased loading capacity of up to 30 kb, which allows carrying of larger genes; and (3) they have an easy production in high titers, which facilitates the translation to human trials and applications. However, as they lack all viral genes, during their production the proteins needed for genome replication, packaging and capsid formation have to be provided by a helper vector [25].

3.3.3 Adenoviral Vector Modifications

The death of Jesse Gelsinger in 1999 during a gene therapy trial revealed one of the critical problems of adenoviruses as vectors, their immunogenicity, which results from the activation of the innate immune response, the cellular immune response and the humoral immune response [26]. To overcome this situation, several modifications were introduced, many of which actually having served as a basis for the development of the different generations of adenoviral vectors described. The wide variety of modifications that were developed had two main goals: (i) overcoming the inflammation response to viral replication and (ii) overcoming the innate immune response, which is dose-dependent. In the same line, the strategies addressing these problems can be organized into two main approaches: genetic modifications and chemical/nonchemical modifications. The genetic modifications are based on the depletion of viral genes and led to the development of the different generations of adenoviral vectors. The chemical and nonchemical modifications are based on the shielding of the viral capsid with different materials (e.g., PEI, polyethylene glycol; HPAM, N-[2-hydroxypropyl]), to prevent its interaction with specific immune receptors. Other modifications were also introduced in adenoviral vectors in order to alter the vector tropism, including (i) capsid pseudotyping, (ii) serotype switching or (iii) binding with antibodies.

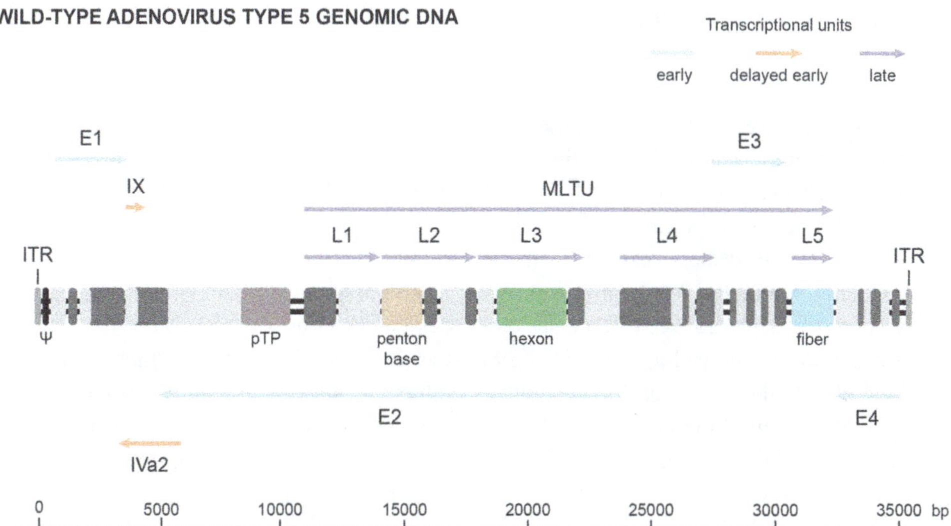

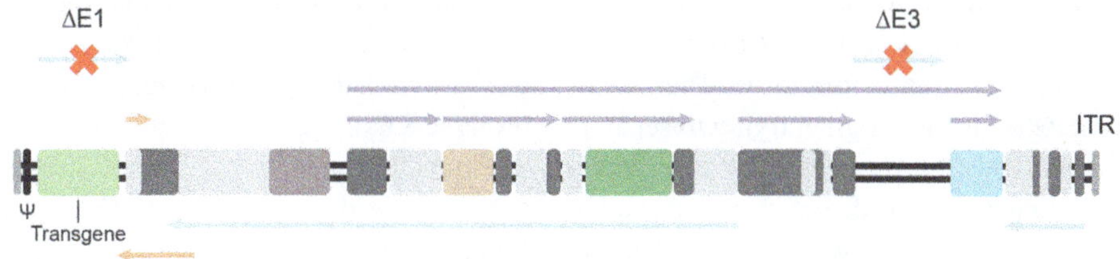

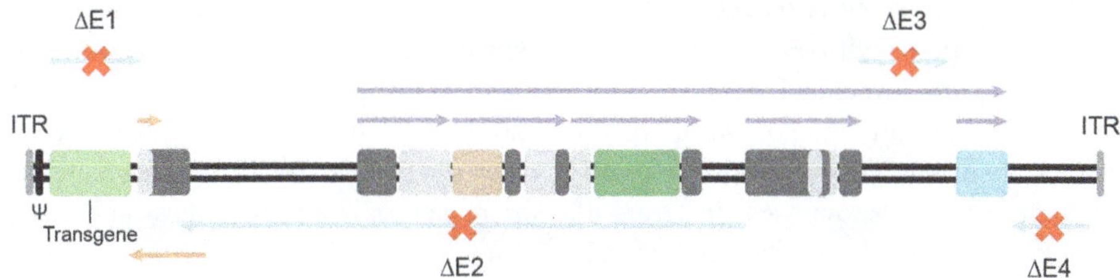

Fig. 3.15 **Adenoviral vector generations** developed for gene therapy applications. The first generation of adenoviral vectors was produced by deleting E1 or both E1 and E3 units, aiming to develop non-replicative vectors and providing cloning capacity for the transgene. The E1-deleted vectors have a cloning capacity of around 5 kb, while the E1- and E3-deleted vectors have a cloning capacity of up to 8 kb. While these vectors fail to replicate *in vitro*, the presence of several other viral genes stimulates a strong immune and inflammatory response. Therefore, a second generation of adenoviral vectors was developed, further deleting E2 and E4 regions, providing a cloning capacity of around 14 kb. However, these second-generation adenoviral vectors still induced some toxicity, also due to the presence of viral genes. So, a third generation of adenoviral vectors was developed based on the complete deletion of the viral genome, except for the ITRs and the packaging signal. These were named adenoviral *gutless* vectors, having a cloning capacity of up to 30 kb and a safer profile, although their production is more complex, as all the viral components are absent.

3.3.4 Adenoviral Vector Production

As mentioned above, the different generations of adenoviral vectors were developed aiming to reduce vector toxicity, mainly by deleting viral genes. However, these genes are crucial for viral replication and thus, in a viral production setting, the producer cells must be provided with the deleted elements. Normally, for the first and second generations of adenoviral vectors, the packaging cell lines used include HeLa, A549, HEK293 and PER.C6 [27]. In the case of *gutless* adenoviral vectors, as they lack all of the viral genes, there is the need to supply all the viral components for replication. This is achieved by the coinfection of the producer cell line with the *gutless* vector and a helper adenovirus (Fig. 3.16) [25].

One important disadvantage in the production of these vectors is the production of a mixed population of viral particles, derived from both the *gutless* vector and the helper virus. Thus, additional strategies were designed trying to separate the viral particles; however, residual contamination of the helper virus is detected in almost all of the production batches, which could lead to potential toxicity events. For example, the use of Cre or other recombinases to excise the packaging signal of the helper virus reduced the contamination by these viruses to 0.1–10%, compared to the *gutless* vectors [28].

Another important limitation of *gutless* vector production is their lower titers compared to previous generations of adenoviruses. In terms of human applications, besides the helper contamination, large-scale production is also very complex, less efficient, and also less safe, as several successive coinfections with the helper virus are needed.

3.3.5 Adenoviral Vectors in Clinical Trials

Since the first gene therapy clinical trial, adenoviral vectors were in the first line as a choice vector for delivering genes. Despite several drawbacks regarding their use, their easy production, versatility and suitability as oncolytic vectors ensure the first place among viral vectors that are most commonly used in gene therapy clinical trials (Table 3.12) [29, 30].

3.4 Adeno-associated Virus (AAV)

Adeno-associated viruses (AAVs) are helper-dependent viruses of the *Dependovirus* genus of the *Parvoviridae* family, which includes virus infecting several mammal species, including humans. AAVs are non-enveloped small virus (18–25 nm), having a capsid with icosahedral symmetry, composed of 60 proteins (Fig. 3.17) [31]. The AAV genome is composed of a single-stranded DNA molecule of around 4.7 kb, comprising two open reading frames (ORFs), *rep* and *cap*, flanked by inverted terminal repeats (ITRs) that form hairpin structures important for replication and packaging events (Fig. 3.18). Three viral promoters have been identified: p5, p19, and p40. The *rep* ORF encodes for four nonstructural proteins involved in replication (Rep78, Rep68, Rep52 and Rep10), whereas the *cap* ORF encodes for three structural proteins of the capsid (VP1, VP2, and VP3) [31, 32].

Over the last years, 12 different AAV **serotypes** have been isolated and identified, and more than 100 variants have been isolated from humans and other primates. The AAV serotype is defined by capsid protein motifs that are identified by different neutralizing antibodies and that provide differences in viral features, such as the type of cells/tissues the virus infect or the main receptors they use to enter the cells (Table 3.13) [33]. Genome organization is conserved across different AAV serotypes, although there are differences in their transcription profiles [31].

3.4.1 Replicative Cycle

The AAV replication cycle starts with the binding of the virus to a specific cell receptors and its internalization through receptor-mediated endocytosis. After cell entry, the virus needs to escape from the endosome, enter the nucleus through the

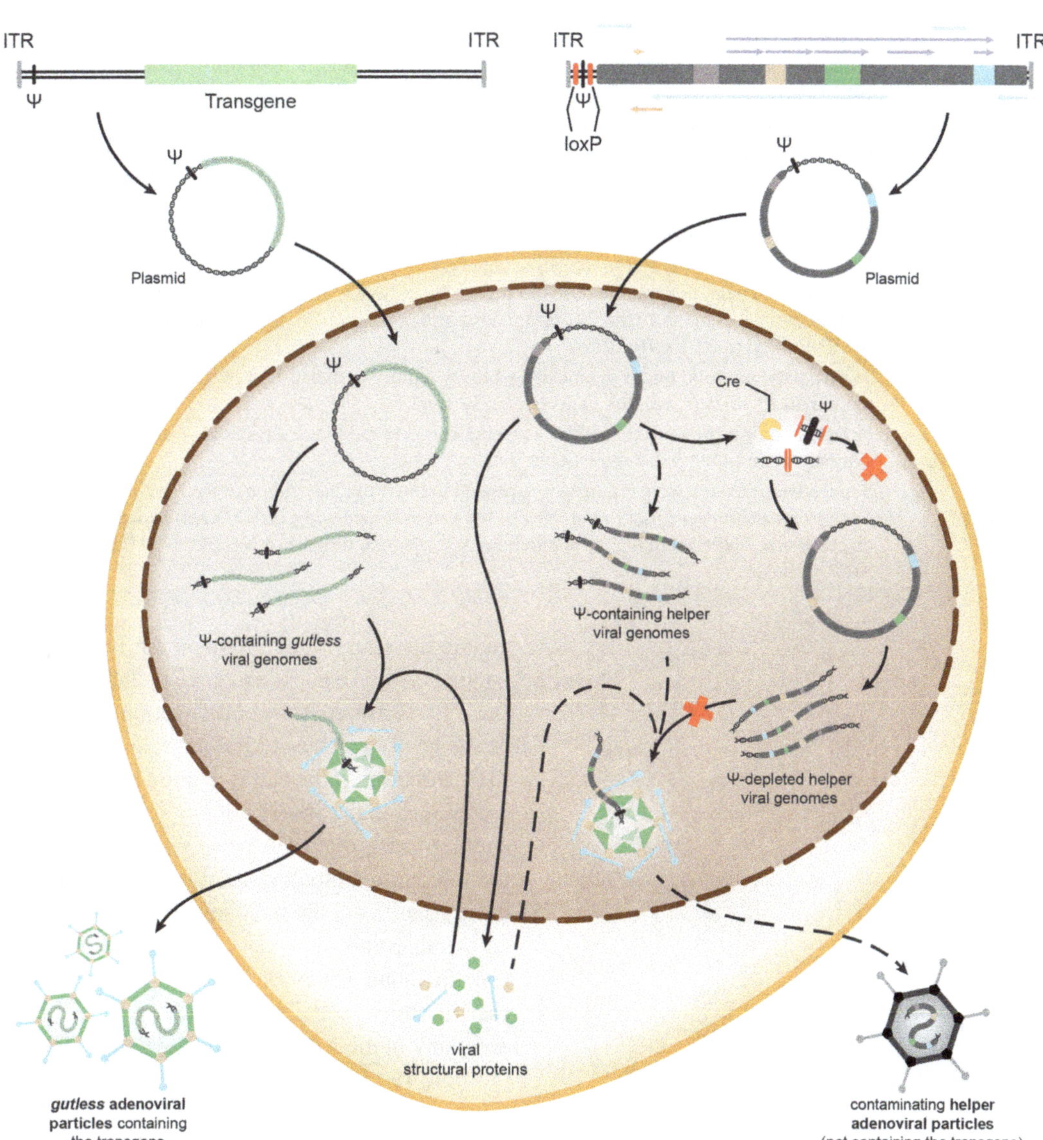

Fig. 3.16 Overview of the production of *gutless* adenoviral vectors using a helper adenoviral vector. As *gutless* adenoviral vectors are devoid of every viral coding unit, their production is based on the transfection of the *gutless* plasmid into the producer cell line, along with the co-transfection of a helper adenoviral plasmid or the infection with a helper adenovirus, containing the entire viral genome. However, this strategy implies that both viruses (*gutless* and helper) are produced, thus complicating the subsequent purification process. Later, an innovation was introduced in this system, which consisted in the removal of the packaging signal (Ψ) of the helper vector by the Cre recombinase, due to the presence of loxP sites in each side of the signal. This reduced the contamination of helper vectors to 0.1– 10%, compared to *gutless* vectors; however, it did not completely prevent their production.

Table 3.12 Example of a gene therapy clinical trial using adenoviral vectors as the delivery system of the therapeutic gene

Study	Rosenthal, E. L., *et al.* (2015) Phase I dose-escalating trial of *Escherichia coli* purine nucleoside phosphorylase and fludarabine gene therapy for advanced solid tumors, *Ann Oncol* 26, 1481–1487
Disease	Advanced solid tumors
Therapeutic gene	*Escherichia coli* purine nucleoside phosphorylase (PNP) combined with fludarabine (fludarabine monophosphate is converted by PNP into fluoroadenine and its phosphorylation disrupts protein synthesis)
Delivery vector	Non-replicative adenovirus (E1 and E3 deleted)
Clinical trial	Phase I dose-escalating trial in the USA: cohorts 1–3 with a constant dose of Ad-PNP (3×10^{11} VP) and increasing fludarabine dose (15 mg/m^2 for cohort 1, 45 mg/m^2 for cohort 2, and 75 mg/m^2 for cohort 3); cohort 4 with a constant dose of fludarabine (75 mg/m^2) and increasing dose of Ad-PNP of up to 2.73×10^{12} VP
Inclusion criteria	Twelve patients, solid tumor diagnosis by biopsy, failure or exhaustion of standard treatments, more than 19 years of age, life expectancy >12 weeks
Type of administration	Intratumoral injection of the Ad-PNP (two injections in the first 2 days) and systemic administration of the fludarabine (days 3–5)
Clinical outcome	Some adverse events were reportedly attributed to the drug, but without dose-limiting toxicity. Tumor regression was observed in most of the patients, achieving 100% in two patients. Better response was observed in higher fludarabine doses, but no difference was observed between Ad-PNP doses.

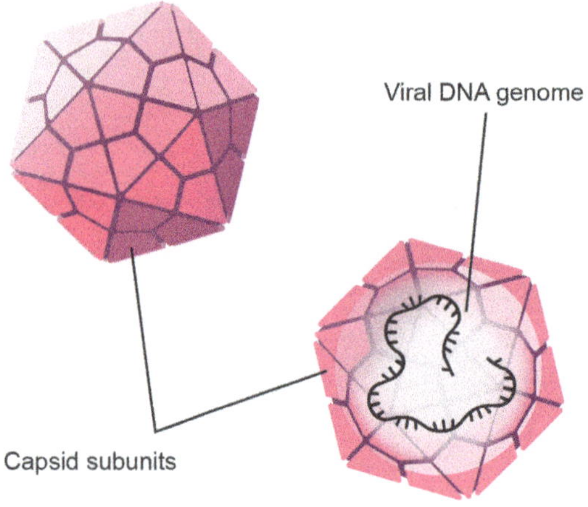

Fig. 3.17 Adeno-associated virus (AAV) structure. AAVs are very simple and small non-enveloped virus, with an icosahedral symmetry capsid composed of 60 proteins, surrounding a single-stranded molecule of DNA.

nuclear pore complex and release its genetic material (uncoating). The second DNA strand is then synthesized and the final transcription step takes place.

The AAV replication cycle is highly dependent on the cellular physiological state and machinery, and infection success is determined by molecular interactions with the host cell at each step of the cycle. If cells are exposed to genotoxic stress (e.g., X-rays) or infected by another virus (e.g., adenovirus or herpesvirus), AAV starts a rapid cycle of replication and release of the newly produced viral particles (Fig. 3.19) [31]. However, if the cell is in good physiological conditions, in the absence of any stressor agent or another virus, the AAV enters a latent stage, and the infection is blocked at multiple steps. Under these conditions, the AAV genome can persist in an extrachromosomal state or integrate in a site-specific manner into the host cell genome. In the case of human cells, this integration occurs preferentially in the AAVS1 site in the q arm of chromosome 19 [34].

3.4.2 From AAV to AAV Vectors

The simplicity of AAVs makes them an outstanding tool as gene therapy vectors. For example, the limited number of viral proteins in the AAV vector reduces the inflammatory response, and the lack of integration reduces the risk of insertional mutagenesis and epigenetic silencing of the transgene (Table 3.14). Moreover, the natural tropism of the different AAV serotypes allows cell-specific targeting, reducing the need for additional engineering. On the other hand, some disadvan-

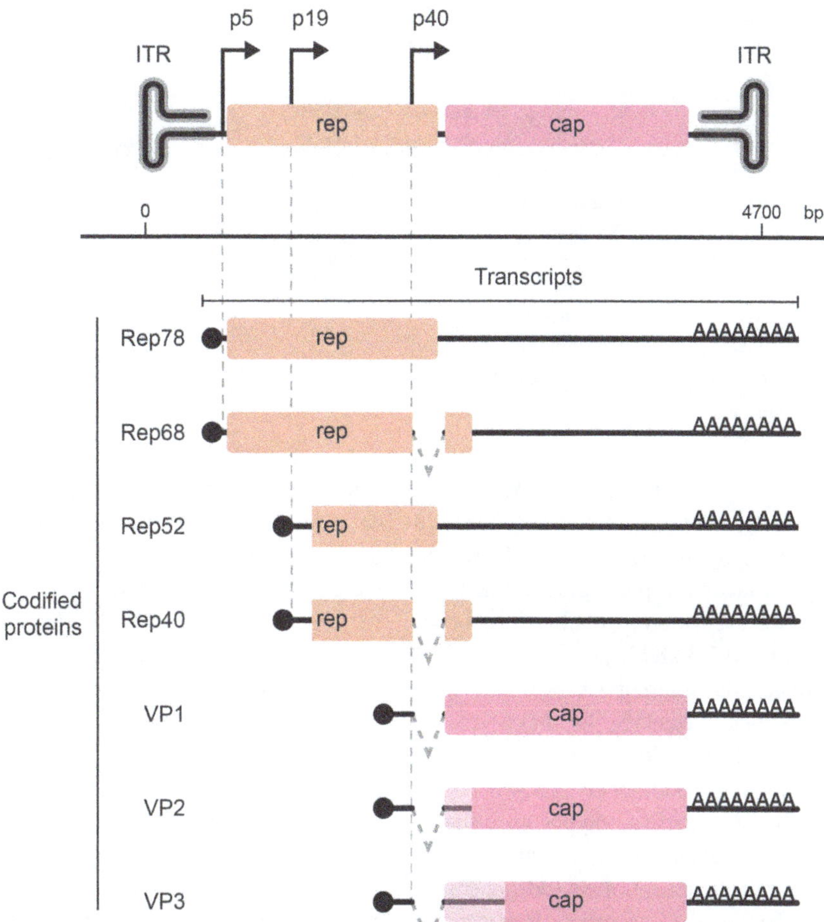

Fig. 3.18 Genome organization of an adeno-associated virus, based on the genome of AAV2. The single DNA strand of 4.7 kb is flanked by two inverted terminal repeats (ITRs), comprising two open reading frames, *rep* and *cap*. The *rep* sequence encodes four non-structural proteins involved in replication (Rep78, Rep68, Rep52 and Rep40), driven by three different promoters (p5, p19, and p40). The *cap* segment encodes three structural proteins of the viral capsid (VP1, VP2, and VP3).

tages of their use include the small cloning capacity, which limits the transgene size and further introduction of tags and promoters.

The size limitation is the main point in designing an AAV vector, imposing that the sequence comprising the ITRs and the transgene to be cloned between them does not exceed 4.9 kb [35]. To overcome this problem, several strategies were developed, such as (i) reducing the transgene size by minimizing the length of regulatory elements or truncating the protein or (ii) splitting the vector system, using fragmented vectors or overlapping dual vectors [36]. However, and despite good advances, there are still some efficacy issues in these strategies, which limit their widespread use.

Another important point to consider in AAV vector design is the delayed transgene expression due to the biology of the AAV single-stranded genome, which requires the synthesis of the complementary strand to be expressed. To overcome this problem, an alternative was envisioned using self-complementary AAV vectors, which package an inverted repeat complementary genome

Table 3.13 Adeno-associated virus serotypes, their origin, and main receptors used for cell entry.

Serotype	Origin	Main receptor	Neutralizing antibodies (%)
AAV1	NHP	Sialic acid (alpha2,3 N-linked and alpha2,6 N-linked)	67
AAV2	Human	Heparan sulfate proteoglycans (HSPGs)	72
AAV3	NHP	Heparan sulfate proteoglycans (HSPGs)	Not known
AAV4	NHP	Sialic acid (alpha2,3 O-linked)	10
AAV5	Human	Sialic acid (alpha2,3 O-linked and alpha2,3 N-linked)	40
AAV6	Human	Sialic acid (alpha2,3 O-linked and alpha2,3 N-linked)	46
AAV7	NHP	Not known	Not known
AAV8	NHP	Laminin receptor (LamR)	38
AAV9	Human	N-linked galactose	47

NHP nonhuman primate

that forms double-stranded DNA, without the requirement for synthesis or base pairing between multiple vector genomes (Fig. 3.20). However, this strategy reduces the packaging capacity of the AAV vector even more, to half of the original length - around 2.3 kb [37].

As mentioned, the different AAV serotypes provide an important advantage to these vectors, allowing the specific transduction of different types of cells without additional modifications. Additionally, some of these serotypes were shown to cross the blood-brain barrier, which opened a new avenue for the treatment of neurodegenerative diseases and the use of less invasive administration routes [38, 39]. In fact, efficient brain transduction was shown in several species such as mice, rats, cats and monkeys, after a systemic administration of AAV, especially for serotype 9.

Despite its promising potential, the systemic administration of AAV still encounters some problems: (i) while neuronal tropism is found in neonatal animals, transduction in adult animals is specially observed in glial cells [38], (ii) a reduction in the transduced cells/volume ratio is observed in aged animals [40], and (iii) the large titers used in preclinical studies suggest that much higher titers will be needed for human clinical trials.

3.4.3 AAV Modifications

Some of the key AAV modifications were developed in order to overcome two of its main disadvantages: the limited cloning capacity and the delayed transgene expression. However, several other modifications were also designed to increase their efficacy and the specificity of AAV vectors and to circumvent some gene delivery barriers.

The presence of AAV neutralizing antibodies in the human population (Table 3.13) limits gene delivery by many natural vectors [41]. A simple form to circumvent this problem in clinical trials is to exclude patients with neutralizing antibodies; however the high percentage of human exposure to different AAV serotypes makes this approach less effective. Several other strategies were tested, such as (i) the use of empty capsids to adsorb anti-AAV antibodies, thus allowing the vector transduction, (ii) the switching of the AAV serotype, or (iii) the engineering of AAV capsids to make them less susceptible to neutralizing antibodies. These latter strategies led to the development of hybrid AAV serotypes involving [42] (i) transcapsidation, in which the genome of one serotype is packaged into the capsid of a different serotype; (ii) adsorption modifications, in which peptides are absorbed on the AAV capsid surface; (iii) mosaic capsids, in which the capsid is packaged with proteins from two AAV serotypes; and (iv) chimeric capsid development, in which capsid proteins are fused with foreign peptide sequences. Some of these strategies were also useful in altering AAV tropism, which can also be achieved through the use of heterologous promoters that increase the specificity of gene

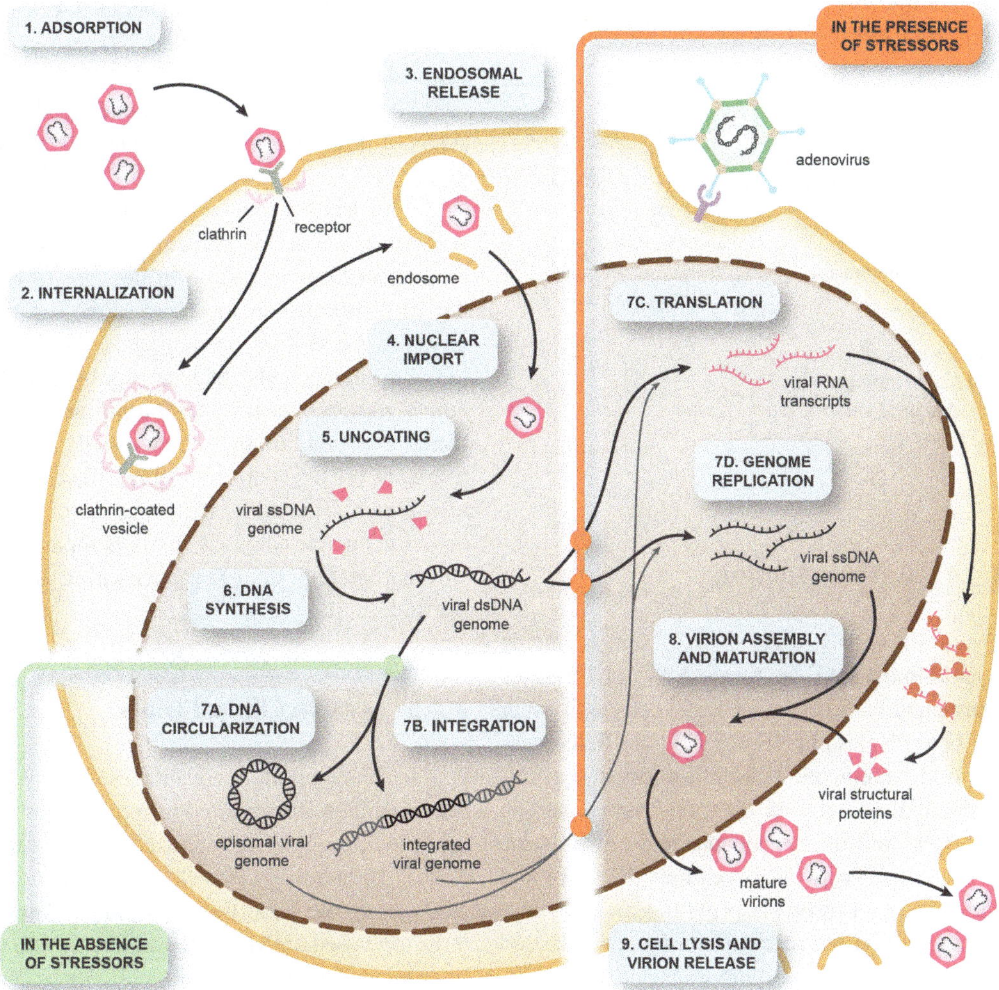

Fig. 3.19 Replicative cycle of an adeno-associated virus, highlighting its main phases. The AAV replicative cycle starts with the binding of the viral particle to particular cell receptors (1) and its internalization through receptor-mediated endocytosis (2). The next steps of the cycle include the endosome escape (3), entry into the nucleus through the nuclear pore complex (4) and uncoating, leading to the release of the single-stranded viral DNA (5). After that, the complementary DNA strand is synthesized, from the ITR's hairpin structures that act as primers for this process (6). The AAV replicative cycle is dependent on the presence of stressors, and if they are absent the double-stranded DNA molecule can circularize and became episomal (7A) or integrate into the cell genome (7B). However, if any given stressor is present (e.g., another virus), the AAV continues its replicative cycle with the translation of viral proteins (7C) and replication of its genome (7D). Finally, the viral components assemble (8) and mature virions are released from the cell.

expression to target cells and ensures high levels of expression of the transgene.

Importantly, several developments were also made trying to overcome the major AAV limitation, which is the limited cloning size, namely by using "overstuffed" and **dual AAV vectors** [43]. In the former option, transgenes exceeding the AAV capacity are packaged as usual. However, this strategy reduces viral titer production, as well as the integrity and functionality of viral particles. In the dual vectors approach, the transgene of interest is divided into 5' and 3' halves, using different methodologies (Fig. 3.21).

Table 3.14 Main advantages and disadvantages of adeno-associated viral vectors for gene delivery.

Advantages	Disadvantages
Transduction of dividing and nondividing cells	Delayed transgene expression
Natural specific tropism	Limited cloning capacity
Stable transgene expression	Complex and time-consuming production
Reduced inflammatory response	Presence of neutralizing antibodies in the human population
Reduced risk of insertional mutagenesis	
Scalable production methods	

3.4.4 AAV Production

The most widely used method for producing **recombinant AAV vectors** (rAAV) involves the transient transfection of HEK293 cells with either two or three plasmids (Fig. 3.22) [44]. In the **two-plasmid system**, one of the plasmids corresponds to the AAV vector with the transgene flanked by the ITRs, and the second plasmid contains the AAV *rep* and *cap* ORFs without the ITRs and also adenoviral proteins providing helper functions. On the **three-plasmid system**, one plasmid is the AAV vector, with the transgene, the ITRs, the promoter and the polyA signal; the second plasmid is the helper plasmid containing adenoviral regions E1, E2, E3 and VA, which contribute to genome replication; and a third plasmid contains AAV *rep* and *cap* genes, which are important for replication and capsid assembly, respectively.

For large-scale productions, the transient delivery of plasmids increases the production cost as a high amount of DNA is required. Moreover, it requires serum-containing media, which could potentiate the toxicity of the viral preparation if not removed completely. Thus, other methods for AAV production integrate one or more components in mammalian cells (mainly HeLa cells) or insect production cell lines (used in this latter case in combination with baculovirus infection). The baculovirus production system was used for the production of AAV1-based gene therapy product Glybera®. These systems are scalable and can use serum-free media; however, the presence of viral components other than AAV still compromises the viral preparation purity.

As for any other viral system, the success of AAV production for clinical use relies on the quality of the purification. In the case of rAAV, the purification steps must ensure the separation of cellular and viral contaminants (protein, lipids and nuclei acids) and the removal of AAV empty capsids. For this, several purification protocols are used, although they usually comprise five steps [45]: (i) cell harvesting, (ii) cell lysis to release AAV particles, (iii) removal of cellular and viral nucleic acids, (iv) particle separation by chromatography and (v) concentration and final sterile filtration.

3.4.5 AAV in Clinical Trials

Due to their important advantages, AAV vectors are being increasingly used in gene therapy clinical trials, targeting different conditions (Tables 3.15, 3.16 and 3.17) [46–48]. The first commercially available gene therapy product (Glybera) in Europe used an AAV1 delivery system to treat lipoprotein lipase deficiency.

3.5 Herpes Simplex Virus

Herpes simplex virus (HSV) is an enveloped virus belonging to the *Herpesviridae* family. Viral particles have around 155–240 nm in size and their structure contains four major elements: (i) a core (electron-opaque), that contains the viral genome; (ii) an icosahedral capsid surrounding the core, with a size of around 125 nm; (iii) an amorphous tegument surrounding the capsid; and (iv) an outer envelope with surface spikes (Fig. 3.23) [49, 50].

The HSV genome consists of a double-stranded DNA chain with approximately 152 kb in length, encoding around 84 genes (Fig. 3.24) [51]. The genome is composed of long (UL) and

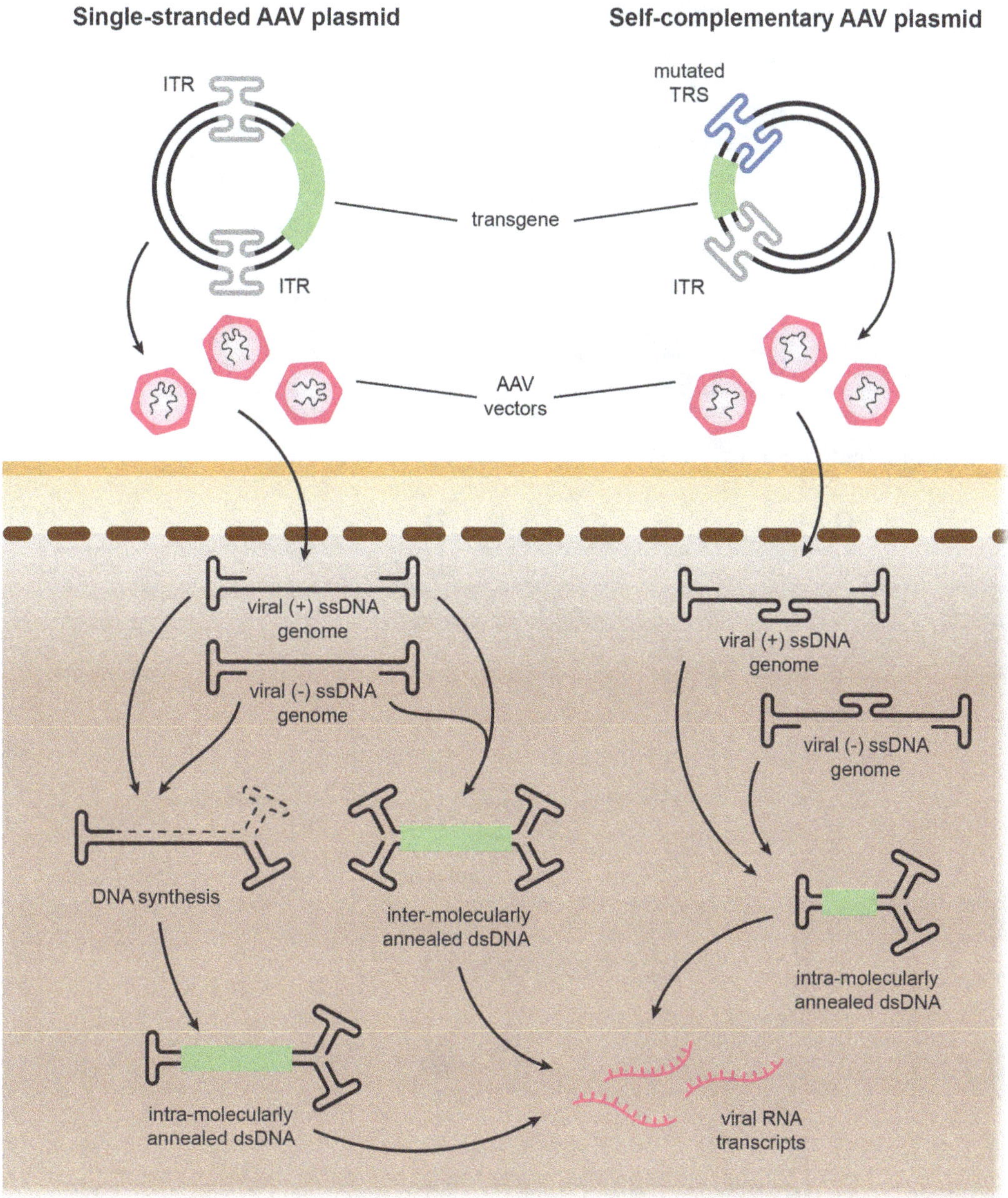

Fig. 3.20 **Adeno-associated virus engineering to produce self-complementary vectors**. Before the start of viral gene expression, common single-stranded AAV vectors have to reach the nucleus and the complementary DNA strand has to be produced. This process leads to a delay and to a reduction in gene expression. An alternative could entail base pairing of two complementary strands delivered by two vectors. However, the process is not very efficient and is highly dependent on cell type. Therefore, a third alternative was engineered to overcome these problems, based on the development of self-complementary AAV (scAAV) vectors, in which both strands are packed in the same molecule (terminal resolution site - TRS). This alternative involves the reduction of the construct size to 2.3 kb, further decreasing the cloning capacity of the AAV. However, it increases the efficiency, speed, and levels of transgene expression.

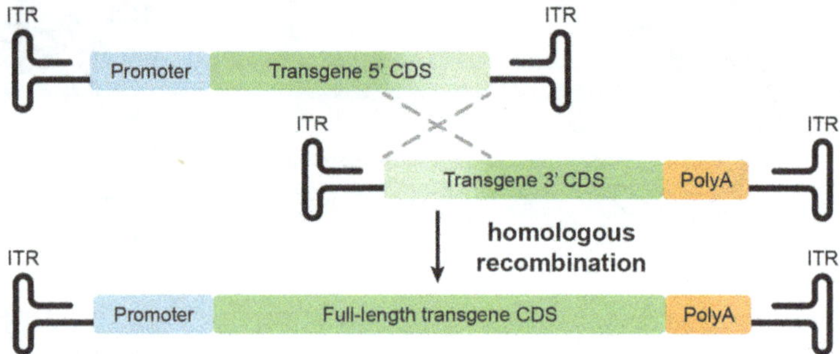

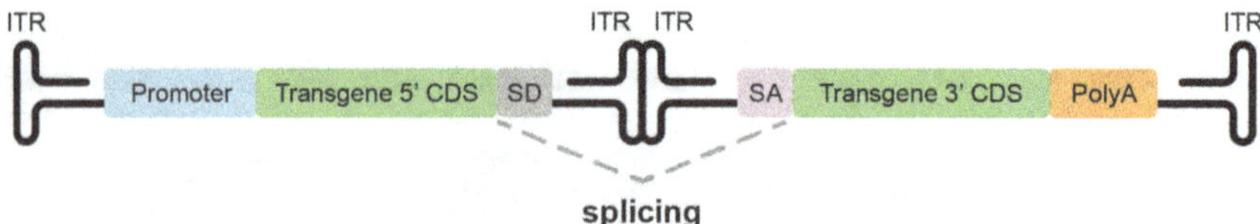

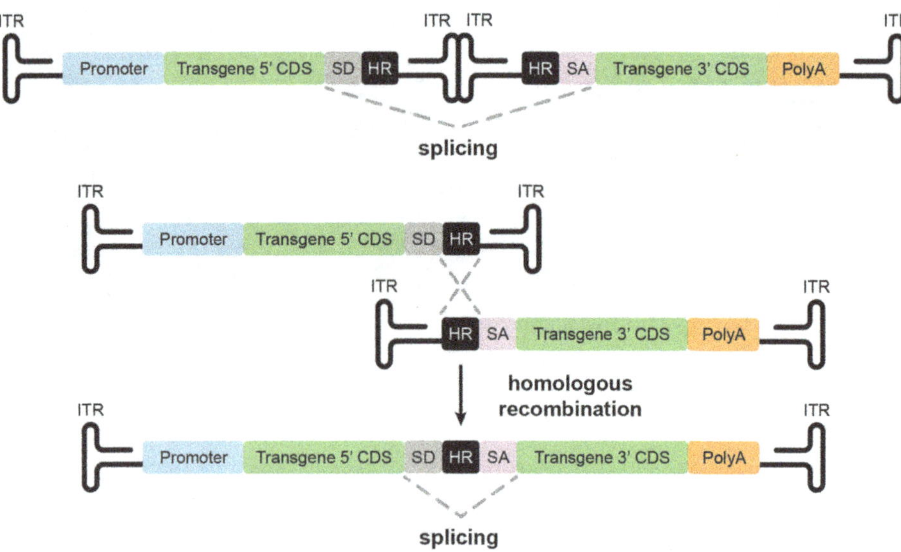

Fig. 3.21 Strategies to improve the cloning capacity of adeno-associated viral vector, based on the development of dual vectors. In the *overlapping* strategy, each vector carries a fragment of the transgene coding sequence (CDS), both containing overlapping parts that are then joined together in the target cell. Despite being used with success in several studies, this strategy requires an extensive process of determining the optimal overlap sequence and can lead to unspecific transgene products. In the *trans-splicing* strategy, two vectors are used, each one containing a different part of the transgene CDS and different splicing sites. The entire expression cassette (promoter, transgene and polyA signal) is joined together in the target cell, through RNA splicing involving those sites. However, this process in not very efficient and can also lead to unspecific transgene products. The *hybrid* vector strategy combines the previous approaches to overcome the main problems found in each one. The vectors contain overlapping sequences of the transgene CDS and different splicing sites, increasing their efficiency compared to the other strategies.

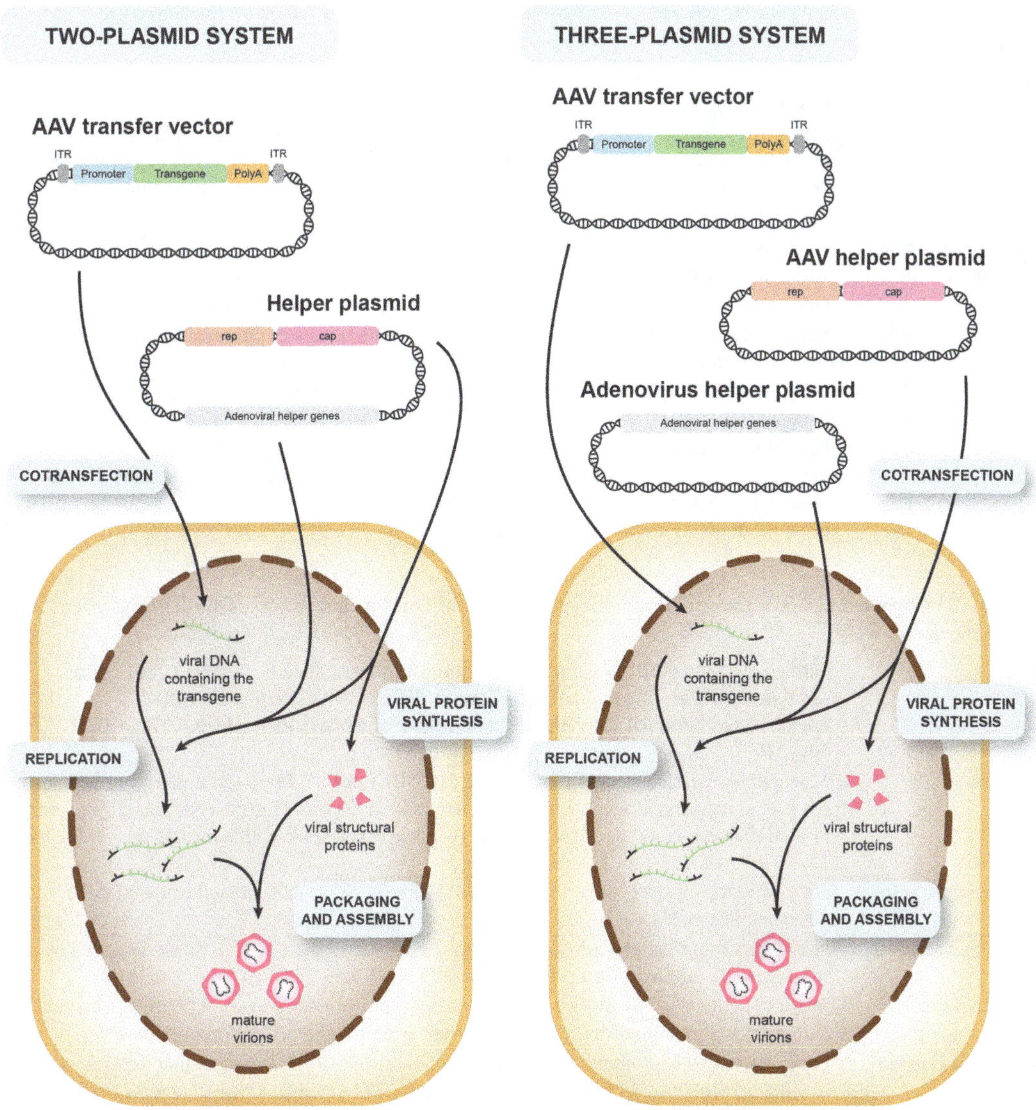

Fig. 3.22 Outline of different strategies to produce recombinant adeno-associated virus, based on the transfection of two or three plasmids. In the two-plasmid system, one of the transfected plasmids corresponds to the AAV vector with the transgene flanked by the ITRs, and the second plasmid contains the AAV *rep* and *cap* ORFs without the ITRs and also adenoviral proteins, providing helper functions. On the three-plasmid system, one of the transfected plasmids is the AAV vector, with the transgene, the ITRs, the promoter and the poly-A signal, the second plasmid is the helper plasmid encoding adenoviral helper proteins, and the third plasmid contains the AAV *rep* and *cap* genes.

short (US) segments, each one flanked by an internal repeat (IR) and a terminal repeat (TR). Each TR is localized at one end of the DNA strand. Another interesting feature of the HSV genome is the classification of its genes as accessory or essential. As the name implies, the **accessory genes** can be deleted without significantly compromising the virus replication, although this may limit the HSV replicative cycle *in vivo*. On the other hand, the deletion of **essential HSV genes** impairs the virus replication cycle.

Table 3.15 Example 1 of a gene therapy clinical trial using adeno-associated viral vectors as the delivery system of the therapeutic gene.

Study	Bartus, R. T., *et al.* (2013) Safety/feasibility of targeting the substantia nigra with AAV2-neurturin in Parkinson patients, *Neurology* 80, 1698–1701
Disease	Parkinson's disease (PD), which is a common neurodegenerative disease characterized by motor impairments, mainly resulting from the progressive degeneration of dopaminergic neurons in the substantia nigra that project axons to the striatum, where dopamine is released
Therapeutic gene	Neurotrophic factor neurturin (NRTN) (CERE-120)
Delivery vector	AAV2 in three doses: 4×10^{11} vg (vector genomes), 5.4×10^{11} vg and 24×10^{11} vg
Clinical trial	Open-label phase I study for safety and feasibility, with a 2-year follow-up in the USA
Inclusion criteria	Six patients with advanced PD in 2 dose cohorts with a 1-month interval between each surgery and a 5-week space between the low-dose and high-dose cohorts
Type of administration	Local delivery of CERE-120 bilaterally in the substantia nigra and in three sites in the putamen
Clinical outcome	No serious adverse events were reported, suggesting good tolerability of the procedure. No increase in AAV antibodies or detectable NRTN antibodies was found in patients' serum. Despite being a safety study, the 2-year follow-up data suggests a decrease in motor behavior abnormalities.

Table 3.16 Example 2 of a gene therapy clinical trial using adeno-associated viral vectors as the delivery system of the therapeutic gene.

Study	Tardieu, M., *et al.* (2014) Intracerebral administration of adeno-associated viral vector serotype rh.10 carrying human SGSH and SUMF1 cDNAs in children with mucopolysaccharidosis type IIIA disease: results of a phase I/II trial, *Hum Gene Ther* 25, 506–516
Disease	Mucopolysaccharidosis type IIIA disease (MPSIIIA or Sanfilippo syndrome type A) is an autosomal recessive disease caused by a mutation in the SGSH gene, encoding a lysosomal protein (heparin-N-sulfamidase). MPSIIIA is a severe disease with a median age of death of 15.4 ± 4.1 years.
Therapeutic gene	An expression cassette with two genes: wild-type SGSH cDNA and wild-type SUMF1 cDNA
Delivery vector	AAV2 genome packaged into a serotype rh.10 capsid (AAVrh.10-hMPS3A) produced by a two-plasmid co-transfection procedure in HEK293T cells (7.2×10^{11} vg/patient)
Clinical trial	Phase I/II clinical study evaluating the intracerebral administration of AAVrh.10-hMPS3A, combined with immunosuppressive treatment, with a 1-year follow-up in France
Inclusion criteria	Four patients with 18 months–6 years of age; onset of MPSIIIA clinical manifestations in the first 5 years of age; SGSH activity in peripheral blood cells <10% of control levels
Type of administration	Local administration of 12 injections (360 μl each), bilaterally targeting the white matter anterior, medial, and posterior to the basal ganglia
Clinical outcome	No adverse events related with the injected product were reported. Also, no increase in the number of infection events was detected, neither any sign of toxicity related with the immunosuppressive drugs. Efficacy results were moderate, with a stabilization in brain atrophy in 2 patients and a slight improvement in behavior, attention, and sleep in 3 patients.

3.5.1 Replicative Cycle

The HSV replicative cycle begins with the viral entry into the host cells, which is mediated by the attachment of viral glycoproteins to the cell surface receptors. After that, the viral envelope fuses with the cellular membrane and releases the tegument-capsid structure into the cell cytoplasm. This structure is then transported to the nucleus (probably in a microtubule-dependent manner), where its genome is released. Inside the nucleus, the transcription of early HSV genes starts immediately, whereas host cell gene transcription and posterior RNA processing events are inhibited, thus promoting viral genome expression. After the start of viral DNA replica-

Table 3.17 Example 3 of a gene therapy clinical trial using adeno-associated viral vectors as the delivery system of the therapeutic gene.

Study	Jacobson, S. G., *et al.* (2012) Gene therapy for Leber's congenital amaurosis caused by RPE65 mutations: safety and efficacy in 15 children and adults followed up to 3 years, *Arch Ophthalmol* 130, 9–24
Disease	Leber congenital amaurosis (LCA) is an autosomal recessive retinal disease caused by mutations in the RPE65 genes, leading to deficiency of the encoded protein
Therapeutic gene	Wild-type RPE65 gene
Delivery vector	AAV2 in five doses: 5.96×10^{10} vg, 11.92×10^{10} vg, 8.94×10^{10} vg, 17.88×10^{10} vg, 7.95×10^{10} vg
Clinical trial	Open label, dose-escalation phase I study, with a 3-year follow-up in the USA
Inclusion criteria	Five cohorts in a total of 15 patients, 11–30 years of age
Type of administration	Subretinal local injection: single in 3 cohorts and two injections in the last 2 cohorts
Clinical outcome	The adverse events reported were only related with the surgery procedure, and no systemic toxicity was detected. An increase in AAV-targeting antibodies was only reported for one patient. In terms of efficacy, visual acuity improved in all the patients, in different levels (major improvements were observed in the eyes with the lower entry acuities)

tion, the expression of late HSV genes begins, producing several structural components of the virus. Viral assembly starts in the nucleus and continues in the cytoplasm with the involvement of the Golgi complex. After the assembly, enveloped viral particles are released through secretory vesicles (Fig. 3.25) [52].

An important feature of HSV is their ability to linger in a latent state, in which the viral genome persists episomally in a stable way, without expression of early or late genes. Only a set of non-translated RNA species (latency-associated transcripts – LATs), without any known function, are synthesized during latency. This HSV feature allows the virus to persist in the host even in the presence of an immune response, and undergo posterior reactivation. The reactivation of the replicative cycle depends on several factors, including the host immune status; however, the molecular events underlying this reactivation are not completely understood.

3.5.2 From HSV to HSV Vectors

At first glance, the use of HSV as vectors for gene therapy could seem inappropriate due to their reported high cytotoxic profile in humans. However, some HSV features make these viruses particularly useful as gene therapy vectors. For example, latent infection, which occurs particularly in neurons, makes HSV attractive as a gene therapy vector targeting the central nervous system (CNS). Other advantages include the large cloning capacity and the reduced risk of insertional mutagenesis, as HSV do not integrate into the host genome (Table 3.18). On the other hand, the high cytotoxicity associated with HSV and stimulation of a strong immune response are important drawbacks for their use in gene therapy. The genetic engineering of HSV, considering its important features, led to the development of three different types of HSV vectors [49]: (1) replication-defective vectors; (2) replication-competent vectors and (3) amplicon vectors (Fig. 3.26).

Replication-Defective Vectors

The replication-defective vectors, as the name implies, are HSV where genes that are essential for replication were deleted or mutated. In HSV it is relatively simple to generate this type of vector, since, although approximately 40 genes are essential, mutation of a single one of them is enough to produce a replication-defective vector [53]. Several combinations of gene deletions have been shown to successfully inhibit virus replication, while preserving some wild-type HSV advantages, like the ability to express transgenes after having established latent infections in

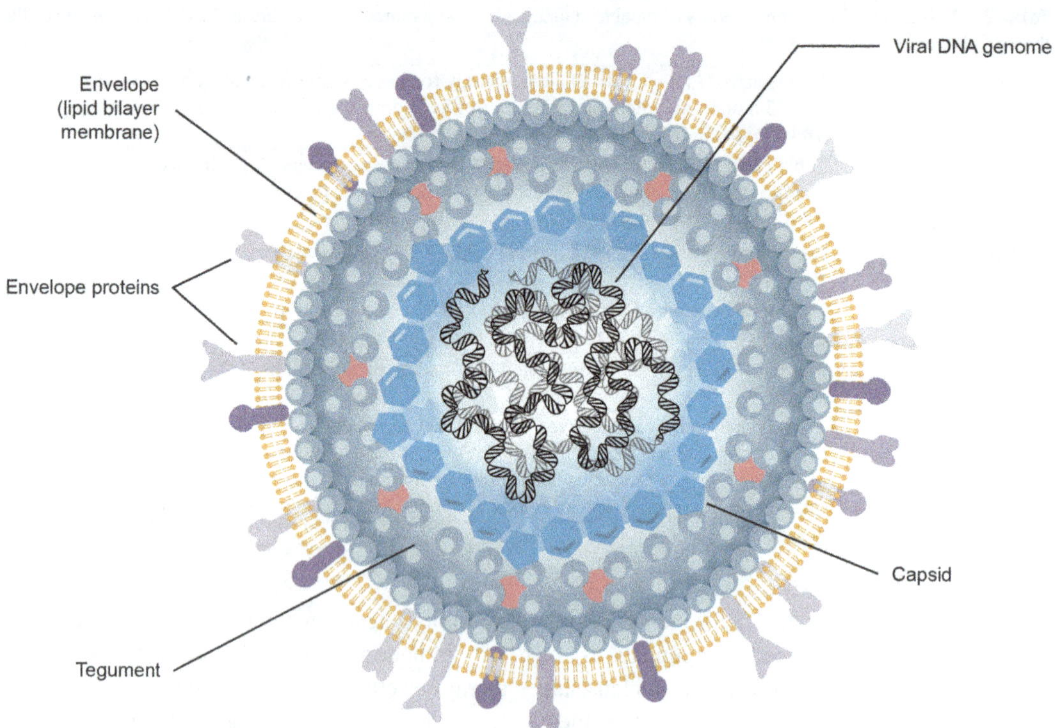

Fig. 3.23 Herpes simplex virus structure highlighting its main components. The virus is composed of a lipid bilayer envelope that contains an amorphous tegument surrounding the protein capsid. The capsid involves the double-stranded DNA molecule that constitutes the viral genome.

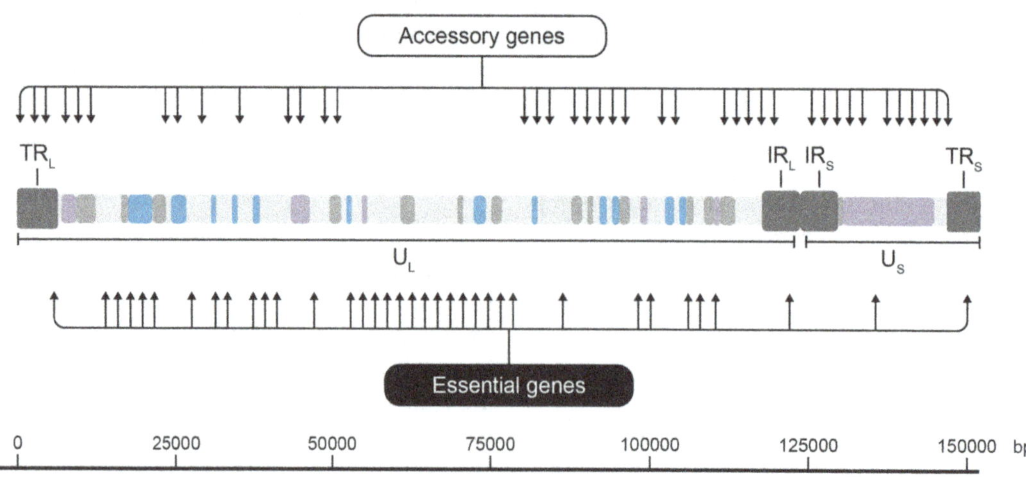

Fig. 3.24 Overview of the genome of the herpes simplex virus, based on HSV-1. The double-stranded DNA molecule is about 152 kb in length and encodes approximately 84 genes. The genome is composed of a long segment (UL) and a short segment (US), separated by an internal repeats (IR). Each side of the molecule is flanked by strand terminal repeats (TR). HSV genes can be divided into two major groups: *essential genes*, which are needed for viral replication, and *accessory genes*, which can be deleted without compromising viral replication.

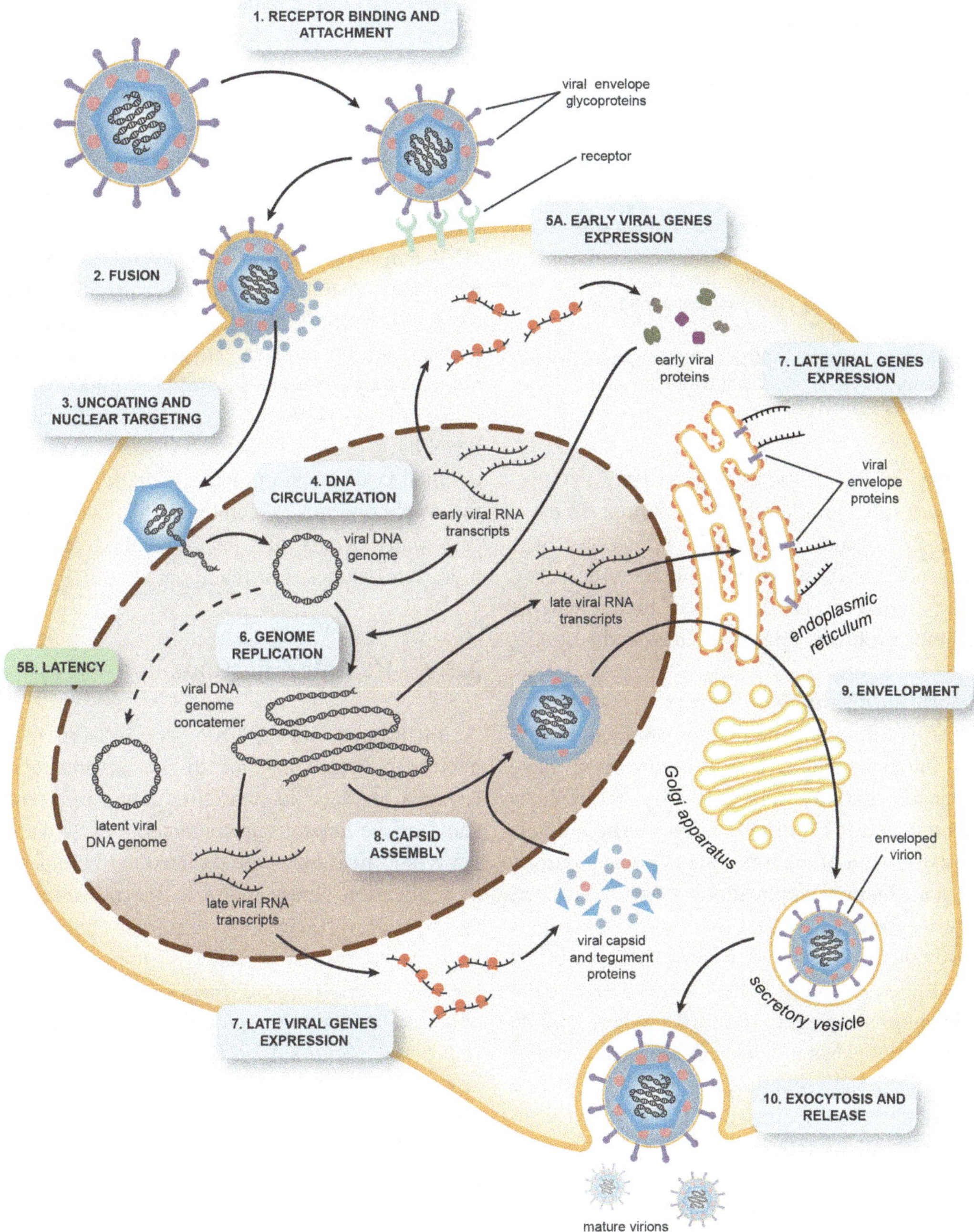

Fig. 3.25 Replicative cycle of a herpes simplex virus. The cycle starts with the attachment of viral glycoproteins of the envelope to the host cell receptors, mediating the viral entry (1). After that, the viral envelope fuses with the cellular membrane and releases the tegument-capsid structure into the cell cytoplasm (2). This structure is transported to the nucleus, where the viral genome enters (3). Although the DNA molecule is linear, inside the nucleus it undergoes circularization (4). Transcription of early viral genes starts immediately, whereas host gene transcription and posterior RNA processing events are inhibited (5A). The HSV can latent within the cell, persisting episomally, without expression of viral genes (5B). If the cycle continues, then the next steps are viral replication (6) and the expression of late genes, producing the structural elements of the virus (7). The capsid assembly starts in the nucleus (8), continues in the Golgi complex (9), and viral particles are finally released by exocytosis (10).

Table 3.18 Main advantages and disadvantages of herpes simplex viral vectors for gene delivery.

Advantages	Disadvantages
Transduction of dividing and nondividing cells	Potentially toxic
High cloning capacity of up to 50 kb	Transient expression of the transgene
Natural tropism to neurons	High levels of pre-existing immunity
Easy production in high titers	Stimulation of a strong immune and inflammatory response
Long-term expression in neurons	Risk of recombination with latent wild-type virus

neurons. However, the associated HSV cytotoxicity was still a problem in these vectors. To overcome this issue, additional engineering was performed in these HSV replication-defective vectors; however, the deletion of toxic elements has some problems related to viral production.

Replication-Competent Vectors

Probably, for most gene therapy applications the ideal nucleic acid delivery vector must have attenuated toxicity and be fully replication-defective, while promoting a long-term and correct expression of the transgene. On the contrary, for gene therapy applications targeting cancer, the use of replicative and toxic virus sometimes constitutes an advantage, especially if the effect is restricted to tumor cells. In this sense, HSV replication-competent vectors, where some nonessential genes are deleted, are used in oncolytic gene therapy, retaining the ability to replicate only in tumor or dividing cells, but with a reduced or inexistent ability to replicate in nondividing cells [54]. This allows the HSV vectors to complete its lytic viral replication cycle in tumor cells and enhances intratumoral distribution, but importantly without producing toxicity problems in other cells. Despite being mostly used for cancer treatment, this type of HSV vectors was also used in studies for CNS disorders.

Amplicon Vectors

HSV amplicon vectors are defective, helper-dependent vectors that carry a concatemeric form of a plasmid DNA, named the amplicon plasmid, instead of the viral HSV genome, while maintaining the structural, immunological, and tropism features of wild-type HSV. The amplicon contains the expression cassette of interest, the HSV-1 origin of replication, and the HSV-1 packaging signal, and thus HSV amplicon vector production is dependent on a helper virus [55]. The HSV amplicon vectors have several advantages as a carrier for gene delivery: (i) a cloning capacity of up to 150 kb; (ii) no viral genes are present, which reduces the associated toxicity; and (iii) they maintain the ability to infect a wide range of cells. However, two important disadvantages have limited their use in gene therapy: (i) the final virus preparation has both the helper and the amplicon virus, which led to the development of toxicity events, and (ii) production of stable high-titer batches is difficult.

3.5.3 HSV Modifications

As mentioned before, HSV vectors were engineered to maintain some of the advantageous features of the wild-type virus while removing some of the deleterious characteristics. Most of HSV modifications were directed to the removal of replication genes and/or to the reduction of toxicity. Nevertheless, several other modifications were tested in these vectors trying to improve/regulate transgene expression, to enhance production protocols and to continue improving vector safety. For most of gene therapy applications, a long-term expression is needed; thus some improvements were made in HSV amplicon vectors to achieve this goal. One strategy involves the development of hybrid amplicon systems, where the transgene capacity of HSV is combined with the potential of AAV to mediate the AAV1 site-specific genomic integration of transgene cassettes surrounded by the AAV ITRs. Another strategy uses transposons to achieve amplicon integration into the host genome.

Several other modifications were developed for oncolytic applications, involving the use of tumor-specific promoters and vector re-targeting

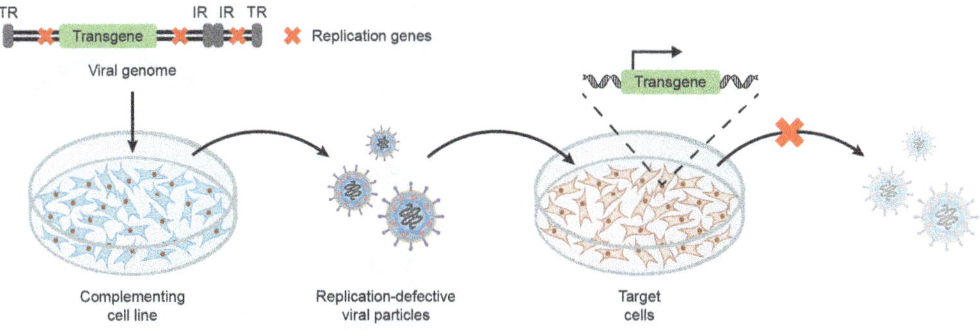

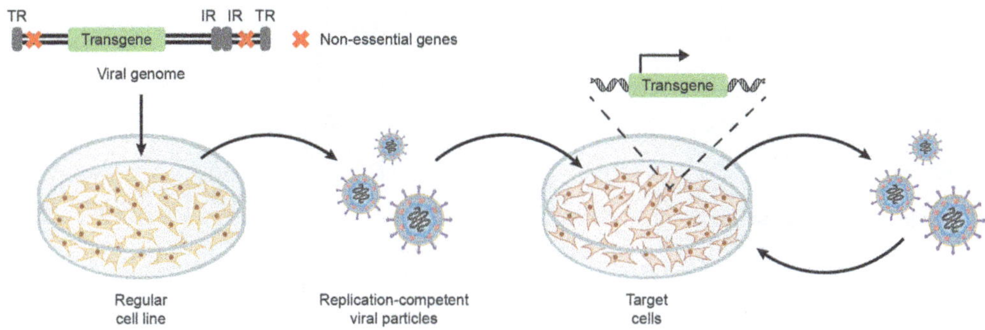

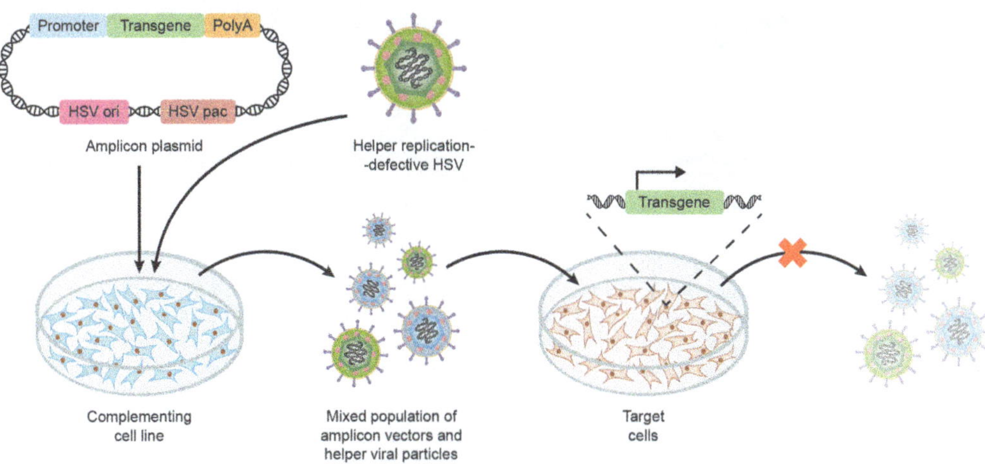

Fig. 3.26 Overview of the different herpes simplex viral vectors used in gene therapy. In the *replication-defective vectors*, some essential genes (or simply one) are deleted, compromising the replication cycle. On the other hand, in the *replication-competent vectors* deletion of viral genes is restricted to accessory genes; these are mainly used for oncolytic gene therapy, as they are replicative. Finally, in the *amplicon vectors* all the HSV genome is deleted, with the exception of the origin of replication and the packaging signal. Although they are safer and have a low toxicity profile, production of amplicon vectors is more complex, as it requires the infection of a replication-defective HSV to provide the viral proteins to the amplicon particles, thus producing a mixed population composed of both viruses.

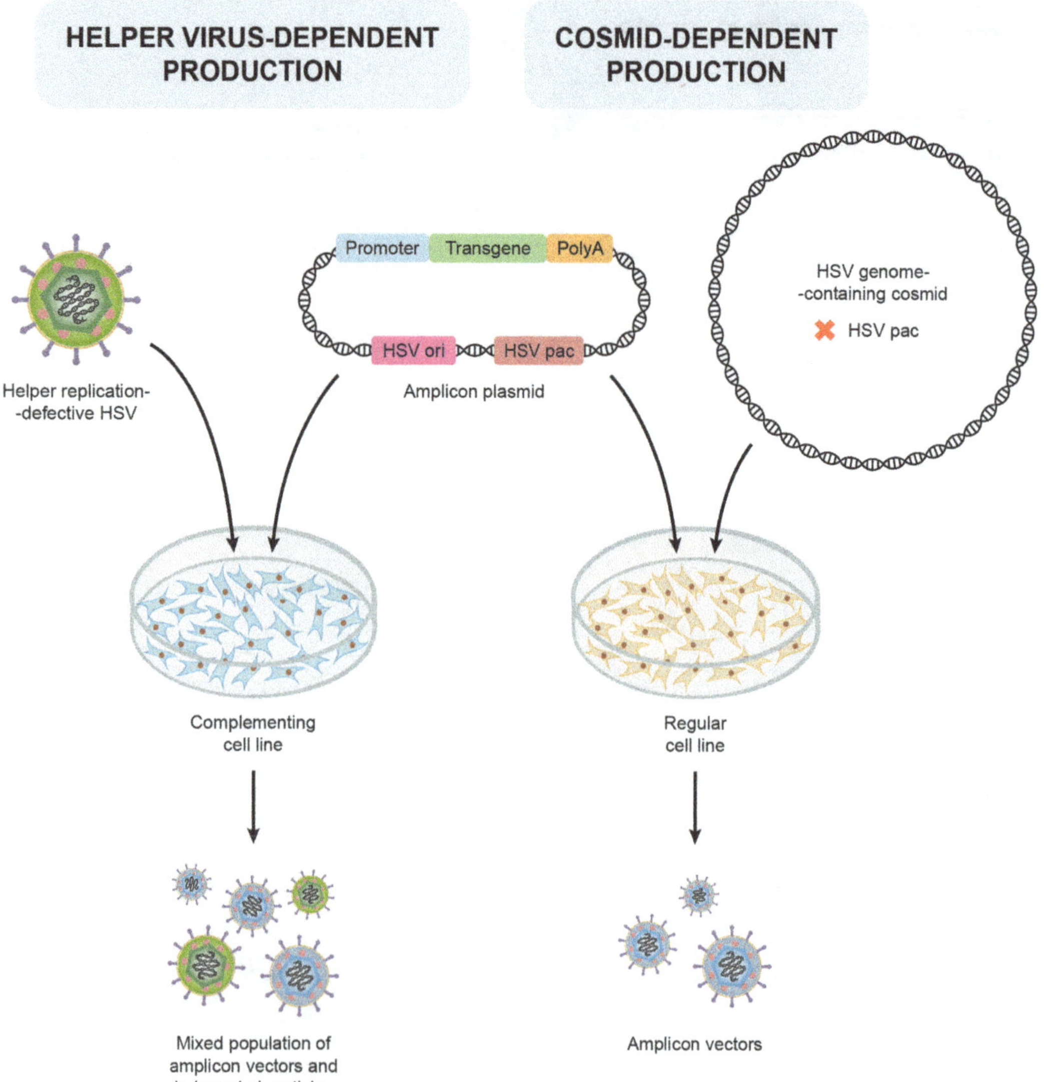

Fig. 3.27 Strategies to produce herpes simplex amplicon vectors. As amplicon vectors lack the viral proteins necessary for the production of functional particles, they have to be provided using different strategies. The producer line can be transfected with the amplicon plasmid and then infected with a helper replication-defective HSV, to provide the required components for viral particle formation. However, the viral batch produced will contain a mix of amplicon and helper viruses. An alternative could be the co-transfection of the amplicon plasmid with overlapping cosmids containing the entire HSV genome (except for the packaging signal - HSV pac). This strategy will provide all the components for the amplicon virus formation, without contamination of additional HSV, thus increasing their safety profile.

as strategies to increase HSV vectors specificity to tumor cells. Also, for cancer therapy, several HSV vectors were "armed" with genes that augment their cytolytic ability, potentiating their therapeutic effect (see Sect. 3.9).

3.5.4 HSV Production

The production of HSV vectors has different needs depending on the type of vector. In the case of HSV replication-competent vectors, several points should be addressed both in the production

Table 3.19 Example of a gene therapy clinical trial using herpes simplex viral vectors as the delivery system of the therapeutic gene.

Study	Markert, J. M., *et al.* (2014) A phase I trial of oncolytic HSV-1, G207, given in combination with radiation for recurrent GBM demonstrates safety and radiographic responses, *Mol Ther* 22, 1048–1055
Disease	Glioblastoma multiforme (GBM)
Therapeutic gene	HSV-1 genes in combination with radiation
Delivery vector	Herpes simplex virus type I (G207, HSV-1 with deletion of essential genes for viral replication in normal cells) in one dose of 1×10^9 pfu (plaque-forming units)
Clinical trial	Open-label, clinical trial phase I, in one site in the USA
Inclusion criteria	Nine patients whose tumor had progressed despite surgery, chemotherapy and radiation therapy
Type of administration	Local, direct inoculation of the G207 vector in multiple sites of the tumor
Clinical outcome	Several adverse events were reported. Six of the nine patients had some improvements for at least one time point, and some displayed longer survival times compared to those normally expected. No HSV was detected in the serum of the patients at any moment of the study. The results suggest that single-dose HSV therapy combined with radiotherapy is safe and has an efficacy potential for GBM treatment.

and in the application steps. In terms of production, considerable safety conditions must be ensured for operators, while, in terms of human applications, it must be guaranteed that no exacerbated toxicity events increase the severity of the illness [56]. On the other hand, while HSV replication-defective vectors have improved safety for human gene therapy, their production is more challenging as the HSV replication components are not present [57]. This production issue is even more complicated in the case of amplicon HSV vectors, where almost all the HSV genes are missing (Fig. 3.27) [58]. The production of these two types of HSV vectors involves steps of *in vitro recombination* of different HSV elements cloned into plasmids, with wild-type HSV virus, and posterior selection and purification of the recombinant vectors. Furthermore, the production of amplicon HSV vectors involves a replication-defective vector as a helper virus, to provide the functions necessary for packaging the amplicon into infectious viral particles. This process, however, poses a problem similar to the one found in the production of *gutless* adenovirus, which is the contamination of the amplicon preparations with the helper virus. Trying to overcome this problem, another process of amplicon HSV vectors was developed. It involves cell transfection of overlapping cosmids, covering the entire HSV genome [59]. Nevertheless, the correct selection of full-recombinant vectors is essential for further production and purification steps, and therefore the method is laborious and currently with low efficiency.

3.5.5 HSV in Clinical Trials

As highlighted in the previous sections, HSV vectors have several advantages for gene therapy applications, especially those related to oncolytic therapy. However, their toxicity and safety issues make their use more complicated than other viral vectors. Data from 2017 revealed that only 3.6% of the vectors used in gene therapy clinical trials are HSV. Nevertheless, in Europe and the USA, a suicide gene therapy with HSV for melanoma was approved, under the commercial name Imlygic® (talimogene laherparepvec). It uses an attenuated HSV-1 vector, presenting a deletion of ICP24.5 and ICP47 genes and the insertion of the granulocyte-macrophage colony-stimulating factor (GM-CSF) gene [60]. This approved therapy and several clinical trials [61] in their final stages highlight the potential of HSV as vectors for oncolytic therapy (Table 3.19).

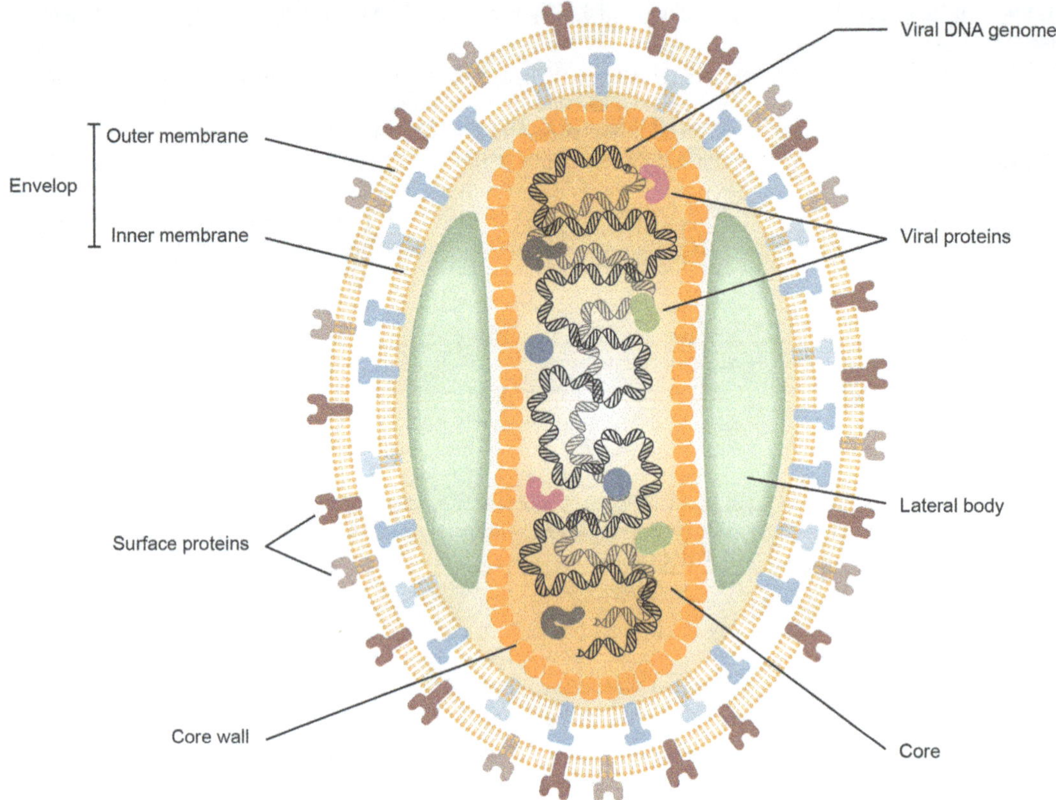

Fig. 3.28 **Vaccinia virus structure and its main components**. Vaccinia viruses are large enveloped virus in a brick shape, formed by two membranes (in their extracellular enveloped virus form). The vaccinia genome, a linear double-stranded DNA molecule, is surrounded by a protein core wall, flanked on two sides by the lateral bodies, which are composed of protein. In the virus core, there are multiple viral proteins that are important for the rapid replicative cycle of the vaccinia virus.

3.6 Vaccinia

Vaccinia virus belongs to the *Poxviridae* family of virus, being widely used and known as the first vaccine for smallpox. Vaccinia virus is among the largest viruses described, with a size of around 250–350 nm, and a brick-like shape. It is an enveloped virus, and its genome consists of linear double-stranded DNA with inverted terminal repeats and a terminal hairpin loop, with approximately 190 kb in length and around 200 genes (Fig. 3.28) [62, 63]. Comparing to the viruses previously described in this chapter, vaccinia has two unique features: (i) its entire life cycle occurs in the cytoplasm, and thus it does not integrate into the host cell genome; and (ii) it exists in two infectious forms: an intracellular mature virus (IMV) and an extracellular enveloped virus (EEV).

3.6.1 Replicative Cycle

Due to the unique characteristics of vaccinia virus, its replication cycle is also unique (Fig. 3.29) [63], relying on its own encoded proteins, which allow a very rapid and efficient cycle (around 10,000 copies of the viral genome are produced within the first 12 hours of infection). Another interesting feature is that vaccinia virus completely shuts down the host cell functions and inhibits the host protein synthesis, inducing a strong cytopathic effect.

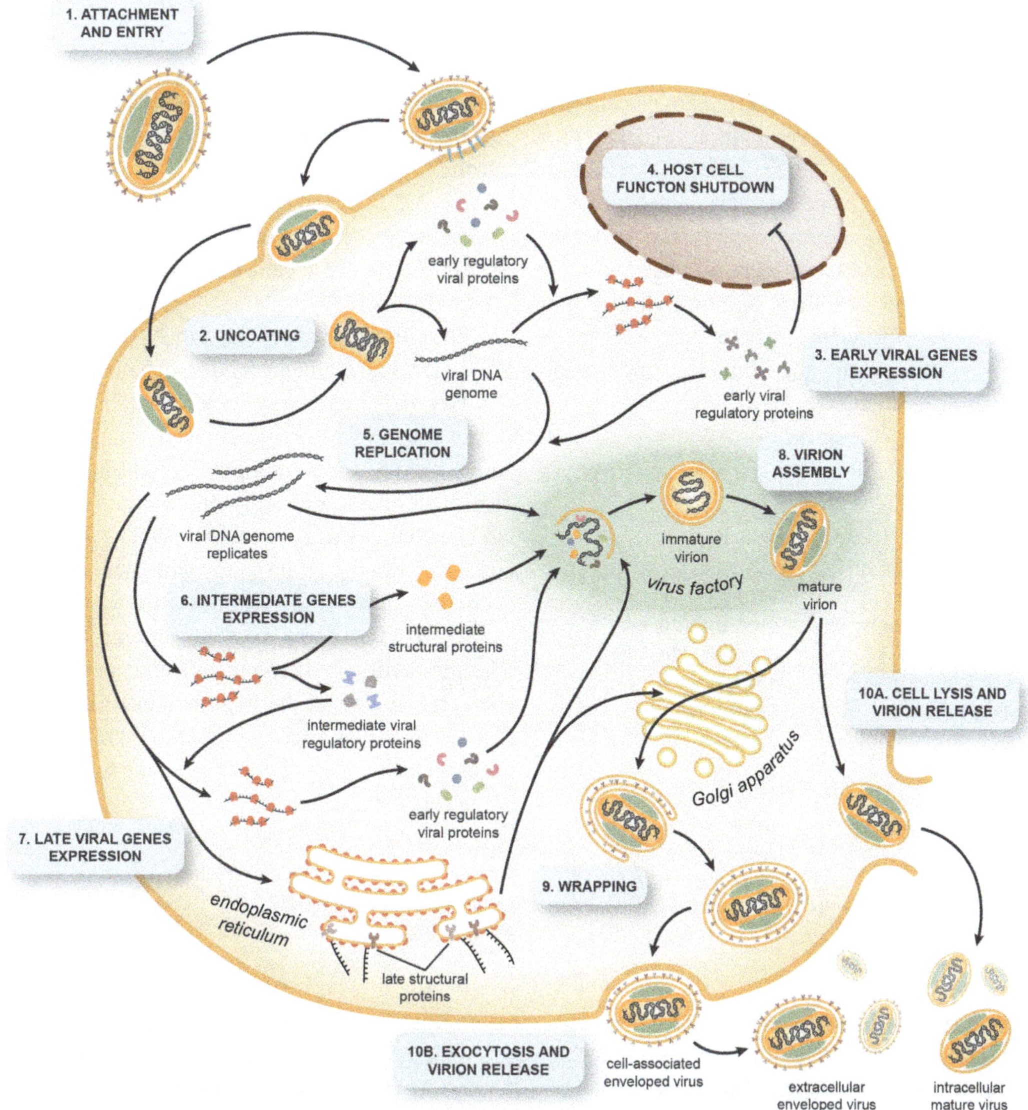

Fig. 3.29 Replicative cycle of the vaccinia virus. The first step is the attachment of the virus to the cell membrane and its entry (1). Inside the cell, viral uncoating occurs, releasing the virus genome and the early regulatory proteins (2). These proteins are crucial for early viral genes expression (3) and the shutdown of the host cell functions (4). After that, in the cytoplasm, the viral genome is replicated (5) and the intermediate genes are expressed, coding for structural proteins (6). This is followed by the late viral genes expression, coding for early regulatory viral proteins (7). All the viral components are assembled (8), and, with the contribution of the Golgi complex, the viral particles are enveloped (9). Finally, the virions are released from the cell as intracellular mature virus, by cell lysis (10A), or as extracellular enveloped virus, by exocytosis (10B).

Mechanisms of cellular entry of vaccinia and other poxvirus are still poorly understood, but as soon as it is in the cytoplasm, it undergoes a process of DNA uncoating and starts the first of three phases of gene expression. The transcription of early genes, which typically encode the proteins needed for viral replication, starts before replication and uses the viral proteins that are encapsulated in the virus. In the second phase, intermediate genes encoding for structural proteins and late transactivators are expressed. The final gene expression phase leads to the production of structural proteins and early transcription factors that are packaged into new viral particles. The replication of the viral genome occurs in the endoplasmic reticulum between the early and intermediate expression phases. The genome and other components are then assembled in Golgi-derived vesicles in the cell cytoplasm. The IMV mature particles then move to the cell surface, where their outer membrane fuses with the cellular membrane, releasing EEV particles.

Table 3.20 Main advantages and disadvantages of vaccinia vectors for gene delivery.

Advantages	Disadvantages
Wide tropism, infecting most mammalian cells	Transient expression of the transgene
High levels of transgene expression	High levels of pre-existing immunity
High cloning capacity of up to 75 kb	Complex genetic engineering
No genome integration	Potential cytopathic effects
Extensive clinical experience	Unknown function of many viral genes
	Difficulty in using heterologous promoters

3.6.2 From Vaccinia Virus to Vaccinia Vectors

Vaccinia virus is not an obvious choice as a vector for gene transfer, as many of its features imply strong disadvantages for gene therapy (Table 3.20). On the other hand, its quick and efficient life cycle makes it very attractive as a vector for oncolytic gene therapy. Another important advantage is its large cloning capacity, of 25 kb without deleting any viral gene and up to 75 kb if some elements are removed [64].

The development of recombinant vaccinia virus vectors (rVVs) is achieved using different methods [63]: (i) homologous recombination taking place inside the cell, although only 0.1% of the new virus will have the transgene; (ii) *in vitro* ligation of the transgene with the vaccinia genome; (iii) use of bacterial artificial chromosome (BAC) to carry the viral genome, allowing the production of recombinant viral genomes in bacteria instead of the mammalian cells used in the two previous approaches; and (iv) recombination of vaccinia amplicons catalyzed by a leporipoxvirus (Shope fibroma virus).

Table 3.21 Example of a gene therapy clinical trial using vaccinia vectors as the delivery system of the therapeutic gene

Study	Mell, L. K., *et al.* (2017) Phase I Trial of Intravenous Oncolytic Vaccinia Virus (GL-ONC1) with Cisplatin and Radiotherapy in Patients with Locoregionally Advanced Head and Neck Carcinoma, *Clin Cancer Res* 23, 5696–5702
Disease	Locoregionally advanced head and neck carcinoma
Therapeutic gene	Vaccinia virus in combination with chemotherapy and radiation
Delivery vector	Vaccinia virus (GL-ONC1, developed as a novel class of immunotherapeutic agent for cancer therapy)
Clinical trial	Single-institution phase I dose escalation clinical trial in the USA
Inclusion criteria	19 patients distributed in 5 cohorts: cohort 1–3 one administration of GL-ONC1 at 3×10^8 pfu, 1×10^9 pfu, and 3×10^9 pfu, respectively; cohort 4: two administrations separated by 5 days at 3×10^9 pfu; cohort 5: four administrations at days 3, 8, 15, and 22 at the same dose of cohort 3 and 4
Type of administration	Intravenous
Clinical outcome	Some adverse events were reported, although the intravenous administration of GL-ONC1 was well tolerated in single and multiple escalating doses. In the follow-up study, 7 deaths and 7 treatment failures were registered, all cases with p16-negative tumors. The remaining 5 treated patients had p16-positive tumors and were all alive and free of disease with median follow-up of 36 months.

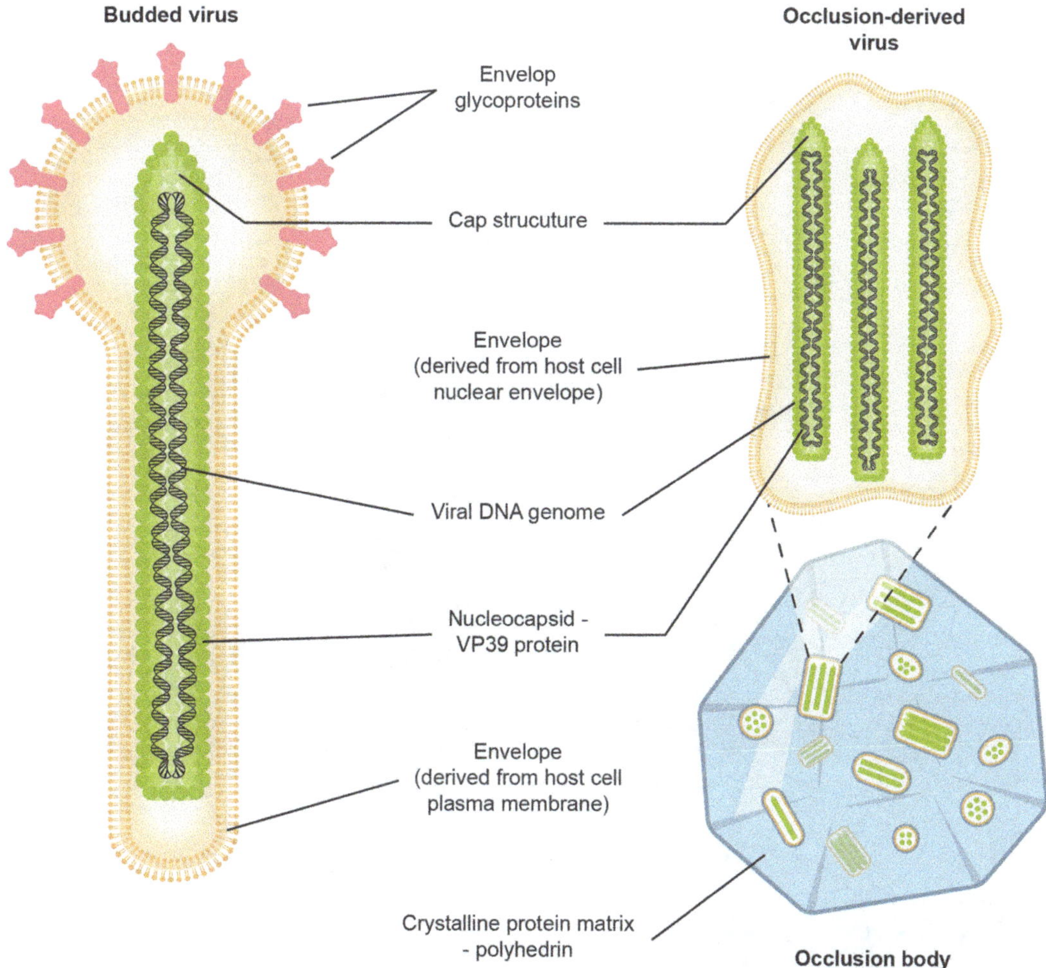

Fig. 3.30 Structure of baculovirus, highlighting the different forms of the virus. Baculoviruses have two forms, the budded virus (BuV) that spreads the infection in the host, and the occlusion-derived virus (ODV) that spreads the infection between hosts. Both forms have a nucleocapsid barrel structure formed by VP39 protein, surrounding a circular double-stranded DNA molecule, and both are enveloped, although the envelope origin is different. The ODV are embedded in a crystalline matrix of proteins forming the occlusion bodies.

3.6.3 Vaccinia Modifications

The complexity of the vaccinia virus increases the difficulty of its genetic engineering, as well as the challenge to develop modifications improving or attenuating some viral features. Nevertheless, in the last years, different studies tried to improve the viral system, for example, using stronger promoters, developing new and better methods for selection of recombinant viruses, and even trying the deletion of some viral genes needed for replication [65, 66].

3.6.4 Vaccinia Vector Production

The classical production of recombinant vaccinia virus involves a recombination step, the selection of the recombinant virus, and a final step of consecutive rounds of amplification of the selected recombinant virus. Alternatively, replication-defective vaccinia vectors can also be produced using a strategy similar to the one employed for *gutless* adenovirus and replication-defective HSV, which utilizes a complementing cell line with the essential viral genes.

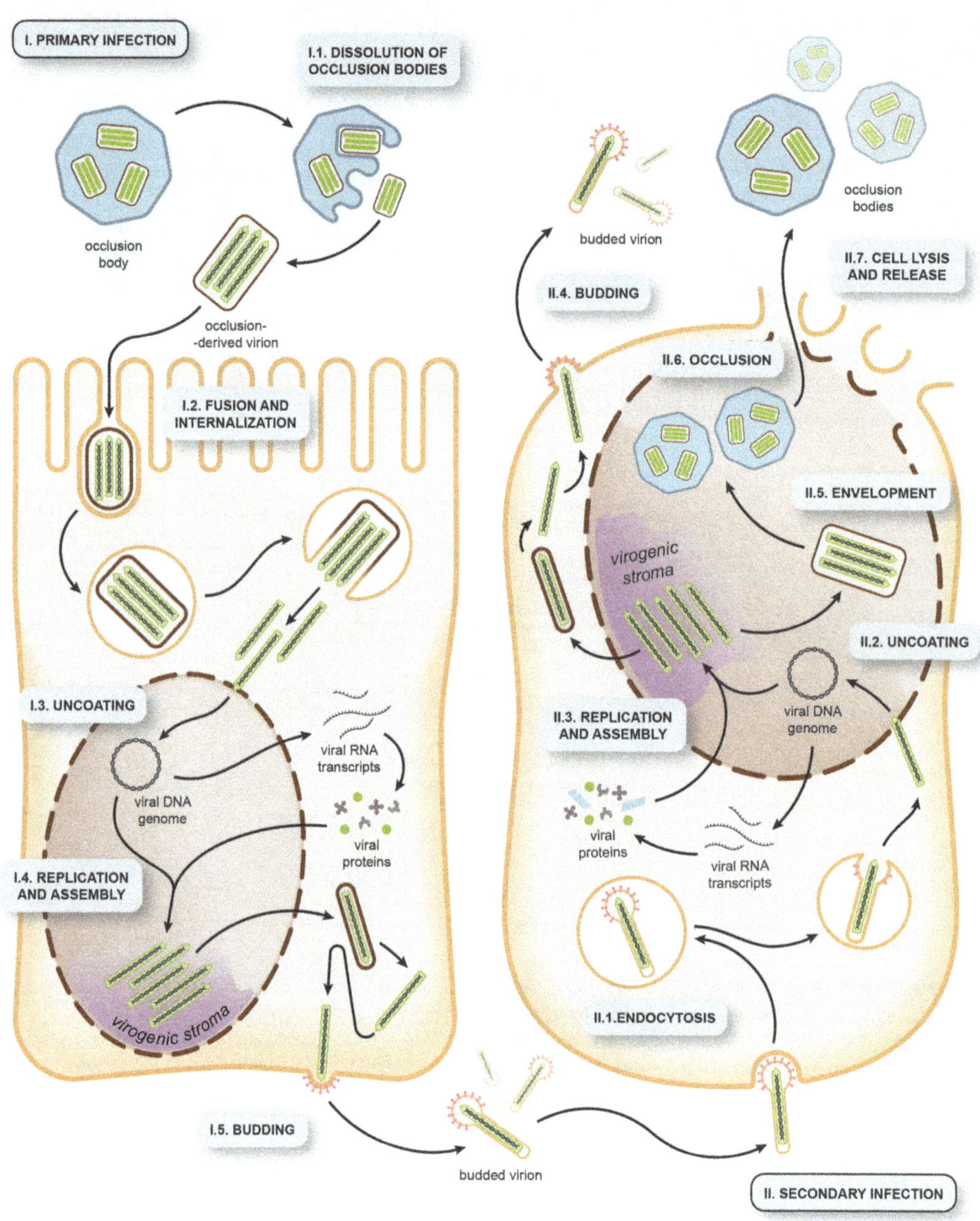

Fig. 3.31 **Replicative cycle of the baculovirus**, highlighting its complexity due to the different forms of the virus. The primary infection (I) is carried out by the occlusion-derived virus (ODV) that dissolves from the occlusion bodies (I.1). After that, ODV fuse with the cell membrane and are internalized (I.2), and in the cytoplasm the virus uncoating occurs (I.3). The next step is the replication of the viral genome and the assembly of the different components of the budded virus (BuV; I.4), which occurs near the cell membrane (I.5). The BuV form is then responsible for the secondary infection (II), entering the cell by endocytosis (II.1), where it is uncoated (II.2), replicated, and assembly of the viral structure takes place (II.3). Some assembled particles form new BuV (II.4), while the ODV are enveloped by the nuclear envelope (II.5), and several particles are grouped into the occlusion bodies (II.6). Finally, the occlusion bodies are released by cell lysis (II.7).

Table 3.22 Main advantages and disadvantages of baculovirus vectors for gene delivery.

Advantages	Disadvantages
Large cloning capacity of up to 38 kb	Transient expression
Easy engineering and production	Becomes inactivated in human serum
No replication in mammalian cells	Unknown long-term effects in mammalian cells
Transduce mitotic and post-mitotic cells	
Low cytotoxicity	
Stable at 4 °C	

3.6.5 Vaccinia in Clinical Trials

Despite the referred limitations, several clinical trials have used vaccinia virus as a vector for gene delivery, especially in oncolytic therapy (Table 3.21) [67].

3.7 Baculovirus

Baculovirus is a general name grouping more than 700 known viruses of the *Baculoviridae* family, that naturally infect insects. The prototype baculovirus is the *Autographa californica* multiple nucleopolyhedrovirus (*Ac*MNPV), which is a large enveloped virus with a rod shape (30–60 × 250–300 nm) [68]. Baculovirus has two forms, the budded virus (BuV) that spreads the infection in the host, and the occlusion-derived virus (ODV) that spreads the infection between hosts (Fig. 3.30) [69]. Both forms have a nucleocapsid barrel structure formed by VP39 protein, but the BuV envelope derives from the host cell membrane, whereas the ODV envelope derives from the nuclear envelope. Moreover, ODV are embedded in a crystalline matrix of proteins to form occlusion bodies (OB). These OB confer protection in the environment but are destroyed when ingested by the host insect, allowing the continuation of the virus replicative cycle. The baculovirus genome is a circular double-stranded DNA molecule, with approximately 180 kb (the *Ac*MNPV has a genome around of 134 kb) and more than 150 open reading frames condensed into a nucleoprotein core with proteins VP39 and VP87 [70, 71].

3.7.1 Replicative Cycle

Baculovirus has a relatively complex replication cycle, mainly due to its two different forms (Fig. 3.31) [72]. The ODV is responsible for the primary infection, which occurs when the host insect larvae ingest the OB, thus releasing the ODV into their midgut. The ODV infect columnar epithelial cells and produce the BuV forms, which then spread the infection throughout the larvae body. The entry of BuV into the insect cells seems mediated by endocytosis, which is also the proposed mechanism for baculovirus entry into mammalian cells (probably through phagocytosis). In the cytoplasm, after endosome escape, the nucleocapsid travels to the nucleus where it releases the genome through the nuclear pore complex. Several studies suggest that the nucleocapsids also enter the nucleus, as some empty nucleocapsids have been found inside the nucleus of infected cells. The baculovirus genome inside the nucleus initiates the transcription cascade for the production of viral mRNAs and nucleocapsid elements. The viral components are assembled in the cytoplasm and after that BuVs are released from the cell.

3.7.2 From Baculovirus to Baculovirus Vectors

Baculovirus have many interesting features that make them very attractive vehicles for gene transfer (Table 3.22). For example, they have a large cloning capacity of up to 38 kb without deleting any viral gene, although, in theory, the size of the nucleocapsid can accommodate up to 100 kb of additional DNA. Another interesting feature is that apparently baculovirus are unable to replicate in mammalian cells, which allows their use without having to delete any essential viral genes. On the side of the disadvantages is the fact that baculovirus vector-mediated expression is transient. Also, the long-term effects of

Table 3.23 Factors that should be taken into account to choose the viral vector

		LV	GammaRV	Ad	AAV	HSV	Vaccinia	Baculovirus
Wild-type virus features	Envelope	Yes	Yes	No	No	Yes	Yes	Yes
	Particle size	80–100 nm	80–100 nm	70–100 nm	18–25 nm	155–240 nm	250–300 nm	30 × 300 nm
	Genetic material	ssRNA	ssRNA	dsDNA	ssDNA	dsDNA	dsDNA	Circular dsDNA
	Genome size	9 kb	7–9 kb	34–43 kb	4.7 kb	152 kb	190 kb	180 kb
	Integration	Yes	Yes	No	No[a]	No	No	No
Recombinant virus features	Transgene capacity	8 kb	7–8 kb	8 kb[b]	4.5 kb	40 kb[c]	25–75 kb	38 kb
	Inflammatory potential	Low	Low	High	Low	High	High	Low
	Immune response	Moderate	Moderate	High	Moderate	High	High	Low
	Duration of expression *in vivo*	Long	Long	Short	Long	Short[d]	Short	Short
	Transduction of quiescent cells	Yes	No	Yes	Yes	Yes	Yes	Yes
	Vector yield	Moderate	Moderate	High	High	High	Moderate	Moderate
	Particle stability	Low	Low	High	High	High	High	High
	Easy engineering	Easy	Easy	Easy	Easy	Difficult	Difficult	Difficult
	Pre-existing host immunity	No	No	Yes	Yes	Yes	Yes	No
	Oncolytic applications	Low	Low	High	Low	Very high	High	Low

[a]It can integrate
[b]It can harbor up to 30 kb in the *gutless* version
[c]It can harbor up to 150 kb in the amplicon version
[d]Long in neurons

baculovirus in mammalian cells are unknown, which could hinder their use for human applications. Nevertheless, the FDA and EMA approval of human vaccine components produced in baculovirus-infected insect cells constitutes an important advance for their future use in gene therapy trials.

3.7.3 Baculovirus Modifications

The pseudotyping of baculovirus envelope was one of the first modifications engineered, aiming to increase the range of cells infected. This included pseudotyping with the widely used VSV-G protein. To increase the specificity of cell transduction, different ligands were also introduced at the baculovirus surface. Baculovirus is also versatile in terms of heterologous promoters, and different tissue-specific or strong expression-driving promoters were introduced with success in these vectors. However, the first studies using recombinant baculovirus *in vivo* failed, probably due to their inactivation by serum components involved in the complement response. Since then, several strategies were designed to overcome this problem, such as using cobra venom factor (CVF) to inhibit complement response and enhance baculovirus vector survival [73].

3.7.4 Baculovirus Production

The baculovirus production process is relatively simple compared to other viral vectors described and can be conducted in low biosafety level facilities. The process involves the generation of recombinant baculovirus in insect cell lines and their posterior concentration and purification to obtain high-titer batches. The construction of recombinant baculovirus can be attained by different methods, including transposition-based and homologous recombination methods.

Importantly, recombinant baculovirus have been used to develop a system to produce other viral vectors, namely AAV and lentivirus. Recombinant AAV particle production in insect cells was achieved by the infection of these cells with different recombinant baculoviruses that produce three AAV capsid proteins and other AAV genes essential for replication and packaging. For the generation of lentiviral vectors a similar strategy was used, with the production relying on four hybrid baculovirus containing the lentivirus genes and their infection in HEK293T cells, thus producing third-generation lentiviral vectors. The baculovirus-based system is able to produce high-titer AAV and lentivirus batches, with a similar *in vitro* and *in vivo* performance to those produced by conventional systems.

3.7.5 Baculovirus in Clinical Trials

The utilization of baculovirus in clinical trials is very limited, although pioneering applications were tested for influenza and lymphoma.

3.8 Choice of the Viral Vector

It is clear from all the descriptions above that the choice of the viral vector used for gene transfer is critical for the success of a particular gene therapy strategy, and it is not an easy decision. Several important factors account for this choice, such as the cargo size needed and the immunogenicity profile of the virus, among many others (Table 3.23). Nevertheless, all the features of a particular viral vector should be taken into consideration when making this selection, since it should reflect several other aspects such as the administration route or the clinical application.

3.9 Oncolytic Virus Applications

As explained before, viral vectors have numerous advantages as carrier systems used to deliver genetic material and currently they are the preferential vector in gene therapy. As mentioned for the HSV, viruses can also be used as oncolytic vectors, an application in which they can act both as carriers and therapy effectors. Oncolytic virus (OV) can be defined as genetically engineered or naturally occurring virus that selectively replicate in and kill cancer cells, without harming normal tissues. Normally, their effect combines tumor-specific cell lysis together with immune stimulation [74]. For this reason, they are now emerging as important agents in cancer treatment. One example is the first oncolytic drug that recently received approval in the USA and Europe, T-Vec® (talimogene laherparepvec), which is a second-generation modified HSV-1 that was approved for the treatment of advanced melanoma.

A wide range of viruses has already been tested as OV for cancer treatment, and they can be categorized into three main groups [75]: (i) wild-type viruses, which are viruses that naturally replicate preferentially in cancer cells and are normally non-pathogenic in humans, which include parvoviruses, myxoma virus (MYXV; poxvirus), Newcastle disease virus (NDV; paramyxovirus), reovirus, and Seneca Valley virus (SVV; picornavirus); (ii) genetically engineered virus to be used as vaccine vectors, including measles virus (MV; paramyxovirus), poliovirus (PV; picornavirus), and vaccinia virus (VV; poxvirus); and (iii) genetically manipulated virus harboring mutations/deletions in genes required for replication in normal cells, but not in tumor cells, which includes adenovirus (Ad), herpes simplex virus (HSV), VV, and vesicular stomatitis virus (VSV; rhabdovirus).

The use of virus in oncolytic therapy has several advantages: (i) they can specifically replicate only in tumor cells, (ii) they can potentiate their effect to neighboring tumor cells, (iii) they enhance sensitivity to conventional therapy and (iv) they have the potential to target dispersed tumor cells. On the other hand, there are also some disadvantages to their use: (i) the immune system can clear the virus before the therapeutic action, (ii) there is a low efficacy in gene transfer *in vivo* and (iii) there is a limited trafficking to some regions of the human body, among other limitations.

This Chapter in a Nutshell

- The main advantage of the viral-based systems used in gene therapy is their high efficiency in delivering transgenes, and because of that they have been extensively used in gene therapy clinical trials.
- On the other hand, the safety profile of viral vectors is one of the main drawbacks for their use, despite the extensive engineering solutions that are continuously being developed. Along with this feature, the limited cloning capacity to introduce transgenes is another disadvantage of their use.
- Lentiviral vectors are mainly based on HIV-1 (retrovirus), which is characterized by the ability to integrate the carried transgene, the transduction of both mitotic and post-mitotic cells and a cloning capacity of around 8 kb.
- Several modifications were made in lentiviral vectors to improve their safety profile, leading led to the development of four generations of vectors. Other improvements include the generation of self-inactivating particles, non-integrative virus and a wide range of pseudotyping options.
- Gamma retrovirus-based vectors are also retrovirus, although their genome structure is simpler than lentivirus. The cloning capacity is around 8 kb; however, they only transduce mitotic cells.
- Adenoviral vectors are based on adenovirus, which are non-enveloped DNA viruses, being the most widely used viral vector in gene therapy clinical trials.
- Depending on the generation, adenoviral vectors can have varying transgene cloning capacity; up to 30 kb in *gutless* adenoviral vectors.
- Adeno-associated viral (AAV) vectors are based on the AAV, that harbors a single DNA molecule comprising two open reading frames. Its cloning capacity is around 4.5 kb.
- One interesting feature of AAV is the existance of a wide range of serotypes with different cellular tropisms, resulting from the presence of capsid protein motifs that are identified by different neutralizing antibodies.
- The herpes simplex viral vectors are mainly based on the HSV-1, which is mainly characterized by a natural tropism to neurons, a high cytotoxic profile and a cloning capacity of up to 150 kb in the amplicon vector version.
- Several modifications were developed to improve HSV vectors' safety profile; how-

ever, the high toxicity of HSV makes it very useful for oncolytic applications.

- Despite their seldom use in gene transfer, vaccinia virus vectors have some interesting advantages that could be explored, such as a quick and efficient replicative cycle, as well as a cloning capacity of 25 kb without the deletion of any viral genes.
- Baculovirus vectors' use in gene therapy is very limited, although the FDA and EMA approval of human vaccine components produced in baculovirus-infected insect cells constitutes an important advance for their future use in gene therapy trials.
- An oncolytic virus can be defined as a genetically engineered or naturally occurring virus that selectively replicates and kills cancer cells, without harming the normal tissues.

Review Questions

1. Which of the following viral vectors induces a potentially high inflammatory profile?
   (a) Adenoviral vectors
   (b) Herpes simplex virus vectors
   (c) Recombinant AAV vectors
   (d) Lentiviral and gamma retroviral vectors
   (e) Gutless adenoviral vectors and herpes simplex virus vectors
2. Which of the following viral vectors has the highest cloning capacity?
   (a) Recombinant AAV vectors
   (b) Gamma retroviral vectors
   (c) Herpes simplex virus vectors
   (d) Second-generation adenoviral vectors
   (e) Lentiviral vectors
3. Which of the following sequences enumerates virus in the order of increasingly difficult genetic engineering?
   (a) AAV, vaccinia, adenovirus
   (b) Vaccinia, baculovirus, lentivirus
   (c) AAV, lentivirus, adenovirus
   (d) Gamma retrovirus, AAV, vaccinia
   (e) Vaccinia, lentivirus, herpes simplex vector
4. Which of the following describes a vector with the ability to transduce post-mitotic cells, has a long-term expression *in vivo*, and has an easy production process?
   (a) Gutless adenoviral vectors
   (b) Lentiviral vectors
   (c) Herpes simplex virus vectors
   (d) Baculovirus vectors
   (e) Gamma retrovirus vectors
5. Which of the following sequences enumerates virus in the order of increasing oncolytic potential?
   (a) Lentivirus, AAV, baculovirus
   (b) AAV, herpes simplex virus, gamma retrovirus
   (c) Gutless adenovirus, lentivirus, herpes simplex virus
   (d) Lentivirus, adenovirus, AAV
   (e) AAV, adenovirus, herpes simplex virus
6. The development of strategies to inactivate retroviral vectors constituted a major advance in the improvement of their safety profile. How was this accomplished?
   (a) Deletion or modification of the U3 region of the LTRs
   (b) Reverse transcriptase inactivation
   (c) Integrase modification
   (d) *gal* and *pol* deletion
   (e) Modification of envelope proteins
7. Lentiviral vectors can lead to transgene integration. Is it possible to revert this viral feature? How?
   (a) No, use AAV as an alternative
   (b) Yes, by having integrase protein without packaging signal
   (c) Yes, by mutating the integrase protein
   (d) No, use non-viral vectors as an alternative
   (e) Yes, by deleting the LTRs
8. Which of the following features better define an oncolytic vector?
   (a) With replication, with transgene integration and with expression in every cell
   (b) Without replication, without transgene integration and with expression in tumor cells
   (c) With replication, hard to produce and with expression in every cell
   (d) Without replication, easy to produce and with expression in tumor cells
   (e) None of the above

References and Further Reading

1. The Journal of Gene Medicine, Gene Therapy Clinical Trials Worldwide. Available at http://www.abedia.com/wiley/
2. Keeler AM, ElMallah MK, Flotte TR (2017) Gene therapy 2017: progress and future directions. Clin Transl Sci 10:242–248
3. Modrow S, Falke D, Truyen U, Schätzl H (2013) Molecular virology. Springer, New York
4. Sakuma T, Barry MA, Ikeda Y (2012) Lentiviral vectors: basic to translational. Biochem J 443:603–618
5. German Advisory Committee Blood (Arbeitskreis Blut), Subgroup 'Assessment of Pathogens Transmissible by Blood' (2016) Human immunodeficiency virus (HIV). Transfus Med Hemother 43:203–222
6. Moore MD, Hu WS (2009) HIV-1 RNA dimerization: it takes two to tango. AIDS Rev 11:91–102
7. Tomás H, Rodrigues A, Alves P, Coroadinha A (2012) Lentiviral gene therapy vectors: challenges and future directions. In: Gene therapy – tools and potential applications. IntechOpen Janeza Trdine, Rijeka, Croatia
8. Kuzembayeva M, Dilley K, Sardo L, Hu WS (2014) Life of psi: how full-length HIV-1 RNAs become packaged genomes in the viral particles. Virology 454–455:362–370
9. Stoltzfus CM (2009) Chapter 1. Regulation of HIV-1 alternative RNA splicing and its role in virus replication. Adv Virus Res 74:1–40
10. Shaw A, Cornetta K (2014) Design and potential of non-integrating lentiviral vectors. Biomedicine 2:14–35
11. Merten OW, Hebben M, Bovolenta C (2016) Production of lentiviral vectors. Mol Ther Methods Clin Dev 3:16017
12. Escors D, Breckpot K (2010) Lentiviral vectors in gene therapy: their current status and future potential. Arch Immunol Ther Exp 58:107–119
13. Dropulic B (2011) Lentiviral vectors: their molecular design, safety, and use in laboratory and preclinical research. Hum Gene Ther 22:649–657
14. Sarkis C, Philippe S, Mallet J, Serguera C (2008) Non-integrating lentiviral vectors. Curr Gene Ther 8:430–437
15. Joglekar AV, Sandoval S (2017) Pseudotyped lentiviral vectors: one vector, many guises. Hum Gene Ther Methods 28:291–301
16. Tiscornia G, Singer O, Verma IM (2006) Production and purification of lentiviral vectors. Nat Protoc 1:241–245
17. Palfi S, Gurruchaga JM, Ralph GS, Lepetit H, Lavisse S, Buttery PC, Watts C, Miskin J, Kelleher M, Deeley S et al (2014) Long-term safety and tolerability of ProSavin, a lentiviral vector-based gene therapy for Parkinson's disease: a dose escalation, open-label, phase 1/2 trial. Lancet 383:1138–1146
18. Cartier N, Hacein-Bey-Abina S, Bartholomae CC, Veres G, Schmidt M, Kutschera I, Vidaud M, Abel U, Dal-Cortivo L, Caccavelli L et al (2009) Hematopoietic stem cell gene therapy with a lentiviral vector in X-linked adrenoleukodystrophy. Science 326:818–823
19. Rodrigues A, Alves P, Coroadinha A (2010) Production of retroviral and lentiviral gene therapy vectors: challenges in the manufacturing of lipid enveloped virus. In: Viral gene therapy. IntechOpen Janeza Trdine, Rijeka, Croatia
20. Maetzig T, Galla M, Baum C, Schambach A (2011) Gammaretroviral vectors: biology, technology and application. Viruses 3:677–713
21. Cavazzana-Calvo M, Hacein-Bey S, de Saint Basile G, Gross F, Yvon E, Nusbaum P, Selz F, Hue C, Certain S, Casanova JL et al (2000) Gene therapy of human severe combined immunodeficiency (SCID)-X1 disease. Science 288:669–672
22. Curiel D (2016) Adenoviral vectors for gene therapy, 2nd edn. Elsevier, London, UK
23. Lee CS, Bishop ES, Zhang R, Yu X, Farina EM, Yan S, Zhao C, Zheng Z, Shu Y, Wu X et al (2017) Adenovirus-mediated gene delivery: potential applications for gene and cell-based therapies in the new era of personalized medicine. Genes Dis 4:43–63
24. Dormond E, Perrier M, Kamen A (2009) From the first to the third generation adenoviral vector: what parameters are governing the production yield? Biotechnol Adv 27:133–144
25. Alba R, Bosch A, Chillon M (2005) Gutless adenovirus: last-generation adenovirus for gene therapy. Gene Ther 12(Suppl 1):S18–S27
26. Capasso C, Garofalo M, Hirvinen M, Cerullo V (2014) The evolution of adenoviral vectors through genetic and chemical surface modifications. Viruses 6:832–855
27. Silva AC, Peixoto C, Lucas T, Kuppers C, Cruz PE, Alves PM, Kochanek S (2010) Adenovirus vector production and purification. Curr Gene Ther 10:437–455
28. Parks RJ, Chen L, Anton M, Sankar U, Rudnicki MA, Graham FL (1996) A helper-dependent adenovirus vector system: removal of helper virus by Cre-mediated excision of the viral packaging signal. Proc Natl Acad Sci U S A 93:13565–13570
29. Wold WS, Toth K (2013) Adenovirus vectors for gene therapy, vaccination and cancer gene therapy. Curr Gene Ther 13:421–433
30. Rosenthal EL, Chung TK, Parker WB, Allan PW, Clemons L, Lowman D, Hong J, Hunt FR, Richman J, Conry RM et al (2015) Phase I dose-escalating trial of Escherichia coli purine nucleoside phosphorylase and fludarabine gene therapy for advanced solid tumors. Ann Oncol 26:1481–1487
31. Weitzman MD, Linden RM (2011) Adeno-associated virus biology. In: Adeno-associated virus: methods and protocols. Methods in molecular biology. New York, USA

32. Samulski RJ, Muzyczka N (2014) AAV-mediated gene therapy for research and therapeutic purposes. Annu Rev Virol 1:427–451
33. Saraiva J, Nobre RJ, Pereira de Almeida L (2016) Gene therapy for the CNS using AAVs: the impact of systemic delivery by AAV9. J Control Release 241:94–109
34. Daya S, Berns KI (2008) Gene therapy using adeno-associated virus vectors. Clin Microbiol Rev 21:583–593
35. Naso MF, Tomkowicz B, Perry WL 3rd, Strohl WR (2017) Adeno-associated virus (AAV) as a vector for gene therapy. BioDrugs 31:317–334
36. Chamberlain K, Riyad JM, Weber T (2016) Expressing transgenes that exceed the packaging capacity of adeno-associated virus capsids. Hum Gene Ther Methods 27:1–12
37. Buie LK, Rasmussen CA, Porterfield EC, Ramgolam VS, Choi VW, Markovic-Plese S, Samulski RJ, Kaufman PL, Borras T (2010) Self-complementary AAV virus (scAAV) safe and long-term gene transfer in the trabecular meshwork of living rats and monkeys. Invest Ophthalmol Vis Sci 51:236–248
38. Bourdenx M, Dutheil N, Bezard E, Dehay B (2014) Systemic gene delivery to the central nervous system using Adeno-associated virus. Front Mol Neurosci 7:50
39. Foust KD, Nurre E, Montgomery CL, Hernandez A, Chan CM, Kaspar BK (2009) Intravascular AAV9 preferentially targets neonatal neurons and adult astrocytes. Nat Biotechnol 27:59–65
40. Polinski NK, Gombash SE, Manfredsson FP, Lipton JW, Kemp CJ, Cole-Strauss A, Kanaan NM, Steece-Collier K, Kuhn NC, Wohlgenant SL et al (2015) Recombinant adenoassociated virus 2/5-mediated gene transfer is reduced in the aged rat midbrain. Neurobiol Aging 36:1110–1120
41. Boutin S, Monteilhet V, Veron P, Leborgne C, Benveniste O, Montus MF, Masurier C (2010) Prevalence of serum IgG and neutralizing factors against adeno-associated virus (AAV) types 1, 2, 5, 6, 8, and 9 in the healthy population: implications for gene therapy using AAV vectors. Hum Gene Ther 21:704–712
42. Choi VW, McCarty DM, Samulski RJ (2005) AAV hybrid serotypes: improved vectors for gene delivery. Curr Gene Ther 5:299–310
43. Ku CA, Pennesi ME (2015) Retinal gene therapy: current progress and future prospects. Expert Rev Ophthalmol 10:281–299
44. Clement N, Grieger JC (2016) Manufacturing of recombinant adeno-associated viral vectors for clinical trials. Mol Ther Methods Clin Dev 3:16002
45. Shin JH, Yue Y, Duan D (2012) Recombinant adeno-associated viral vector production and purification. Methods Mol Biol 798:267–284
46. Bartus RT, Baumann TL, Siffert J, Herzog CD, Alterman R, Boulis N, Turner DA, Stacy M, Lang AE, Lozano AM et al (2013) Safety/feasibility of targeting the substantia nigra with AAV2-neurturin in Parkinson patients. Neurology 80:1698–1701
47. Tardieu M, Zerah M, Husson B, de Bournonville S, Deiva K, Adamsbaum C, Vincent F, Hocquemiller M, Broissand C, Furlan V et al (2014) Intracerebral administration of adeno-associated viral vector serotype rh.10 carrying human SGSH and SUMF1 cDNAs in children with mucopolysaccharidosis type IIIA disease: results of a phase I/II trial. Hum Gene Ther 25:506–516
48. Jacobson SG, Cideciyan AV, Ratnakaram R, Heon E, Schwartz SB, Roman AJ, Peden MC, Aleman TS, Boye SL, Sumaroka A et al (2012) Gene therapy for leber congenital amaurosis caused by RPE65 mutations: safety and efficacy in 15 children and adults followed up to 3 years. Arch Ophthalmol 130:9–24
49. Glorioso JC (2014) Herpes simplex viral vectors: late bloomers with big potential. Hum Gene Ther 25:83–91
50. Liu F, Zhou ZH (2007) Comparative virion structures of human herpesviruses. In: Arvin A, Campadelli-Fiume G, Mocarski E, Moore PS, Roizman B, Whitley R, Yamanishi K (eds) Human herpesviruses: biology, therapy, and immunoprophylaxis. Cambridge University Press, Cambridge
51. Manservigi R, Argnani R, Marconi P (2010) HSV recombinant vectors for gene therapy. Open Virol J 4:123–156
52. Lachmann R (2004) Herpes simplex virus-based vectors. Int J Exp Pathol 85:177–190
53. Marconi P, Argnani R, Epstein AL, Manservigi R (2009) HSV as a vector in vaccine development and gene therapy. Adv Exp Med Biol 655:118–144
54. Shen Y, Nemunaitis J (2006) Herpes simplex virus 1 (HSV-1) for cancer treatment. Cancer Gene Ther 13:975–992
55. Epstein AL (2009) Progress and prospects: biological properties and technological advances of herpes simplex virus type 1-based amplicon vectors. Gene Ther 16:709–715
56. Argnani R, Lufino M, Manservigi M, Manservigi R (2005) Replication-competent herpes simplex vectors: design and applications. Gene Ther 12(Suppl 1):S170–S177
57. Berto E, Bozac A, Marconi P (2005) Development and application of replication-incompetent HSV-1-based vectors. Gene Ther 12(Suppl 1):S98–S102
58. Epstein AL (2005) HSV-1-based amplicon vectors: design and applications. Gene Ther 12(Suppl 1):S154–S158
59. Zhou F, Gao SJ (2011) Recent advances in cloning herpesviral genomes as infectious bacterial artificial chromosomes. Cell Cycle 10:434–440
60. Pol J, Kroemer G, Galluzzi L (2016) First oncolytic virus approved for melanoma immunotherapy. Oncoimmunology 5:e1115641
61. Markert JM, Razdan SN, Kuo HC, Cantor A, Knoll A, Karrasch M, Nabors LB, Markiewicz M, Agee BS, Coleman JM et al (2014) A phase 1 trial of oncolytic HSV-1, G207, given in combination with radiation for recurrent GBM demonstrates safety and radiographic responses. Mol Ther 22:1048–1055

62. Harrison SC, Alberts B, Ehrenfeld E, Enquist L, Fineberg H, McKnight SL, Moss B, O'Donnell M, Ploegh H, Schmid SL et al (2004) Discovery of antivirals against smallpox. Proc Natl Acad Sci U S A 101:11178–11192
63. Guo ZS, Bartlett DL (2004) Vaccinia as a vector for gene delivery. Expert Opin Biol Ther 4:901–917
64. Mastrangelo MJ, Eisenlohr LC, Gomella L, Lattime EC (2000) Poxvirus vectors: orphaned and underappreciated. J Clin Invest 105:1031–1034
65. Wyatt LS, Xiao W, Americo JL, Earl PL, Moss B (2017) Novel nonreplicating vaccinia virus vector enhances expression of heterologous genes and suppresses synthesis of endogenous viral proteins. MBio 8:e00790-17
66. Hodge JW, Poole DJ, Aarts WM, Gomez Yafal A, Gritz L, Schlom J (2003) Modified vaccinia virus ankara recombinants are as potent as vaccinia recombinants in diversified prime and boost vaccine regimens to elicit therapeutic antitumor responses. Cancer Res 63:7942–7949
67. Mell LK, Brumund KT, Daniels GA, Advani SJ, Zakeri K, Wright ME, Onyeama SJ, Weisman RA, Sanghvi PR, Martin PJ et al (2017) Phase I trial of intravenous oncolytic vaccinia virus (GL-ONC1) with cisplatin and radiotherapy in patients with locoregionally advanced head and neck carcinoma. Clin Cancer Res 23:5696–5702
68. Rohrmann GF (2013) Baculovirus molecular biology, 3rd edn. National Center for Biotechnology Information, Bethesda
69. Au S, Wu W, Pante N (2013) Baculovirus nuclear import: open, nuclear pore complex (NPC) sesame. Viruses 5:1885–1900
70. van Oers MM, Vlak JM (2007) Baculovirus genomics. Curr Drug Targets 8:1051–1068
71. Chen YR, Zhong S, Fei Z, Hashimoto Y, Xiang JZ, Zhang S, Blissard GW (2013) The transcriptome of the baculovirus Autographa californica multiple nucleopolyhedrovirus in Trichoplusia ni cells. J Virol 87:6391–6405
72. Ghosh S, Parvez MK, Banerjee K, Sarin SK, Hasnain SE (2002) Baculovirus as mammalian cell expression vector for gene therapy: an emerging strategy. Mol Ther 6:5–11
73. Ono C, Ninomiya A, Yamamoto S, Abe T, Wen X, Fukuhara T, Sasai M, Yamamoto M, Saitoh T, Satoh T et al (2014) Innate immune response induced by baculovirus attenuates transgene expression in mammalian cells. J Virol 88:2157–2167
74. Lawler SE, Speranza MC, Cho CF, Chiocca EA (2017) Oncolytic viruses in cancer treatment: a review. JAMA Oncol 3:841–849
75. Goldufsky J, Sivendran S, Harcharik S, Pan M, Bernardo S, Stern RH, Friedlander P, Ruby CE, Saenger Y, Kaufman HL (2013) Oncolytic virus therapy for cancer. Oncolytic Virother 2:31–46

Barriers to Gene Delivery

4

As mentioned before, the choice of a particular vector for gene delivery is a critical point in the success of gene therapy. From the details presented in the previous chapters concerning the different possible vectors, important conclusions emerge. First, an ideal vector should combine several features, ensuring both safety and efficacy. Second, there are advantages and disadvantages for each of the described vectors, and the choice between them is very much dependent on the target disease/condition. All in all, a correct selection is a crucial point for the success of a particular gene therapy.

In order to ensure the success of a particular therapeutic approach, vectors need to overcome several barriers. These barriers can be grouped into three main categories: extracellular barriers, intracellular barriers and technical barriers (Fig. 4.1). It is clear that, for non-viral systems, overcoming most of the barriers is especially challenging, and many of them could explain the lower gene delivery efficiency non-viral vectors display when compared to viral vectors. Nevertheless, some of the barriers also pose an obstacle for gene delivery mediated by viral vectors, for example, some physiological barriers or the immune response (Table 4.1). Additionally, one important point is that some barriers may not be present or be relevant in specific gene therapy applications. For example, extracellular barriers do not apply to *ex vivo* gene therapy, and overcoming endothelial barriers is not a very relevant issue if local delivery to cells/tissue is performed.

4.1 Extracellular Barriers

4.1.1 Unspecific Interactions

This barrier to gene delivery refers to nonspecific interactions of the vectors with molecular or cellular components that could reduce their specificity and efficiency. It is particularly important and relevant if a systemic administration route is used, affecting mainly the non-viral chemical vectors, such as liposomes or polymers [1].

Physicochemical characteristics such as the charge or size of the vectors can lead to nonspecific interactions with serum protein or enzymes (vectors with a positive charge can interact with negatively charged proteins and enzimes, for example), thus resulting in a decrease of efficiency. Additionaly, vectors may-interact nonspecifically with cell surface molecules such as glycosaminoglycans, also leading to a decrease in their efficiency. The interaction of these non-viral vectors with some plasma compounds leads to their opsonization, causing aggregation, dissociation, and degradation of the DNA-vector complex. Furthermore, the presence of degradative enzymes in the bloodstream can lead to the degradation of the DNA, if it is not well protected inside the vector. In the case of viral vectors, systemic delivery can also be inefficient if the vector is not able to evade immune detection.

The extracellular matrix poses another barrier, as it provides additional conditions for unspecific

C. Nóbrega et al., *A Handbook of Gene and Cell Therapy*,
https://doi.org/10.1007/978-3-030-41333-0_4

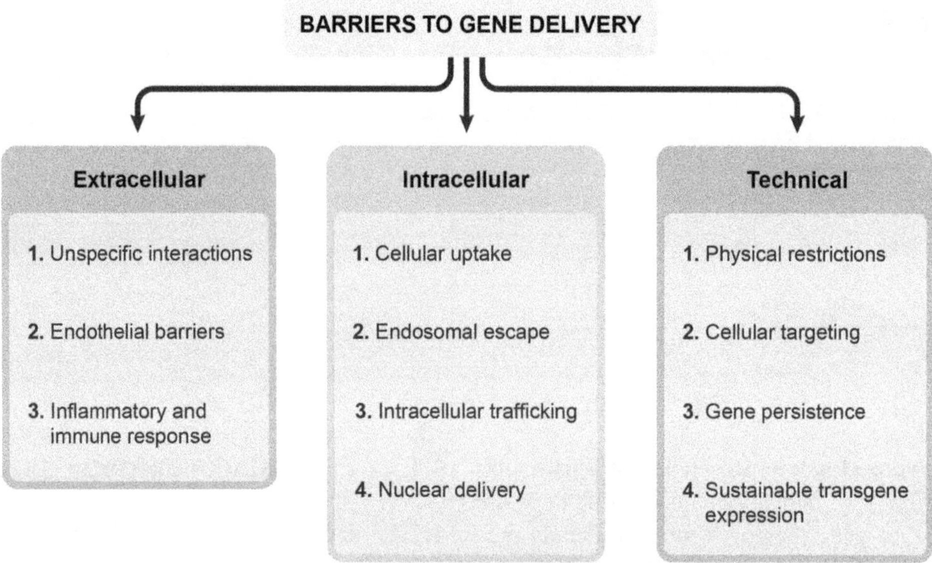

Fig. 4.1 Overview of the main barriers to gene delivery, categorized in extracellular, intracellular, and technical barriers.

Table 4.1 Relevance of the different barriers to gene delivery, for viral and non-viral vectors.

	Viral vectors	Non-viral vectors
Unspecific interactions	+	+++
Endothelial barriers	+++	+++
Inflammatory and immune response	+++	+
Cellular uptake	+	+++
Endosomal escape	+	+++
Intracellular trafficking	+	+++
Nuclear delivery	+	+++
Physical restrictions	+	+++
Cellular targeting	+++	+++
Gene persistence	+++	+++
Sustainable gene expression	+++	+++

+++ Barrier more relevant

interactions, especially for non-viral vectors. The extracellular matrix is a complex and dense network of molecules that fill the spaces between cells. Thus, the vectors can interact in a nonspecific manner with the matrix components, preventing their binding to the cellular membranes and further entry into (or uptake by) the cells.

Different modifications of chemical non-viral vectors were developed to minimize the occurence of unspecific interactions, such as protecting the DNA from nuclease activity by using cationic liposomes (e.g., DOTAP:DOPE) or using PEGylated lipids and polymers (lypids and polymers combined with polyethylene glycol - PEG) to prevent aggregation of the vectors [2].

4.1.2 Endothelial Barriers

If a systemic delivery route is used for gene delivery, another set of barriers appears. The endothelium functions as a major barrier at the interface between the blood and the tissues, by limiting the entry of plasma, cells and molecules from the circulation into the organ parenchyma [3]. Thus, it is clear that both viral and non-viral vectors must be able to cross this barrier in order to reach the target cells/tissue.

For example, the blood-brain barrier (BBB) constitutes an important and efficient endothelial barrier to gene delivery in the central nervous system (CNS), and even most of the viral vectors are unable to cross it. The BBB is a highly restricting barrier, regulating the movement of molecules, ions and cells between the blood and the CNS, thus tightly controlling CNS homeostasis (Fig. 4.2). It is now known that the loss of the restrictive proprieties of the BBB underlies the pathogenesis and progression of several neurological diseases [4].

Despite being restrictive, there are diverse pathways that transport molecules across the BBB (Fig. 4.2), which can be explored to facilitate the entry of the vector (and the therapeutic gene) into the CNS. For example, adding an anti-transferrin receptor antibody to the liposome surface allows particles to cross the BBB, due to the presence of transferrin receptors at the surface of endothelial cells and a receptor-mediated phenomenon of transcytosis [5]. Another strategy includes the incorporation of a short peptide derived from rabies virus glycoprotein (RVG-9r) into the liposomes. RVG-9r has a known tropism for the CNS, thus providing the ability for the liposomes to cross the BBB [6]. As already mentioned in the previous chapter, some adeno-associated virus (AAV) serotypes are also able to cross the BBB and can thus be used in the systemic delivery of genes targeting the CNS [7].

4.1.3 Inflammatory and Immune Response

The immune response probably constitutes one of the major barriers to gene delivery, and it targets both non-viral and viral vectors. In a gene therapy context, the reaction of the immune system to exogenous molecules and their carriers is divided into two main responses: innate immunity and adaptive immunity (Fig. 4.3) [8].

The innate immune response occurs shortly after gene delivery and consists of a rapid secretion of proinflammatory cytokines and chemokines and a nonspecific global stimulation of the immune system. This secretion of proinflammatory factors can be restricted to the administration site, if local delivery is used, or be more widespread in the case of systemic delivery.

The adaptive immune response occurs later, with the recruitment of antigen-presenting cells (APCs) that process and expose antigens, the activation of lymphocytes) and the production of proinflammatory cytokines. It includes a cell-mediated response component, involving T-cell (CD4+ and CD8+) and natural killer (NK)-cell recruitment, and a humoral response component mediated by B-cells, that is characterized by the production of neutralizing antibodies specific to the vector or transgene-derived antigens.

Despite being relevant for both types of vectors, the activation of the immune response is particularly important if gene delivery is mediated by viral vectors [9]. For example, adenoviral and herpes simplex vectors elicit a strong immune response, both innate and adaptive. On the contrary, AAV and lentiviral vectors do not elicit a strong innate response, although AAV can lead to an adaptive immune response due to the presence of neutralizing antibodies for the different serotypes [10]. The immune response against the viral vector or to the transgene contributes to a decrease of gene transfer efficiency, to the elimination of the therapy or the treated cells over time and even to the production of massive inflammation reactions. The death of Jesse Gelsinger was caused by an exacerbation of the innate immunity triggered by the injection of heavy doses of adenoviral vectors [11].

Being a crucial barrier to gene delivery, several strategies were developed and tested to circumvent the immune response, including lowering the vector doses, delivering the gene to immune-privileged sites (e.g., the eye), modifying the genome and structure of the vector or performing immune suppression [12].

4.2 Intracellular Barriers

4.2.1 Cellular Uptake

The entry of the carrier vector into the target cell, crossing the plasma membrane, constitutes the first intracellular barrier to gene delivery. Despite being a relevant barrier for both types of vectors, cellular internalization is a problem mainly for non-viral vectors.

Cellular membranes serve as selective barriers to the entry of molecules into cells. Small compounds can diffuse across the membrane passively, whereas the entry of polar and large molecules is achieved through transporter pro-

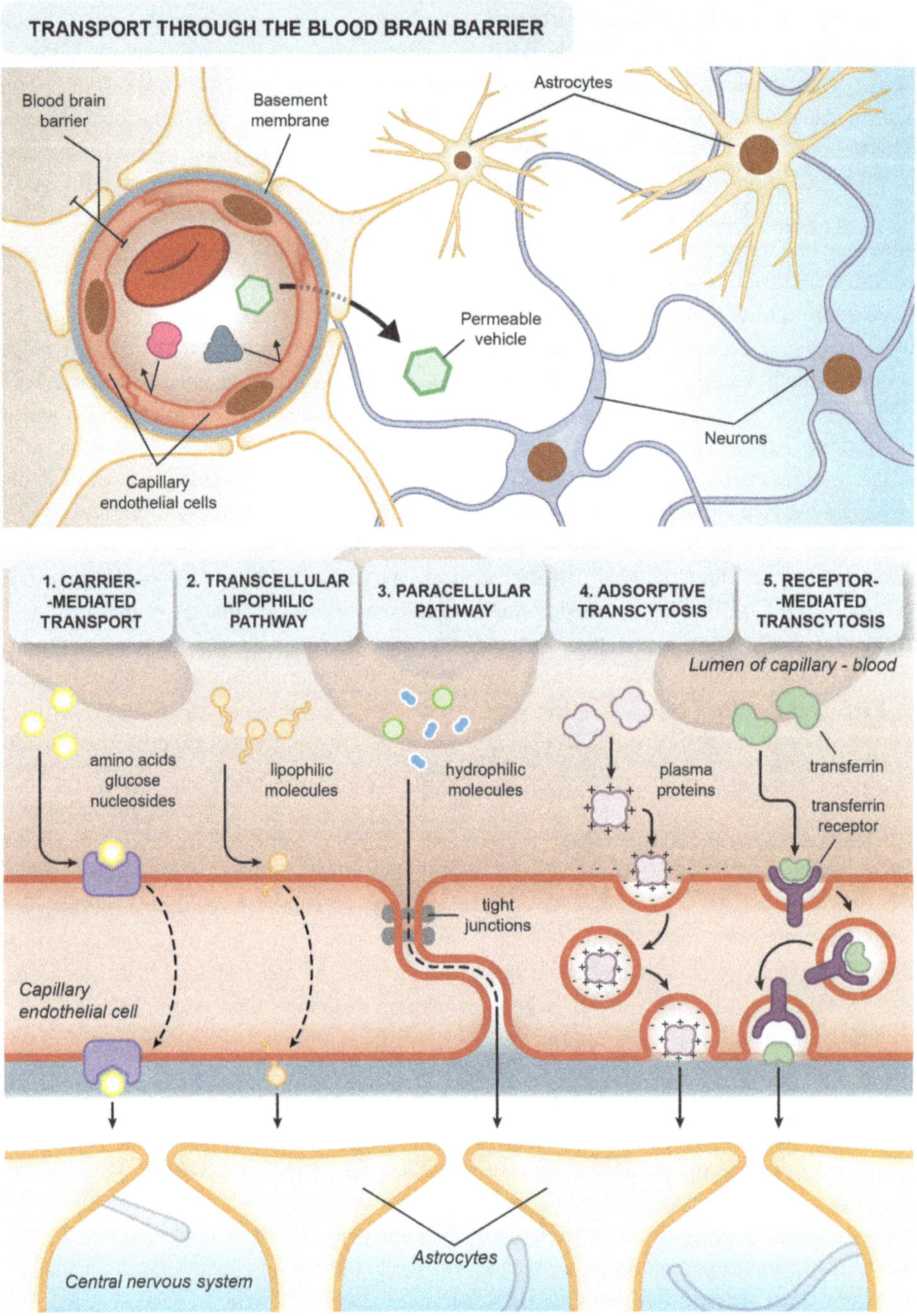

Fig. 4.2 Main pathways for molecule transport across the blood-brain barrier (BBB). The BBB is a highly restricting barrier formed by capillary endothelial cells and astrocytes, that regulates the movement of molecules, ions and cells between the blood and the central nervous system. The capillary basement membrane contacts with the processes of astrocytes, and these cells interact with neurons. Since the BBB is selective, several mechanisms exist in order to transport specific molecules across it. The *carrier-mediated transport* machineries (e.g., Glut1) are responsible for the transportation of small endogenous molecules, such as amino acids and glucose (1). The *transcellular lipophilic pathway* corresponds to the passive diffusion of small lipid-soluble substrates (2). The *paracellular pathway* is primarily used by small hydrophilic molecules (3). *Adsorptive transcytosis* is a low-capacity route for the receptor-independent passage of some molecules, such as albumin (4). *Receptor-mediated transcytosis* is responsible for the transportation of large molecule, through receptor-mediated endocytosis/transcytosis (5).

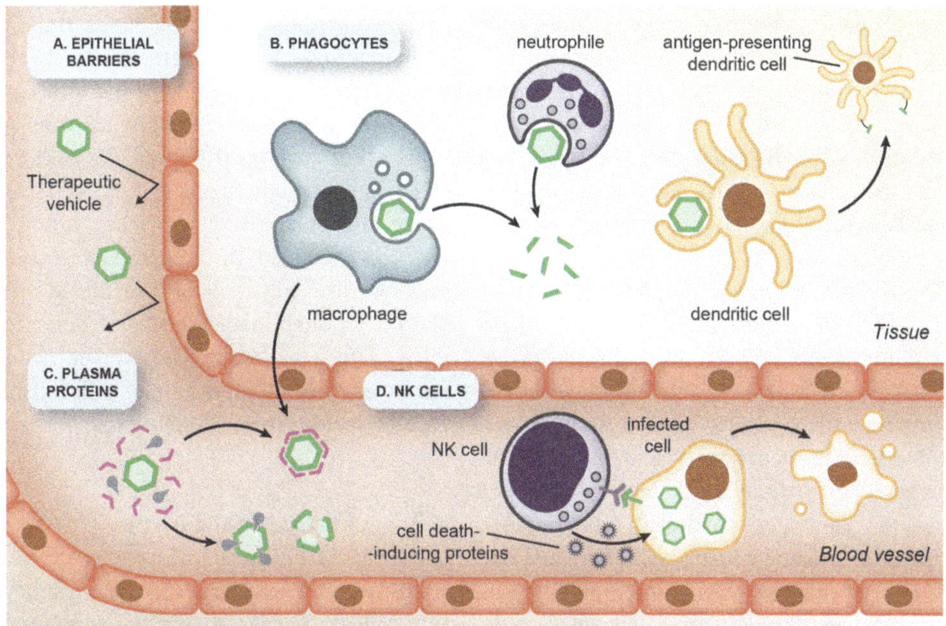

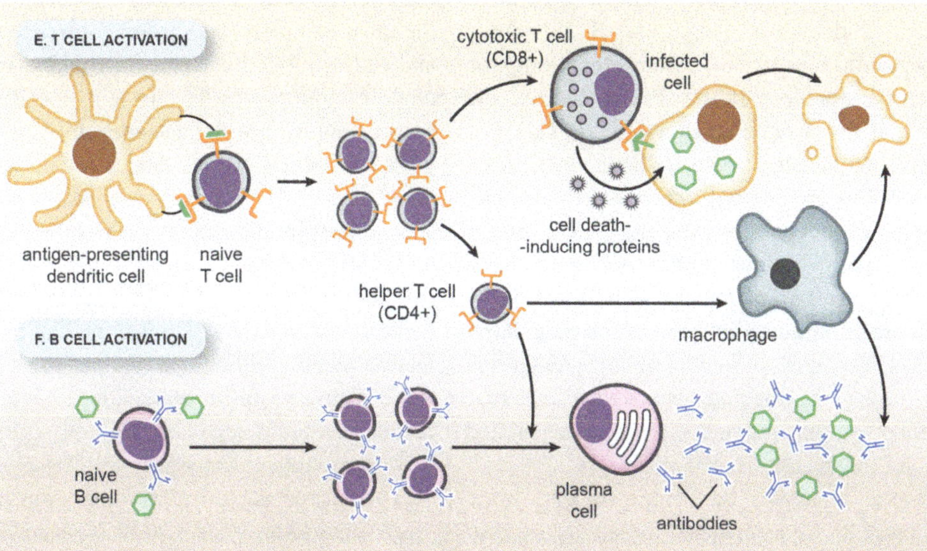

Fig. 4.3 Innate and adaptive immunity, as a response to exogenous molecules delivered in gene therapy. The innate immune response occurs shortly after gene delivery and involves several players. The *epithelial barriers* (A) act as physical and chemical obstacles, preventing the passage of the molecules from circulation to the tissue. If molecules overcome this barrier, different phagocytic cells, or *phagocytes* (B), will engulf the exogenous molecules and destroy them. Macrophages, for example, recognize exogenous molecules, engulfing and destroying them through phagocytosis. They can also release cytokines to signal and recruit other cells. Dendritic cells are specialized antigen-presenting cells, that along with macrophages act as a bridge between the innate immune system and the adaptive immune system. Certain plasma proteins (C) can lead to the opsonization of the carrier molecules and/or to their destruction. Finally, the *natural killer (NK) cells* (D) destroy cells that were modifyed by the exogenous molecules, upon recognition by specific receptors. The adaptive immune response occurs later, with the recruitment of antigen-presenting cells carrying the antigens, the *activation* of both CD4+ and CD8+ *T cells* (E), and the production of proinflammatory cytokines. Another component of this system is the *activation* of *B-cells* (F), which express antibodies on their cell surface. B-cell are activated when they encounter their target antigen and then quickly divide in order to produce either memory B-cells or effector B-cells, which differentiate into antibody-segregating plasma cells.

teins or by the formation of vesicles. In the case of large molecules, their entry occurs through different mechanisms collectively named **endocytosis**. This process actively transports molecules into the cells through vesicles, and functions as a way for the cells to process and destroy dangerous particles (like viruses or bacteria) and reshape the membrane surface. In a very broad sense, endocytosis can be divided into two main steps: (i) binding to the membrane, followed by membrane deformation, and (ii) intracellular sorting. The endocytic vesicle cargo includes membrane proteins and their extracellular ligands, which after internalization are involved in different processes such as cell signaling or nutrient uptake.

Several endocytic pathways have been described so far, but the main, well-characterized, pathways are phagocytosis, micropinocytosis, clathrin-mediated endocytosis, caveolae-mediated endocytosis and clathrin- and caveolin-independent endocytosis (Fig. 4.4) [13]. These pathways differ in several features, including the size and morphology of the vesicles formed, the cargo and the proteins involved (Table 4.2). For example, in phagocytosis, very large vesicles (up to 1000 nm) are formed, whereas in caveolae-mediated endocytosis the vesicles have a size of around 50 nm. The different entry pathways and their features are extremely important in the development of non-viral vectors for gene delivery. For example, anionic polymers enter the cells via caveolae-mediated endocytosis, whereas neutral or cationic polymers use the clathrin- and caveolin-independent mechanisms [14].

Several strategies and improvements were developed trying to enhance the ability of non-viral vectors to penetrate the membrane, such as conjugating the vectors with cell-penetrating peptides (CPPs), which are short peptides able to carry cargoes and cross the membrane without affecting its structural integrity [15]. Other strategies are based on the use of ligands (e.g., transferrin), that are coupled to the vector and bind to particular cell surface receptors (e.g., transferrin receptor). This facilitates entry because the vector is internalized when the ligand undergoes receptor-mediated endocytosisis [16].

In the case of viral vectors, the natural cellular entry mechanisms used by viruses can be explored or altered. Viral-mediated entry into the cells requires specific interactions between host cell receptors and proteins of the viral envelope or capsid, often named viral attachment proteins (VAP). For enveloped viruses, the cellular entry is accomplished by fusion with the cellular membrane. For non-enveloped viruses, normally, the entry is mediated by a specific mechanism of endocytosis, such as clathrin-mediated endocytosis [17]. In order to alter the cellular entry characteristics of a particular viral vector, the system can be engineered for example, with capsid/envelope molecules from another virus.

4.2.2 Endosomal Escape

Endo-lysosomes are probably the most important intracellular barrier to the delivery of genetic material to cells. Endocytosis mechanisms are the most common route for nanoparticle uptake, and following cellular entry the endocytic vesicles become accessible to early endosomes, before the fusion with late endosomes and finally with lysosomes, where their content is degradated (Fig. 4.4). The lysosome, as an organelle specialized in digestion, with a low pH and hydrolytic enzymes, is an important hurdle to be overcome by vectors entering the cells. Of course, the specific endocytic pathway mediating vector internalization will be determinant to its fate, as some pathways favor degradation, while others favor endosomal accumulation and sorting in a non-degradative manner [18].

This barrier is particularly important for non-viral vectors, as viruses have evolved efficient systems for endosomal escape and release. For example, and as mentioned above, the cellular entry of enveloped viruses is mediated by fusion with the cellular membrane, directly delivering the internal part of the virus into the cytoplasm and thus avoiding the endo-lysosome pathway. Non-enveloped viruses escape the endosome by lysing the vesicle membrane or by generating a pore that allows their release [19].

The understanding of the mechanisms of viral and bacterial escape is very important in the development of improvements that allow non-viral vectors to escape the endo-lysosomal path-

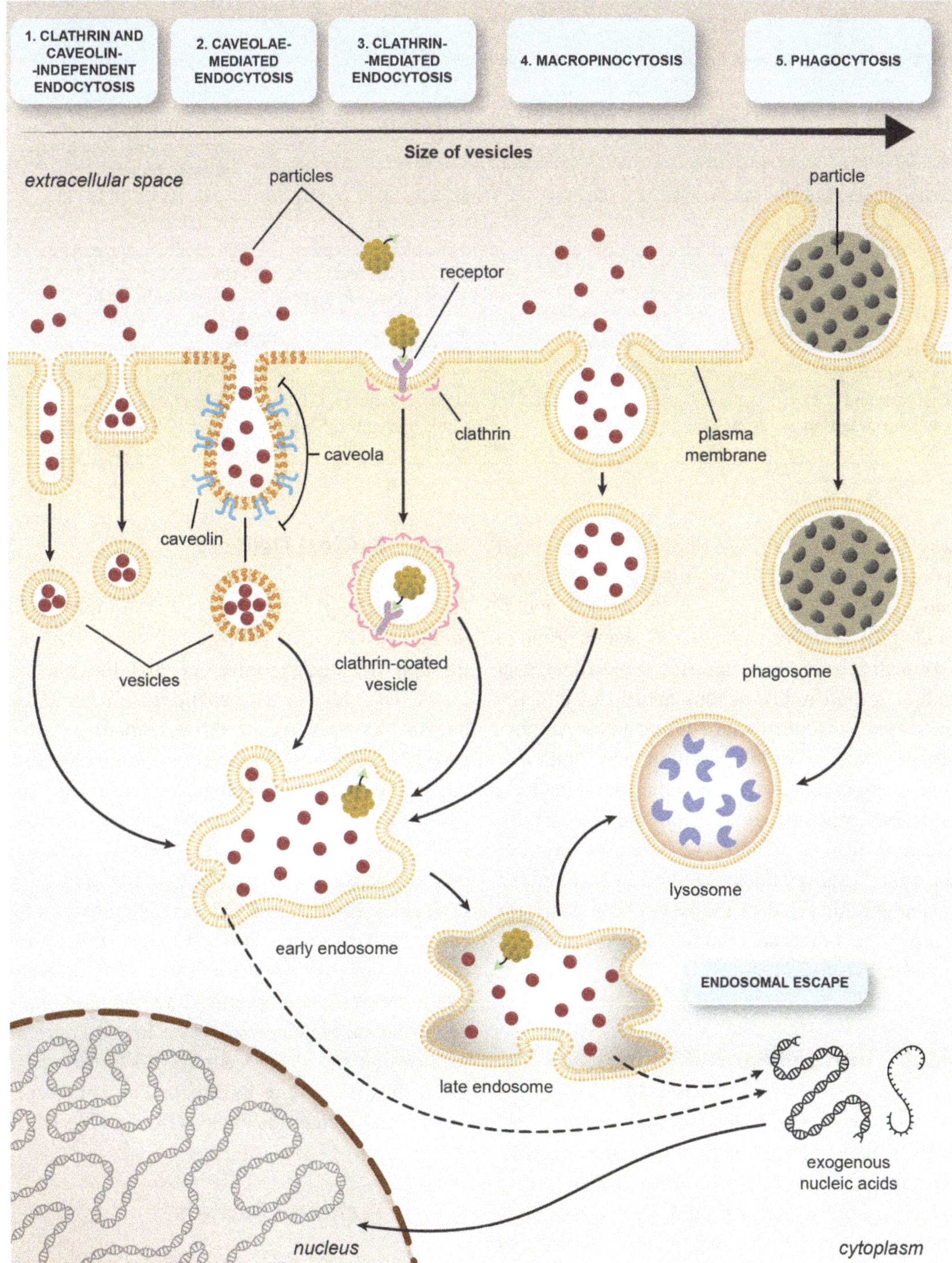

Fig. 4.4 Overview of the different endocytosis mechanisms. Cellular membranes act as selective barriers to the entry of external molecules. Small compounds can passively diffuse across the membrane, whereas polar and large molecules enter through transporters or through vesicles. The latter case involves different mechanisms collectively called endocytosis. In the case of *clathrin and caveolin-independent endocytosis*, molecules enter the cell through invagination of the cellular membrane, independently from the clathrin and caveolin proteins (1). This mechanism is used by some neutral and cationic polymers to enter the cells. The formation of endocytic vesicles in *caveolae-mediated endocytosis* and *clathrin-mediated endocytosis* depends on caveolin (2) and clathrin, respectively (3). *Macropinocytosis* (4), involves engulfment and uptake of large amounts of fluids, particles and membranes portions. Finally, *phagocytosis* (5) is a cellular entry mechanism for very large particles (≥0.5 μm), which are engulfed by sizeable plasma membrane protrusions.

Table 4.2 Main features of the different endocytosis pathways

	Morphology	Vesicle size	Cargoes	Proteins implicated
Phagocytosis	Cargo shaped	1000 nm	Pathogens, apoptotic remains	Actin, dynamin, IQGAP1, amphiphysin1, Rho kinase, adhesion proteins
Macropinocytosis	Disarranged	500 nm	Fluid phase markers, tyrosine kinase receptor	Actin, PAK1, PI3K, Ras, Src, HDAC6
Clathrin-mediated endocytosis	Vesicular	100 nm	Tyrosine kinase receptor, protein-coupled receptors, transferrin receptor, anthrax toxin	Dynamin, clathrin, AP2, epsin, SNX9, synaptojanin, actin, amphiphysin and others
Caveolae-mediated endocytosis	Vesicular/ tubovesicular	50 nm	CTxB, SV40, GPI-linked proteins	Dynamin, caveolins, PTRF, src, PKC, actin
Clathrin- and caveolin-independent endocytosis	Tubular/ ring-like	50–80 nm	Fluid phase markers, CTxB, GPI-linked proteins, SV40, Integrins	ARHGAP10, actin, GRAF1, other GRAFs, CLICs, dynamin

way. For example, the use of binding agents such as cationic amphiphilic peptides (AMPs) seems capable of creating pores in the vesicle membrane [20]. Another strategy uses agents prone to protonation (e.g., histidine-rich molecules), that induce a high influx of ions and water into the endosome, leading to its rupture [21]. Another strategy that is widely used in non-viral vectors development is their coupling with bacteria and virus proteins/peptides that are relevant for escaping from the endo-lysosome pathway. For example, viral peptides like the HA2 subunit of hemagglutinin (HA) or the HIV-1 trans-activator gene product (Tat) are commonly incorporated in non-viral vectors [19].

4.2.3 Intracellular Trafficking

After escaping the endosome, the vectors must deliver their genetic material to the nucleus. This intracellular transport barrier is also important, especially for non-viral vectors, as they must ensure the trafficking to the nucleus and protect the genetic material from degradation by cytoplasmic nucleases. The incorporation of specific molecules such as polyethylenimine (PEI) in non-viral vectors allows safe intracellular transport of the DNA cargo through the microtubules to the nucleus.

4.2.4 Nuclear Delivery

After reaching the nucleus, the vectors must finally ensure the passage of the transgene through the nuclear envelope and the nuclear pores. The entry of the genetic material into the nucleus seems to occur through three possible routes [22]: (i) incorporation during mitosis, when the nuclear envelope is disrupted; (ii) by diffusion through the nuclear pore complex (NPC), for molecules smaller than 10 nm; and (iii) by active transport through the NPC, for molecules smaller than 25 nm. For postmitotic cells, the first option is not possible; thus, viral and non-viral vectors must ensure gene delivery using the other two options. One common strategy to enhance nuclear delivery is to add a nuclear localization signal/sequence (NLS) to the transgene, thus promoting its attachment to proteins that naturally enter the nucleus through the NPC.

4.3 Technical Barriers

4.3.1 Physical Restrictions

It is clear from the previous description of several barriers that size matters; the total vector size is a crucial factor, with larger particles having more difficulty in overcoming extracellular and intracellular barriers. In the development of non-viral vectors, the size is particularly important, as it can determine the cellular entry mechanism and, therefore, the subsequent intracellular trafficking steps. Size is also important in the case of viral vectors, considering their limited cloning capacity. This capacity can limit the size of the transgene to be used and conditionate the type of viral vector that is chosen.

4.3.2 Cellular Targeting

Another important technical barrier in gene therapy, which affects both viral and non-viral vectors, is cellular targeting, which concerns the ability to ensure a correct and specific targeting of the transgene to the desired cells. Proper targeting will ensure a higher efficacy of the therapy and, importantly, will also contribute to reduce adverse effects and toxicity events related to off-target delivery. The local administration of the vectors can overcome this problem to a certain extent; however, for many applications, this approach is too invasive and therefore not an option. In the case of viral vectors, some instances of natural tropism can be explored, although this strategy can be largely restricted to a limited number of cells/tissue types. As already mentioned, for non-viral vectors specific ligands can be incorporated at the surface, thus directing gene delivery to the target cells.

4.3.3 Gene Persistence

The main goal of gene therapy is to ensure, whenever possible, a onetime solution therapy. Thus, the difficulty in maintaining gene expression over time is a technical barrier that should be overcome to ensure the success of the therapy. This point is even more critical in dividing cells, where exogenous episomal DNA could be lost during cell division. However, in nondividing cells, this point may also be relevant, as gene expression could be lost over time, for example due to epigenetic repression. The possibility that some viral vectors offer a long-term gene expression due to genome integration seems the solution for this barrier, although this possibility is not so straightforward as (i) there is the chance of insertional mutagenesis caused by the transgene, (ii) the transgene expression levels may be very different (higher or lower) than the physiological requirements, and (iii) loss of gene expression due to different repression mechanisms can occur over time.

4.3.4 Sustainable Gene Expression

As highlighted above, it is important to ensure persistence of the expression of the inserted gene. Furthermore, it is also important to guarantee that the inserted transgene is transcribed and translated in the amounts required for it to elicit a therapeutic effect, while avoiding toxicity/saturation events. Moreover, it would be favorable that transgene expresison could be turned off in case adverse effects were observed, using, for example, the systems for controlling gene expression that were already described (see further details in Chap. 1).

This Chapter in a Nutshell

- To ensure a correct and efficient delivery of genes, the vector must overcome several extracellular and intracellular barriers. Moreover, several technical barriers must also be considered in gene therapy.
- The main extracellular barriers to gene delivery include unspecific molecular interactions, the existence of endothelial barriers and the immune response.
- The intracellular barriers to gene delivery include limitations of cellular uptake, endosome-lysosome escape, intracellular trafficking and nuclear delivery.
- In terms of technical barriers, physical restrictions, cellular targeting limitations, and the difficulty to guarantee gene persistence and sustainable gene expression are among the most important challenges to gene delivery.

Review Questions

1. Why is the immune response a particularly important extracellular barrier when using viral vectors to gene delivery?
   (a) Because viral vectors enter the cell using membrane receptors
   (b) Because viral proteins elicit a strong immune response
   (c) Because viral vectors only activate the innate immune response
   (d) Because viral proteins elicit immune suppression
   (e) Because viral vectors cannot deliver genes into immune-privileged sites
2. Which of the following strategies is not used to enhance the ability of non-viral vectors to escape the endo-lysosome pathway?
   (a) Binding with agents that create pores in the endocytic vesicles
   (b) Use of agents prone to protonation
   (c) Use of neutralizing antibodies
   (d) Use of viral peptides
   (e) Use of bacterial proteins
3. The delivery of a transgene using a non-viral vector with 50 nm to nondividing cells failed to reach the nucleus. Which of the following features could explain the failure?
   (a) The use of non-viral vectors is exclusive to dividing cells
   (b) The nuclear pore complex does not allow particles of the size used
   (c) The gene was not integrated in the cell genome
   (d) The gene did not have a nuclear localization signal
   (e) The gene size was more than 250 kb
4. Which of the following is not considered a technical barrier to gene delivery?
   (a) Physical restrictions
   (b) Unspecific interactions
   (c) Cellular targeting
   (d) Gene persistence
   (e) Sustainable gene expression

References and Further Reading

1. Schatzlein AG (2003) Targeting of synthetic gene delivery systems. J Biomed Biotechnol 2003:149–158
2. Ramamoorth M, Narvekar A (2015) Non viral vectors in gene therapy- an overview. J Clin Diagn Res 9:GE01–GE06
3. Feletou M (2011) The endothelium: part 1: multiple functions of the endothelial cells-focus on endothelium-derived vasoactive mediators. Morgan & Claypool, San Rafael
4. Daneman R, Prat A (2015) The blood-brain barrier. Cold Spring Harb Perspect Biol 7:a020412
5. Johnsen KB, Burkhart A, Melander F, Kempen PJ, Vejlebo JB, Siupka P, Nielsen MS, Andresen TL, Moos T (2017) Targeting transferrin receptors at the blood-brain barrier improves the uptake of immunoliposomes and subsequent cargo transport into the brain parenchyma. Sci Rep 7:10396
6. Conceicao M, Mendonca L, Nobrega C, Gomes C, Costa P, Hirai H, Moreira JN, Lima MC, Manjunath N, Pereira de Almeida L (2016) Intravenous administration of brain-targeted stable nucleic acid lipid particles alleviates Machado-Joseph disease neurological phenotype. Biomaterials 82:124–137
7. Bourdenx M, Dutheil N, Bezard E, Dehay B (2014) Systemic gene delivery to the central nervous system using Adeno-associated virus. Front Mol Neurosci 7:50
8. Chaplin DD (2010) Overview of the immune response. J Allergy Clin Immunol 125:S3–S23
9. Nayak S, Herzog RW (2010) Progress and prospects: immune responses to viral vectors. Gene Ther 17:295–304
10. Boutin S, Monteilhet V, Veron P, Leborgne C, Benveniste O, Montus MF, Masurier C (2010) Prevalence of serum IgG and neutralizing factors against adeno-associated virus (AAV) types 1, 2, 5, 6, 8, and 9 in the healthy population: implications for gene therapy using AAV vectors. Hum Gene Ther 21:704–712
11. Steinbrook R (2008) The Gelsinger case. In: Emanuel EJ (ed) The Oxford textbook of clinical research ethics. Oxford University Press, Oxford
12. Sack BK, Herzog RW (2009) Evading the immune response upon in vivo gene therapy with viral vectors. Curr Opin Mol Ther 11:493–503
13. Mayor S, Pagano RE (2007) Pathways of clathrin-independent endocytosis. Nat Rev Mol Cell Biol 8:603–612
14. Perumal OP, Inapagolla R, Kannan S, Kannan RM (2008) The effect of surface functionality on cellular trafficking of dendrimers. Biomaterials 29:3469–3476
15. Gupta B, Torchilin VP (2006) Transactivating transcriptional activator-mediated drug delivery. Expert Opin Drug Deliv 3:177–190

16. Lee KM, Kim IS, Lee YB, Shin SC, Lee KC, Oh IJ (2005) Evaluation of transferrin-polyethylenimine conjugate for targeted gene delivery. Arch Pharm Res 28:722–729
17. Doms M (2016) Basic concepts: a step-by-step guide to viral infection. In: Katze MG, Law GL, Korth MJ, Nathanson N (eds) Viral pathogenesis: from basics to systems biology. Elsevier, Amsterdam
18. Jones CH, Chen CK, Ravikrishnan A, Rane S, Pfeifer BA (2013) Overcoming nonviral gene delivery barriers: perspective and future. Mol Pharm 10:4082–4098
19. Varkouhi AK, Scholte M, Storm G, Haisma HJ (2011) Endosomal escape pathways for delivery of biologicals. J Control Release 151:220–228
20. Huang HW, Chen FY, Lee MT (2004) Molecular mechanism of peptide-induced pores in membranes. Phys Rev Lett 92:198304
21. Moreira C, Oliveira H, Pires LR, Simoes S, Barbosa MA, Pego AP (2009) Improving chitosan-mediated gene transfer by the introduction of intracellular buffering moieties into the chitosan backbone. Acta Biomater 5:2995–3006
22. Cohen S, Au S, Pante N (2011) How viruses access the nucleus. Biochim Biophys Acta 1813:1634–1645

5 Stem Cells and Tissue Regeneration

Stem cells are non-differentiated and unspecialized cells characterized by their ability of self-renewing by proliferating through symmetric division, which maintains the pool of stem cells. On the other hand, these cells can also undergo asymmetric division, resulting in two different cells, one stem cell and another cell that becomes compromised into one cell lineage and gives rise to mature specialized cells. Stem cells are present in all stages of human development, from the zygote to adulthood. During embryogenesis, they are responsible for tissue and organ formation. Although in adults they are present in much smaller numbers, they are still present in specific regions, *germinative niches*, with the important role of tissue maintenance and regeneration. However, in adults the tissue regeneration ability of these cells in some organs like the brain is very limited.

Stem cells can be classified according to their source (origin) and potency, i.e., the range of different types of cells they can give rise to upon differentiation (Fig. 5.1). Thus, according to **potency**, stem cells can be classified as totipotent, pluripotent, multipotent, and unipotent. **Totipotent** stem cells can, upon differentiation, originate all types of cells, from the three germinal layers (endoderm, mesoderm, and ectoderm), and originate also the support structures necessary for embryo development (placenta and umbilical cord). **Pluripotent** stem cells can originate all the cells of the three germinal layers upon differentiation, but cannot give rise to embryo development support structures. **Multipotent** stem cells are compromised to one of the germinal layers, only giving rise to cells of that layer. Finally, **unipotent** stem cells will only originate one type of cell [1].

Concerning the source of these cells, they can be classified in accordance with (i) the stage of development of the organism from which they were isolated, as adult, fetal and embryonic; (ii) the type of tissue from which they were isolated, as bone marrow mesenchymal stromal cells (BM-MSCs), adipose tissue mesenchymal stromal cells (AD-MSCs) and the umbilical cord blood mesenchymal stromal cells (UC-MSCs) [1]; or (iii) the manner these cells were generated, as induced pluripotent stem cells (iPSC) or chemically induced pluripotent stem cells (CiPSC). These different types of stem cells have been widely explored in cell replacement therapy applications, and the choice of the type of cells to be implemented is dependent on the therapeutic purpose and on the tissue to be regenerated.

Additionally, stem cells have also been used to develop new experimental models that serve as human *in vitro* models of organs that are very difficult to access, such as the brain. Moreover, with the establishment of human iPSC, cultures of human iPSC-derived neurons

C. Nóbrega et al., *A Handbook of Gene and Cell Therapy*,
https://doi.org/10.1007/978-3-030-41333-0_5

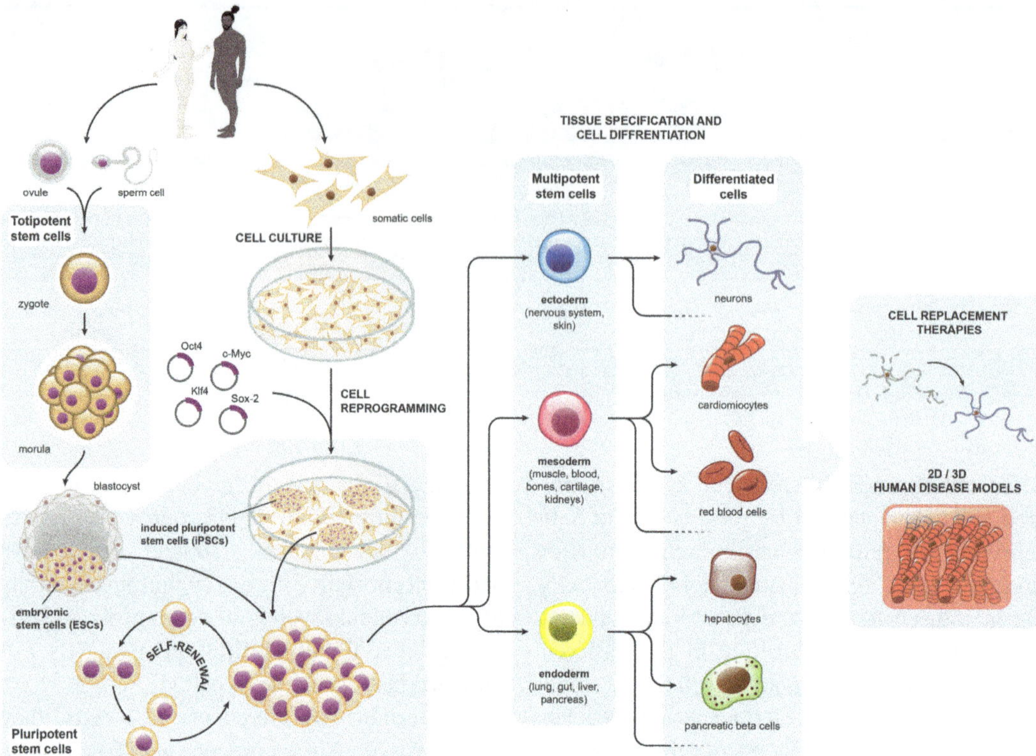

Fig. 5.1 Different sources of stem cells available for stem cell-based therapies. Stem cells are non-differentiated and unspecialized cells that can be classified in accordance with their source (origin) and potency, i.e., the range of different types of cells they can give rise to upon differentiation. *Totipotent stem cells* can originate all the different types of cells of the organism, plus the supporting structures required for embryo development; *pluripotent stem cells* also originate all the different types of cells present in the organism but are not capable of giving rise to the structures supporting embryo development; and *multipotent stem cells* are compromised into one of the germinal layers and therefore can only originate cells from the ectoderm, mesoderm or endoderm. Totipotent cells are present in the morula after fertilization, while pluripotent stem cells can be obtained by reprogramming somatic cells into *induced pluripotent stem cells* or can be obtained in the inner cell mass of the blastocyst (*embryonic stem cells*). Pluripotent stem cells and can be induced into multipotent stem cells and more specialized cells through *tissue specification and cell differentiation* procedures, serving as *2D and 3D human disease models* and sources of cells to be implemented in *cell replacement strategies*.

have been modeled *in vitro*, and also 3D mini-human brains, with the organoid technology [1].

This chapter describes different types of stem cells available to be used in cell replacement therapeutic strategies, with a special focus on adult and fetal neural stem cells (NSC), embryonic stem cells (ESC) and iPSC, and discusses new cell-based tools and models, such as organoids and 3D bioprinting. Although many of the examples provided concern aspects related to therapy and modeling of the nervous system, stem cell technology can be applied to the vast diversity of organs, tissues and cell types of the organism.

5.1 Adult and Fetal Neural Stem Cells

Presently, it is widely known that human adult tissues have niches of stem cells that are responsible for tissue maintenance and repair. The adult brain is no exception, presenting niches of NSC, in specific regions such as the subventricular zone (SVZ), the subgranular zone (SGZ) in the dentate gyrus of the hippocampus, the external germinal layer of the cerebellum, the striatum, the substantia nigra and the cortex [2], which are activated in trauma episodes [3]. However, the number of stem cells available and the repair potential of these cells is very different depending on the tissue where they are present. Unfortunately, the repair ability of the adult brain is very limited, which could be in part explained by the low plasticity of the adult brain. Consequently, and at the present, the most efficient neuroregeneration results described were obtained through the transplantation of NSC, instead of only relying on the activation of the niches of neural stem cells.

Adult NSC have the capacity for self-renewal, which maintains the pool of stem cells, and are also multipotent cells that, upon differentiation, originate cells of the nervous system, namely neurons, astrocytes and oligodendrocytes [1, 4]. These cells can be isolated from the adult nervous system (in the niches previously enumerated) and also from fetal brains. Adult NSC can be easily and safely maintained and expanded for months in chemically defined culture media, which offer a renewable source of NSC that can be used as *in vitro* models and also as a source of cells to be used for brain transplantation experiments [3]. Upon transplantation, these cells can trigger several repair mechanisms, namely cell replacement and neuroprotection. **Cell replacement** is achieved through the substitution of the dead neuronal cells, given that the transplanted cells have the ability to migrate to the injured brain region and differentiate into new cells. **Neuroprotection** is mediated by two mechanisms: (i) production of neurotrophic factors, like BDNF, GDNF, and NGF [4, 5], which will increase the survival and performance of the host neurons and glia, and (ii) positive modulation of the immune system, reducing neuroinflammation [4, 6, 7], which is associated with brain injury and neurodegenerative diseases and that actively contributes to neuronal death [8]. Although NSC can positively modulate brain injuries and neurodegenerative diseases through both neuroprotection and cell replacement, it remains unclear whether these cells become functionally integrated into the host neuronal network in an extension that undoubtedly contributes to the improvements triggered by NSC transplantation. Several reports of NSC transplantation into adult brains indicate a small extension of integration of new neurons [9].

The results of the transplantation of human fetal tissues into patients with Parkinson's disease are particularly noteworthy, namely, fetal mesencephalic substantia nigra obtained from a 13-week-old fetus that dramatically improved both rigidity and dyskinesia symptoms [10], and also the deep brain transplantation of fetal brain tissue that restored local dopamine synthesis and storage, thus alleviating the disease motor symptoms [11–13]. However, in some cases, transplantation led to more heterogeneous results, as in some patients with Huntington's disease, the benefits from the grafts seem to be temporary [1].

Nevertheless, the use of human fetal tissues for clinical applications is associated with important ethical limitations (Table 5.1), given that the tissues are obtained from fetuses of spontaneous abortions [13] and the use of human fetus tissues for research purposes is highly controversial and even illegal in some countries [3]. An additional problem is the high number of cells/tissues required for human transplantation, as a single patient may require cells from three to eight donors [3], which increases the difficulty to obtain enough donors to perform a clinical study with several patients. Furthermore, the different donors' source may trigger immunological problems and compromise the success of the intervention. Moreover, cells used for transplantation have to be characterized regarding their safety profile, namely for contamination with infectious agents, genetic instability (normal karyotype) due to their extensive expansion and endotoxin levels [13]. This increases the complexity of the transplantation process with fetal neural stem cells.

Overall, despite some encouraging results using fetal stem cells, the important ethical questions, associated and some less promising studies, make them still not the ideal source of cells for broader therapeutic applications.

5.2 Embryonic Stem Cells

Human embryonic stem cells (ESC) are undifferentiated pluripotent stem cells, derived from the inner cell mass of the blastocyst of preimplantation-stage supernumerary or nonviable embryos, obtained from *in vitro* fertilization

Table 5.1 Main advantages and disadvantages of the different stem cells used in cell therapy.

	Cell potency	Sources	Main problems	Main advantages
Fetal/ adult stem cells	Multipotent	Fetus and adult brain	Ethical concerns; prone to immune rejection; difficult reproducibility from batch to batch	Some degree of specialization, which could make differentiation easier
ESC	Pluripotent	Embryos	Ethical concerns; prone to immune rejection	High cell proliferation rate; ability to originate all different types of cells
iPSC	Pluripotent	Cell reprogramming	Safety aspects are under evaluation	High cell proliferation rate; ability to originate all different types of cells; devoid of ethical concerns

[14]. ESC have an unlimited self-renewal ability and, upon differentiation, they can originate cells from all three primary germ layers. They are capable of dividing without differentiating for a prolonged period in culture and can be propagated indefinitely, maintaining a normal karyotype and without undergoing senescence [15]. In fact, there is evidence that ESC have significantly improved DNA repair capacity and telomerase activity, which along which particular mitochondrial and epigenetic changes [16] may explain the lack of senescence observed in ESC, as compared to somatic cells.

The remarkable genetic stability, the unlimited ability to proliferate and the ability to give rise to any type of specialized cell make ESC the ideal source for cell-based therapies. A remarkable study conducted in a patient with Parkinson's disease, who was transplanted with human cells derived from embryonic ventral mesencephalon, demonstrated that these cells could survive up to 24 years after the transplantation event. Moreover, the authors observed major motor improvement, recovery of striatal dopaminergic function, graft-derived dopaminergic reinnervation of the putamen, and no evidence of immune response [17].

Despite the promising results obtained with ESC so far [1, 11, 17, 18], their use in clinical practice is involved in high ethical controversy (Table 5.1), due to the fact that the generation of ESC lines require the destruction of human embryos, even thoughsupernumerary or non-viable human embryos have been used. These ethical problems have placed an enormous pressure in the development of alternative human pluripotent stem cell sources, such as the induced pluripotent stem cells.

5.3 Stem Cells Generated Through Cell Reprogramming

Although cell reprogramming only became an attractive research field in 2006, with Yamanaka and collaborators' breakthrough work [19], the reprogramming concept is prior to that. In fact, already in 1958, somatic cell nuclear transfer (SCNT), which entails the transfer of the nucleus of a somatic cell to the cytoplasm of an enucleated oocyte, was used to reprogram cells [20]. Cell fusion, involving the fusion of the adult cell to be reprogrammed with an embryonic stem cell, had also been previously performed [21, 22]. Nevertheless, given the technical complexity of these strategies, the process described by Yamanaka, involving the generation of induced pluripotent stem cells from somatic cells through the expression of defined factors, has become much more widely used, being quickly adopted and adapted by several labs, definitively boosting the cell reprogramming research field (Fig. 5.2).

5.3.1 Induced Pluripotent Stem Cells (iPSC)

Fully differentiated somatic cells can be reprogrammed into a pluripotent state, which enables them to originate other cell types. These cells that were reprogrammed and induced into a pluripotent state are designated as **induced pluripotent stem cells** (iPSC) [19]. However, cell reprogramming does not necessarily require that cells go through the pluripotent state; it is possible to directly reprogram cells from one differentiated

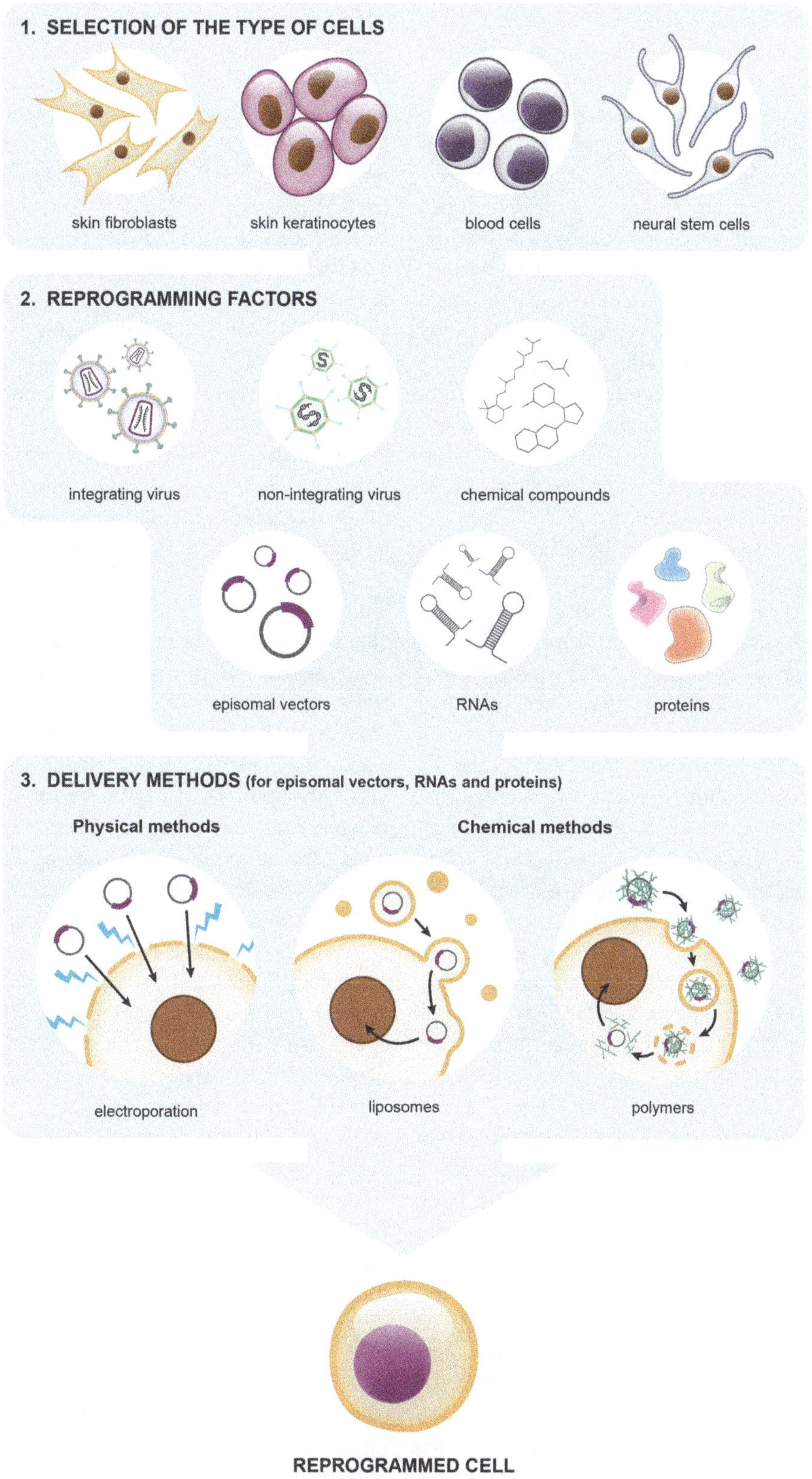

1. SELECTION OF THE TYPE OF CELLS
skin fibroblasts
skin keratinocytes
blood cells
neural stem cells
2. REPROGRAMMING FACTORS
integrating virus
non-integrating virus
chemical compounds
episomal vectors
RNAs
proteins
3. DELIVERY METHODS (for episomal vectors, RNAs and proteins)
Physical methods
Chemical methods
electroporation
liposomes
polymers
REPROGRAMMED CELL

Fig. 5.2 Overview of the cell reprogramming process, highlighting the different cells that can be used as starting material, as well as the diversity of the reprogramming factors and methods to deliver them into the cells.

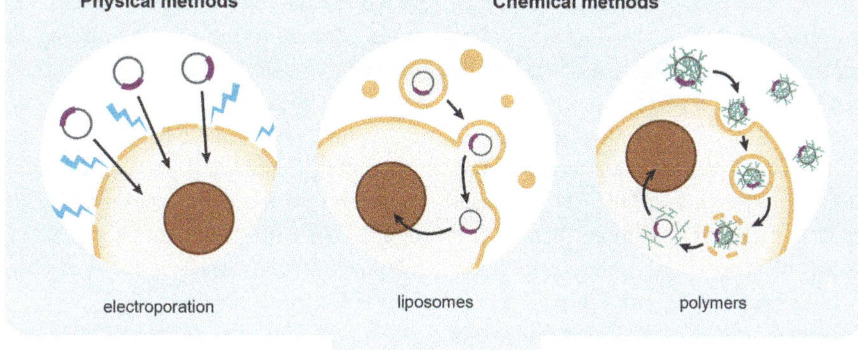

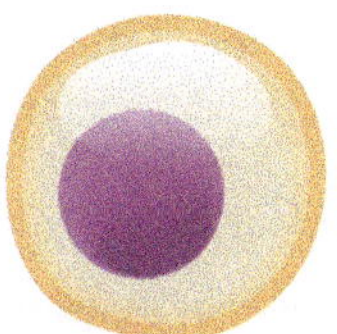

lineage to another differentiated lineage (e.g., from fibroblasts to neurons), without going back to an intermediate stem cell state, the so-called direct reprogramming or transdifferentiation [21, 23].

In many aspects, iPSC are very similar to natural pluripotent ESC, namely regarding their morphology, cell potency, immortal growth, high cell proliferation rate, stem cell markers (as pluripotency markers) and chromatin methylation patterns (related to pluripotency). Besides these phenotypic characteristics, both cell types also share the ability to originate **embryoid bodies** (multicellular structures similar to embryonic development), to form **teratomas** (tumors containing the three germ layers observed during embryogenesis) and to originate viable chimeras, which proves them able to give rise to all tissues [15, 23].

Cell reprogramming can be achieved by the forced expression of four reprogramming factors [19] important in ESC function, Sox-2, Klf4, Oct4, and c-Myc, which will drive cells into a pluripotent state. Other alternative reprogramming factor combinations have been identified and tested, such as Sox-2, Oct4, Lin28 and Nanog [15, 22, 24]. Additionally, these factors can be replaced and/or aided by small molecules that boost cell pluripotency, improving the efficiency and safety of the process.

The four reprogramming factors described by Yamanaka play important roles in stem cell pluripotency, in the control of cell proliferation and in apoptosis. Octamer-binding transcription factor 4, **Oct4**, also known as POU5F1, represses genes related to cell differentiation; therefore, it is responsible for preserving cell pluripotency [22, 23]. Sex-determining region Y-box 2, **Sox-2**, is a mitotic bookmarking transcription factor that remains bound to chromosomes during mitosis and regulates phenotypic maintenance of stem cells at the mitosis M-G1 transition. It has been demonstrated that the absence of this embryonic stem cell transcriptional factor at the M-G1 transition impairs pluripotency maintenance. The V-myc avian myelocytomatosis viral oncogene homolog, **c-Myc**, is not essential for cell reprogramming, although this proto-oncogene increases the process efficiency by increasing the proliferation rate. However, the overexpression of this gene increases the levels of the p53 protein, which may drive cells into apoptosis. Nevertheless, this effect is compensated by the Krüppel-like factor 4, **Klf4**, which increases the p21 protein levels, decreasing the proliferation rates and p53 levels, thereby reducing apoptosis risk [22, 23, 25].

Reprogramming of differentiated cells into iPSC has been extensively studied in the last decade; consequently, several strategies and protocols have been developed to induce cell pluripotency. The most common method uses retroviral delivery of the reprogramming factors, which enables the genomic integration of the factors important for pluripotency. However, lentiviruses and retroviruses integrate into the genetic material randomly, thereby increasing the probability of endogenous oncogenes activation and of triggering insertional mutagenesis and teratoma formation. To overcome this issue, excisable systems have been used, like the Cre-loxP and PiggyBac transposon system that allows the transgenes excision after the reprogramming is done, as well as non-integrative viruses, such as adenovirus and Sendai virus. Additionally, non-viral strategies such as the use of liposomes, polymers and nucleofection to deliver DNA plasmids containing the reprogramming factors can also be used [22, 23]. Besides the reprogramming factors, it is possible to obtain iPSC through the delivery of (i) synthetic RNA encoding the reprogramming factors, (ii) proteins and miRNA that target the pluripotency genes, and (iii) small molecules that will act in cellular pathways involved in pluripotency, namely in DNA methylation and chromatin condensation [15, 23].

Besides the choice of the delivery method/vehicle, another important aspect to consider when deciding on the reprogramming protocol is the type/source of cells to be reprogrammed. In fact, the efficiency and kinetics of reprogramming are different for different types of cells. For example, it has been reported that keratinocytes have a higher reprogramming efficiency as compared to fibroblasts and that immature cells are easier to be reprogrammed when compared to differentiated cells [24]. Finally, the cell growth culture conditions, namely, the supporting matrix, culture medium composition and factors or small molecules used, will, naturally, influence cell characteristics and the reprogramming efficiency and success.

5.3.2 Chemically Induced Pluripotent Stem Cells (CiPSC)

One of the major concerns regarding the safety of iPSC and their derived cells is that they have been generated through the introduction of genomic modifications, frequently using viral vectors. Therefore, to avoid these genomic modifications, one approach that gained relevance in the last years is the use of small chemical compounds to trigger cell pluripotency. Small molecules have been used in combination with reprogramming factors to improve reprogramming efficiency, but they can completely replace the reprogramming factors. Chemical reprogramming has several advantages over genetic manipulation, namely, that (i) it allows better regulation over protein function through a tight control of the drug concentration used; (ii) it permits a better temporal control, given that drugs can interact with their targets quickly and their effect is reversed/ended by removing the drug; and (iii) it is a simpler and cheaper strategy, as compared to the use of viral vectors [15, 21]. Nevertheless, small molecules can trigger off-target effects affecting proteins with similar conformation [15], and thus the safety of this strategy is also under evaluation.

5.3.3 Epigenetic, Metabolic and Cellular Pathway Modulation by Small Molecules to Induce Pluripotency

The small molecules that have been used in reprogramming protocols are modulators of epigenetics, metabolism, cell signaling, apoptosis and cell senescence [26], which are processes and pathways that are manipulated during cell reprogramming.

Epigenetics refers to gene expression control mechanisms that are independent of the DNA sequence. The two main epigenetic modulator classes are DNA methylation and histone acetylation/methylation. **DNA methylation** inhibits transcription by blocking the access of transcription factors to the gene promoters. Silent genes (in terms of transcription), such as pluripotency genes, frequently have hypermethylated promoters. Thus, the reactivation of these hypermethylated genes is required for an efficient/complete reprogramming. **Histones** are structural proteins that bind DNA into chromatin; they are also epigenetic regulators because their modification alters the degree of chromatin condensation, dictating the access to genes and consequently their transcription. DNA expression is modulated through the methylation and acetylation states of the histones [15, 23, 26]. The three major classes of small molecules that act as epigenetic regulators are (i) DNA methyltransferases (DNMT) inhibitors, such as RG108 and 5-aza-2′-cytidine (decitabine), which prevent DNA methylation mediated by DNMT enzymes; (ii) histone deacetylase (HDAC) inhibitors, such as suberoylanilide, trichostatin A and valproic acid, which avoid the removal of acetyl groups from lysine residues of histones performed by HDAC enzymes; and (iii) histone methyltransferase (HMT) inhibitors, such as BIX-01294, which prevent histone methylation mediated by these enzymes, repressing transcription.

The modulation of cell signaling has also been explored in several protocols, and it was demonstrated that Rho kinase, glycogen synthase kinase 3 (GSK-3), transforming growth factors (TGF) receptors and mitogen-activated protein kinase (MAPK) are capable of enhancing cell reprogramming [26]. Moreover, the use of BIX-01294, an HMT inhibitor, and BayK8644, an L-type calcium channel agonist, in combination with Oct3/4 and Klf4 triggers the transition of fibroblasts into iPSC [21, 26]. The use of HDAC inhibitors reduces the propensity for cell differentiation, increasing the self-renewing capacities and leading to an enhancement of the reprogramming efficiency. One known example of these types of molecules is valproic acid, commonly used in reprogramming protocols to replace the proto-oncogene c-Myc [23, 26], aiming to reduce the risk of tumor formation. Vitamin C also improves reprogramming efficiency; it induces DNA methylation, reducing cellular differentiation tendency and senescence [15].

Notably, researchers have described the generation of iPSC with only seven small molecules [26]: VC6TFZ (valproic acid), CHIR99021 (an inhibitor of GSK-3 kinases), 616452 (RepSox), tranylcypromine, forskolin, 2-methyl-5-hydroxytryptamine (2-Me-5HT) and D4476.

5.3.4 Direct Reprogramming Mediated by Reprogramming Factors and/or Small Molecules

The process of pluripotency induction has been raising safety concerns related to the potential development of tumors by pluripotent stem cells. Therefore, several alternative protocols of cell reprogramming in which cells do not pass through a pluripotent state have been developed. The cell reprogramming process, in which a somatic cell acquires the identity of a different somatic cell (e.g., a fibroblast becomes a neuron) without going through a pluripotent intermediate stage, is designated by **direct reprogramming** [27, 28].

In the case of neurons, for example, direct reprogramming has been accomplished through the following procedures:

(i) Forced expression of neuronal transcription factors such as Ascl1, Brn2 and Myt1l [29];
(ii) Chemical manipulation of cellular pathways regulating neuronal fate, through which chemical-induced neurons (CiN) can be obtained using seven molecules - valproic acid, CHIR99021, RepSox, forskolin, SP600125, GO6983 and Y27632 [26];
(iii) Forced expression of specific miRNA, such as miR-9/9∗ and miR-124 [30];
(iv) The downregulation of RNA-binding proteins PTB/nPTB [31] or p16–p19 [27, 32].

On the other hand, it is also possible to directly convert somatic cells into NSC that later can be induced to differentiate into neurons. Direct chemically induced NSC (CiNSC) are possible to obtain with only three small compounds: valproic acid, CHIR99021 and RepSox (an inhibitor of TGF pathways) [26]. Thus, there are several approaches to obtain neuronal cells by reprogramming somatic cells. Nevertheless, the obtained cells reflect the process used; in particular, epigenetic memory is retained from the cells of origin. Therefore, more studies assessing the impact of these epigenetic differences in the safety and functionality of the cells are mandatory.

5.3.5 iPSC and iPSC-Derived Neural Progenitors' Issues

The very low reprogramming efficiency of somatic cells into iPSC has been pointed out as a problem; however, once iPSC colonies are formed, they rapidly expand originating a large number of cells. Thus, compared with other questions, such as safety, this is a minor problem. Additional important issues concerning iPSC use include (i) the safety of the reprogramming process, that many times employs integrating viral vectors, and how this can lead to tumor development, and (ii) the epigenetic differences between iPSC and ESC and how the observed differences in iPSC affect long-term safety, namely the genomic stability of iPSC-derived cells used for cell replacement purposes [21]. These are important questions that hopefully will be answered in the years to come by clinical trials using iPSC-derived cells.

The reversion of differentiated cells into pluripotent stem cells able to originate any type of differentiated cells of the human body has revolutionized the field of cell-based research, providing new sources of cells for fundamental and applied research. It enables the generation of patient-specific tissues for cell and organ replacement therapies. Moreover, it enables the generation of human disease models for many rare diseases lacking appropriate models, which will allow the study of these diseases and the screening of new therapies [22]. Examples of new disease models that have been galvanized by iPSC are the organoids and 3D bioprinting, which are detailed and discussed in the following sections.

5.4 Brain in a Dish: From Organoids to 3D Bioprinting

5.4.1 Organoids

Organoids are three-dimensional (3D) cellular clusters that self-organize themselves into structures that resemble mini-organs and somehow exhibit similar organ functionality and architecture [33, 34]. These structures are obtained from pluripotent stem cells or adult stem cells that, under the influence of organ-specific biochemical cues and with an appropriate supporting matrix, will self-organize into the organ-specific micro-anatomy and organ-specific differentiated cell types [35]. The stem cells originating the organoids will persist, sustaining the 3D structure with the ability for self-renewal and for differentiation into new specialized cells.

Stem cells have an intrinsic ability to assemble into 3D structures composed of different cell types. As the behavior of stem cells is controlled by their microenvironment, this process can be guided and aided with the use of a matrix, like Matrigel, and suitable exogenous factors. These factors, frequently morphogens (BMP, Wnt, Shh), small molecules (retinoic acid) and growth factors (FGF2, FGF8), are used to control several signaling pathways involved in the proliferation, migration, and differentiation of stem cells, in order to originate the desired tissue-mimicking organoid [33, 35]. Laminin, fibronectin, collagen and Matrigel are some examples of physical supportive components used in organoids; nevertheless, Matrigel is the most commonly used matrix. Matrigel is a commercially available cell culture product obtained from reconstituted basement membrane extracted from Engelbreth-Holm-Swarm (EHS) mouse sarcoma, which is a tumor rich in extracellular matrix proteins. This matrix composition is approximately 60% laminin, 30% collagen IV and 8% entactin (a protein that contributes to the structural organization of the extracellular matrix). It also contains heparan sulfate proteoglycan (perlecan), TGF-β, epidermal growth factor, insulin-like growth factor, fibroblast growth factor, tissue plasminogen activator and matrix metalloproteinases, as indicated by the suppliers. The concentration of the different factors in the Matrigel, previously enumerated, is below the concentrations usually used in cell differentiation and patterning processes. Thus, besides the structural support, these extracellular matrices also provide some supplementation of signaling cues of basement membrane ligands to sustain cell survival, migration and attachment, resulting in the organoid formation. Naturally, there are more elaborated scaffolds available, such as (i) decellularized matrices (obtained through a process involving the removal of the cells from an organ/tissue leaving the extracellular matrix scaffold of the original organ/tissue); (ii) 3D scaffolds produced by biomaterial customization, with synthetic polymers incorporating specific compounds/signaling factors to control the location and time at which stem cells will be presented to a specific stimulus; and (iii) nanoparticles loaded with signaling factors that can be incorporated into the scaffolds slowly releasing the factors or releasing them upon a specific stimulus [33]. The combination of the exogenously added factors, with the endogenous factors produced by the stem cells and the matrix, creates a dynamic environment that temporally and spatially controls the different cells that are originated and how they are assembled, giving rise to the organoid formation.

Organoids are powerful *ex vivo* models for tissue morphogenesis and organogenesis studies given that they recapitulate several biological and biochemical parameters, including the spatial organization of organ-specific cells, cell-cell and cell-matrix interactions, biochemical composition and physiological functions within the organoid [33, 35]. These systems are also able to recapitulate many important properties and steps of the organogenesis, being far more complex and therefore more challenging than 2D cell cultures (Table 5.2). The major advantages of organoids are the following: (i) their composition and 3D organization resemble the living organ to be studied; (ii) there is an increasing easiness in the manipulation protocols; (iii) they can be obtained from different sources of cells (adult NSC, ESC and iPSC); and (iv) they can be expanded for a long time without genomic instability. Thus, they have been used as disease models and also to study and manipulate tissue development and regeneration. Organoids allow performing experiments that are not possible to be conducted in patients and originated experimental models for rare diseases, namely neurologic and neurodegenerative diseases, that lacked appropriate experimental models [35, 36].

However, despite the very encouraging features of this system, organoids have some limitations, such as lacking immune cells and a robust vascularization. This makes the study of inflammatory reactions impossible and thwarts the total comprehension of pathways under the influence of the immune system, such as neurogenesis. The lack of vascularization is also an important problem, causing the death of the organoid's core cells (in bigger organoids) as a consequence of nutrient deprivation and lack of proper oxygenation [37]. Nevertheless, some groups are now working trying to overcome these limitations and yield more relevant and significant organoid models.

Table 5.2 Comparison of the main characteristics of 2D cell cultures and 3D organoids.

	2D cultures	3D organoids
Homogeneity and purity	Relatively homogenous, mostly with one or a few types of cells	More heterogeneous, with different types of cells
Culture period	Shorter	Longer
Cell transplantation	Yes	Not reported yet (more complex and challenging)
Cost	Low	Moderate to high
Easiness to set up	Easy to moderately challenging	Very challenging
Reproducibility	High	Low
Vascularization	No	Possible
Structural organization	Poor	Good – more closely resembles the *in vivo* tissue/organ
Genomic instability	High	Low

5.4.2 Three-Dimensional (3D) Bioprinting

3D bioprinting aims the construction of living, functional organs to be used in reconstructive medicine by printing living cells and matrices in a 3D structure that resembles the shape of the organ to be generated [38, 39]. Despite sounding like science fiction, this research field is more than one decade old and has been evolving faster in the last years, involving different fields such as mechanical engineering, cellular biology, systems biology and robotics. Despite the advances, the synchronous printing of living cells, matrix components and soluble factors (such as growth factors and guiding cues for cell proliferation, migration and differentiation) is an understandably complex process. This process comprises different steps, including obtaining a 3D image of the organ to be printed, the conversion of the image to an informatics printing pattern, the selection of the *bioink* composition (cells, matrix and soluble compounds) and, of course, the actual printing procedure, culturing, and finally the organ quality control [38–40].

Combining cells with biomaterials can be done through two main processes: top-down and bottom-up printing. In the top-down process, the cells are homogeneously dispersed into the biomaterials and shaped to resemble the desired organ. The major flaw of this process is that real organs are not homogenous in their composition as they have different types of cells and extracellular matrices in different regions. This problem has been explored in bottom-up printing, which constructs the organ layer by layer, using different types of bioinks in parallel, with different types of cells, hydrogels and soluble compounds. This process takes advantage of computer-controlled high-resolution cell deposition that allows performing this cell distribution process at a microscale level. This will hopefully result in site-specific cell differentiation [38–40].

The three main bio-components in 3D bioprinting are cells, hydrogel and soluble compounds. The types of cells to be used are selected depending on the organ to print; now with iPSC, it is much easier to derive the different types of cells that compose an organ. Concerning the hydrogel to implement, there are two main groups of hydrogels available: natural and synthetic polymers. Some natural polymers currently used are components of the extracellular matrix (laminin, collagen and fibronectin), while others include chitosan, alginate, hyaluronic acid and nanocellulose. Synthetic polymers such as poloxamer and polyethylene glycol (PEG) diacrylate and PEG methacrylate are also used. Hydrogels should provide a stable environment for the cells, have low cytotoxicity, and should promote proliferation and differentiation. Other features such as the mechanical properties of the hydrogel, including stiffness, pore size and mechanical strength, will also influence the tissue architecture and cell activity [38, 39, 41].

Bioprinting also offers the possibility of constructing micro-organs in microchips, allowing recording and controlling microenvironmentl changes in real time [38]. The integration of 3D tissue-engineered organs with microfluidic systems, the so-called "organ-on-a-chip", allowed the design of chips comprising different chambers with different cell types that represent the different tissues or organs communicating through microfluidic networks. The transport of fluids between these chambers mimics the blood flow, and the fluid flow generates mechanical forces that recapitulate the *in vivo* microenvironment experienced by cells. Generally, these chips integrate biosensors with microscopy-based readouts allowing a real-time study of concentration gradients and cellular processes in a microscale base, which is more cost-effective and thus reduces cell culture costs [42]. These highly controllable "organ-on-a-chip" devices allow the performance of tissue morphogenesis and organogenesis studies and also the performance of high-throughput toxicological studies to test new drugs and therapies. Nevertheless, this very promising and complex innovation has several challenges to be overcome, namely the biological control of the different types of cell populations to guarantee the correct positioning and functioning, and the efficient vascularization of the organ through the inclusion of functional vessels capable of being integrated in, and supplied by, the "circulatory system" [39, 40].

This Chapter in a Nutshell

- Cell replacement therapy seeks to replace missing or functionally impaired cells in order to restore tissue and organ functions lost in pathological processes or trauma.
- Stem cells are non-differentiated and unspecialized cells characterized by their ability of self-renewing by proliferating through symmetric division, which maintains the pool of stem cells.
- Stem cells can be classified according to their source (origin) and potency, i.e., the range of different types of cells that they can give rise to upon differentiation.
- Embryonic stem cells (ESC) are undifferentiated pluripotent stem cells derived from the inner cell mass of the blastocyst, having an unlimited self-renewal ability and the capacity to originate cells of all three primary germ layers upon differentiation.
- Induced pluripotent stem cells (iPSC) are derived from somatic cells that have been reprogrammed back into an embryonic-like pluripotent state.
- Organoids are three-dimensional (3D) cellular clusters that self-organize themselves into structures that resemble mini-organs and exhibit similar organ functionality and architecture.
- 3D bioprinting aims the construction of living, functional organs to be used in reconstructive medicine by printing living cells and matrices in a 3D structure that resembles the shape of the organ to be generated.

Review Questions

1. Stem cells are non-differentiated and unspecialized cells characterized by their ability to self-renew, maintain the pool of stem cells and differentiate into mature specialized cells. Which of the following sentences is correct?
   (a) There are different types of stem cells, namely the pluripotent neural stem cells and embryonic stem cells.
   (b) Pluripotent stem cells can be isolated from the inner cell mass of the blastocyst.
   (c) Pluripotent stem cells can be obtained by transdifferentiation.
   (d) All of the above
   (e) None of the above
2. Neural stem cells can be obtained from:
   (a) Isolation from fetal and adult brains
   (b) Direct chemical induction using valproic acid, CHIR99021 and RepSox
   (c) Neural induction of pluripotent stem cells
   (d) All of the above
   (e) None of the above
3. Cell reprogramming enables obtaining new types of cells to be used as human cell models and as a source of cells for cell-based therapies. Choose the option that best applies.
   (a) Somatic cell nuclear transfer (SCNT) and cell fusion are strategies that allow cell reprogramming
   (b) To obtain induced pluripotent stem cells (iPSC) it is mandatory to use reprogramming factors, by delivering them with either viral or non-viral systems
   (c) Through direct reprogramming, it is possible to obtain differentiated cells such as neurons or multipotent stem cells like NSC
   (d) Answers (a) and (c)
   (e) Answers (b) and (c)
4. Fully differentiated somatic cells can be reprogrammed into a pluripotent state. The resulting cells are designated as induced pluripotent stem cells (iPSC), which are able to originate other cell types. Which of the following sentences is correct?
   (a) iPSC are very similar to the natural pluripotent ESC, namely in terms of morphology, cell potency, immortal growth and ability to originate embryoid bodies and to form teratomas
   (b) Several factors dictate the reprogramming efficiency, namely the type/source of cells to be reprogrammed
   (c) The four reprogramming factors described by Yamanaka (Sox-2, Klf4, Oct4 and c-Myc) play important roles in stem cell pluripotency and control cell proliferation and apoptosis
   (d) The reprogramming efficiency of somatic cells into iPSC is very low
   (e) All of the above
   (f) None of the above
5. The development of new disease models and platforms such as organoids and 3D bioprinting has been galvanized by the discovery of iPSC. Which of the following sentences is correct?
   (a) Organoids are three-dimensional (3D) cellular clusters that self-organize themselves into structures that resemble mini-organs, which are very promising to study, for example, pathways under the influence of the immune system, such as neurogenesis
   (b) Organoids can be obtained from different sources of cells (aNSC, ESC and iPSC) and have been used as disease models and also to study and manipulate tissue development and regeneration
   (c) Chitosan, alginate, hyaluronic acid, nanocellulose, poloxamer and polyethylene glycol (PEG)-conjugated synthetic polymers are examples of compounds used in 3D bioprinting
   (d) Answers (a) and (c)
   (e) Answers (b) and (c)
   (f) All of the above

References and Further Reading

1. Mendonca LS, Onofre I, Miranda CO, Perfeito R, Nobrega C, de Almeida LP (2018) Stem cell-based therapies for polyglutamine diseases. Adv Exp Med Biol 1049:439–466
2. Rolfe A, Sun D (2015) In: Kobeissy FH (ed) Brain neurotrauma: molecular, neuropsychological, and rehabilitation aspects, CRC Press/Taylor & Francis: Boca Raton
3. Pluchino S, Zanotti L, Deleidi M, Martino G (2005) Neural stem cells and their use as therapeutic tool in neurological disorders. Brain Res Brain Res Rev 48(2):211–219
4. Mendonca LS, Nobrega C, Hirai H, Kaspar BK, Pereira de Almeida L (2015) Transplantation of cerebellar neural stem cells improves motor coordination and neuropathology in Machado-Joseph disease mice. Brain 138(Pt 2):320–335
5. Blurton-Jones M, Kitazawa M, Martinez-Coria H, Castello NA, Muller FJ, Loring JF, Yamasaki TR, Poon WW, Green KN, LaFerla FM (2009) Neural stem cells improve cognition via BDNF in a transgenic model of Alzheimer disease. Proc Natl Acad Sci U S A 106(32):13594–13599
6. Pluchino S, Zanotti L, Brambilla E, Rovere-Querini P, Capobianco A, Alfaro-Cervello C, Salani G, Cossetti C, Borsellino G, Battistini L et al (2009) Immune regulatory neural stem/precursor cells protect from central nervous system autoimmunity by restraining dendritic cell function. PLoS One 4(6):e5959
7. Pluchino S, Zanotti L, Rossi B, Brambilla E, Ottoboni L, Salani G, Martinello M, Cattalini A, Bergami A, Furlan R et al (2005) Neurosphere-derived multipotent precursors promote neuroprotection by an immunomodulatory mechanism. Nature 436(7048):266–271
8. Lyman M, Lloyd DG, Ji X, Vizcaychipi MP, Ma D (2014) Neuroinflammation: the role and consequences. Neurosci Res 79:1–12
9. Drago D, Cossetti C, Iraci N, Gaude E, Musco G, Bachi A, Pluchino S (2013) The stem cell secretome and its role in brain repair. Biochimie 95(12):2271–2285
10. Madrazo I, Leon V, Torres C, Aguilera MC, Varela G, Alvarez F, Fraga A, Drucker-Colin R, Ostrosky F, Skurovich M et al (1988) Transplantation of fetal substantia nigra and adrenal medulla to the caudate nucleus in two patients with Parkinson's disease. N Engl J Med 318(1):51
11. Lindvall O, Sawle G, Widner H, Rothwell JC, Bjorklund A, Brooks D, Brundin P, Frackowiak R, Marsden CD, Odin P et al (1994) Evidence for long-term survival and function of dopaminergic grafts in progressive Parkinson's disease. Ann Neurol 35(2):172–180
12. Piccini P, Brooks DJ, Bjorklund A, Gunn RN, Grasby PM, Rimoldi O, Brundin P, Hagell P, Rehncrona S, Widner H et al (1999) Dopamine release from nigral transplants visualized in vivo in a Parkinson's patient. Nat Neurosci 2(12):1137–1140
13. Ishii T, Eto K (2014) Fetal stem cell transplantation: past, present, and future. World J Stem Cells 6(4):404–420
14. Ilic D, Ogilvie C (2017) Concise review: human embryonic stem cells-what have we done? What are we doing? Where are we going? Stem Cells 35(1):17–25
15. Jung DW, Kim WH, Williams DR (2014) Reprogram or reboot: small molecule approaches for the production of induced pluripotent stem cells and direct cell reprogramming. ACS Chem Biol 9(1):80–95
16. Zeng X, Rao MS (2007) Human embryonic stem cells: long term stability, absence of senescence and a potential cell source for neural replacement. Neuroscience 145(4):1348–1358
17. Li W, Englund E, Widner H, Mattsson B, van Westen D, Latt J, Rehncrona S, Brundin P, Bjorklund A, Lindvall O et al (2016) Extensive graft-derived dopaminergic innervation is maintained 24 years after transplantation in the degenerating parkinsonian brain. Proc Natl Acad Sci U S A 113(23):6544–6549
18. Ilic D, Devito L, Miere C, Codognotto S (2015) Human embryonic and induced pluripotent stem cells in clinical trials. Br Med Bull 116:19–27
19. Takahashi K, Yamanaka S (2006) Induction of pluripotent stem cells from mouse embryonic and adult fibroblast cultures by defined factors. Cell 126(4):663–676
20. Gurdon JB, Elsdale TR, Fischberg M (1958) Sexually mature individuals of Xenopus laevis from the transplantation of single somatic nuclei. Nature 182(4627):64–65
21. Anastasia L, Pelissero G, Venerando B, Tettamanti G (2010) Cell reprogramming: expectations and challenges for chemistry in stem cell biology and regenerative medicine. Cell Death Differ 17(8):1230–1237
22. Singh VK, Kumar N, Kalsan M, Saini A, Chandra R (2015) Mechanism of induction: Induced Pluripotent Stem Cells (iPSCs). J Stem Cells 10(1):43–62
23. Baranek M, Belter A, Naskret-Barciszewska MZ, Stobiecki M, Markiewicz WT, Barciszewski J (2017) Effect of small molecules on cell reprogramming. Mol BioSyst 13(2):277–313
24. Brouwer M, Zhou H, Nadif KN (2016) Choices for induction of pluripotency: recent developments in human induced pluripotent stem cell reprogramming strategies. Stem Cell Rev 12(1):54–72
25. Deluz C, Friman ET, Strebinger D, Benke A, Raccaud M, Callegari A, Leleu M, Manley S, Suter DM (2016) A role for mitotic bookmarking of SOX2 in pluripotency and differentiation. Genes Dev 30(22):2538–2550
26. Biswas D, Jiang P (2016) Chemically induced reprogramming of somatic cells to pluripotent stem cells and neural cells. Int J Mol Sci 17(2):226
27. Drouin-Ouellet J, Pircs K, Barker RA, Jakobsson J, Parmar M (2017) Direct neuronal reprogramming for disease modeling studies using patient-derived neurons: what have we learned? Front Neurosci 11:530
28. Velasco I, Salazar P, Giorgetti A, Ramos-Mejia V, Castano J, Romero-Moya D, Menendez P (2014) Concise review: generation of neurons from somatic

cells of healthy individuals and neurological patients through induced pluripotency or direct conversion. Stem Cells 32(11):2811–2817
29. Vierbuchen T, Ostermeier A, Pang ZP, Kokubu Y, Sudhof TC, Wernig M (2010) Direct conversion of fibroblasts to functional neurons by defined factors. Nature 463(7284):1035–1041
30. Victor MB, Richner M, Hermanstyne TO, Ransdell JL, Sobieski C, Deng PY, Klyachko VA, Nerbonne JM, Yoo AS (2014) Generation of human striatal neurons by microRNA-dependent direct conversion of fibroblasts. Neuron 84(2):311–323
31. Xue Y, Qian H, Hu J, Zhou B, Zhou Y, Hu X, Karakhanyan A, Pang Z, Fu XD (2016) Sequential regulatory loops as key gatekeepers for neuronal reprogramming in human cells. Nat Neurosci 19(6):807–815
32. Sun CK, Zhou D, Zhang Z, He L, Zhang F, Wang X, Yuan J, Chen Q, Wu LG, Yang Q (2014) Senescence impairs direct conversion of human somatic cells to neurons. Nat Commun 5:4112
33. Yin X, Mead BE, Safaee H, Langer R, Karp JM, Levy O (2016) Engineering stem cell organoids. Cell Stem Cell 18(1):25–38
34. Kretzschmar K, Clevers H (2016) Organoids: modeling development and the stem cell Niche in a dish. Dev Cell 38(6):590–600
35. Fatehullah A, Tan SH, Barker N (2016) Organoids as an *in vitro* model of human development and disease. Nat Cell Biol 18(3):246–254
36. Huch M, Koo BK (2015) Modeling mouse and human development using organoid cultures. Development 142(18):3113–3125
37. Kelava I, Lancaster MA (2016) Dishing out mini-brains: current progress and future prospects in brain organoid research. Dev Biol 420:199
38. Mandrycky C, Wang Z, Kim K, Kim DH (2016) 3D bioprinting for engineering complex tissues. Biotechnol Adv 34(4):422–434
39. Gao G, Huang Y, Schilling AF, Hubbell K, Cui X (2018) Organ bioprinting: are we there yet? Adv Healthc Mater 7(1): 1701018.
40. Collins SF (2014) Bioprinting is changing regenerative medicine forever. Stem Cells Dev 23(Suppl 1):79–82
41. Ji S, Guvendiren M (2017) Recent advances in bioink design for 3D bioprinting of tissues and organs. Front Bioeng Biotechnol 5(23)
42. Bhise NS, Ribas J, Manoharan V, Zhang YS, Polini A, Massa S, Dokmeci MR, Khademhosseini A. (2014) Organ-on-a-chip platforms for studying drug delivery systems. J Control Release. 190:82–93

6 Gene Therapy Strategies: Gene Augmentation

As it was already mentioned in the first chapter of this book, but will also be made clear throughout the remaining chapters, the term gene therapy encapsulates different approaches and applications based on the use and delivery of different nucleic acids. The most straightforward and perhaps most simple strategy for a gene-based therapy consists in adding a new protein-coding gene, which is an approach called **gene augmentation therapy** (Fig. 6.1). For monogenic recessive disorders in which the causative mutated is nonfunctional, the therapeutic gene to be delivered will be the normal wild-type form of the gene. The delivery of a correct copy of the gene is expected to restore the production of the defective or missing protein and thus revert the disease phenotype. This type of gene augmentation therapy is often also referred to as **gene replacement therapy**. As already discussed, this rationale was behind the first gene therapy clinical trial performed and is also at the basis of the majority of the gene therapy clinical trials performed so far.

However, for monogenic dominant diseases or more complex multigenic diseases, this approach would not be effective. For these diseases, the use of gene therapy strategies based on gene silencing or editing would be more suitable (these will be discussed in more detail in Chaps. 7 and 8). Nevertheless, gene augmentation strategies can still be used in these more complex diseases, namely by using protein-coding genes that improve cellular function and homeostasis, contributing to an improvement of the disease phenotype, although in theory it will not cure the disease or solve its molecular cause. For example, genes codifying for growth factors, cytokines or autophagy-activating proteins have been proven successful as therapeutic agents in several complex diseases. This type of gene therapy is often generally referred to as **gene addition therapy**.

Two important considerations must be made about gene augmentation therapy strategies. First, it is important to take into account the size of the gene to be added, as this will conditionate the choice of vector and therefore the efficiency and safety of the therapy. According to the last annotation of the human genome, on average, a human gene has ~29 kb, which makes packaging entire genes unfeasable for most of the viral vectors commonly used in gene therapy. One form to circumvent this problem is to use only the cDNA, which on average has ~2.5 kb and therefore can be packaged in any viral vector. However, for larger genes, packaging the cDNA in viral vectors will still be virtually impossible. For example, Duchenne muscular dystrophy is caused by mutations in the *DMD* gene, which codifies the dystrophin protein. The gene has more than 220 kb and its cDNA around 14 kb, which makes it unsuitable, for example, to be packaged into an AAV or a lentiviral vector. A second important consideration to be made is related with the expression levels of the therapeutic gene. This question was

C. Nóbrega et al., *A Handbook of Gene and Cell Therapy*,
https://doi.org/10.1007/978-3-030-41333-0_6

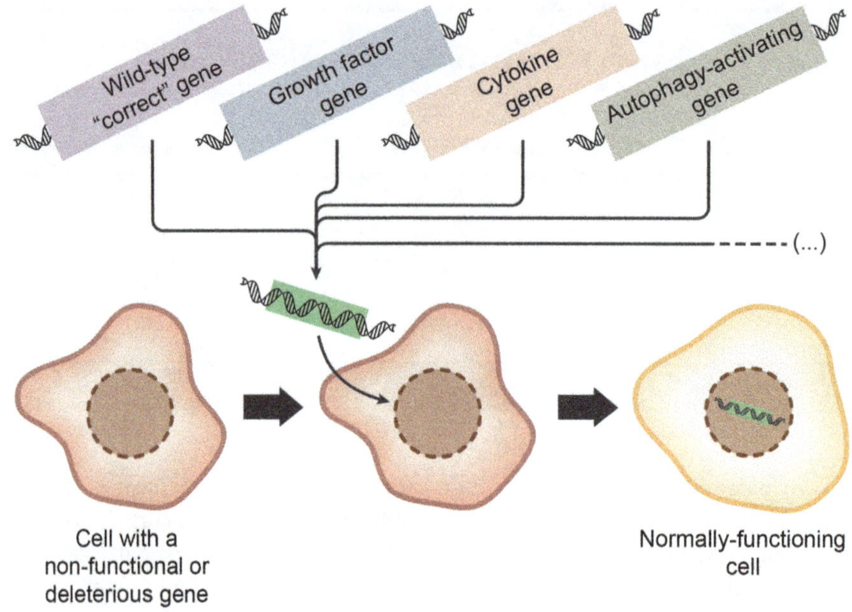

Fig. 6.1 **Gene augmentation therapy**, which refers to the introduction of an additional functional gene, aims at compensating a pathological mutation or improving the homeostasis of a compromised cell.

already discussed in the first chapter as one of the important aspects that need to be adressed in gene therapy applications. Nevertheless, this question is particularly sensitive in gene augmentation therapy strategies, as the aim is to produce correct/therapeutic levels of protein expression.

The following sections develop important ideas and concepts related to gene replacement and gene addition therapy, providing examples of both strategies.

6.1 Gene Replacement

Gene replacement therapy can be applied both *ex vivo* and *in vivo*. In fact, several currently approved gene therapy products use gene replacement therapy, such as Strimvelis® (*ex vivo*) and Luxturna® (*in vivo*). Additionally, there is an extensive list of studies using gene replacement therapy for different diseases both in preclinical and clinical settings. Therefore, it is probable that other gene replacement therapies will be approved in the near future [1].

The single-dose gene replacement therapy (AVXS-101) for spinal muscular atrophy (SMA), based on intravenous delivery by AAV9 and with spectacular results, is an example of a therapy that reached the market very recently [2]. SMA constitutes a group of genetic disorders characterized by the degeneration of anterior horn cells and subsequent muscle atrophy and weakness. There is a wide range of clinical manifestations, different degrees of severity and variances in life expectancy, which led to the classification of different clinical types of SMA (Table 6.1) [3]. The incidence of SMA is of 1 in 10,000 live births, whereas the prevalence of the carrier state is of 1 in 54 [4]. Around 95% of SMA cases are caused by homozygous deletion of the *survival motor neuron 1 (SMN1)* gene [5]. In healthy human subjects there are two forms of the *SMN* gene: *SMN1* and *SMN2* (Fig. 6.2). Both genes encode for the SMN protein, but the *survival motor neuron 2 (SMN2)* gene bears a nucleotide substitution in an exon splicing enhancer that results in the exclusion of exon 7 during transcription. Thus, the protein formed is truncated and non-functional, being rapidly degraded. However, the exon 7 exclusion is not complete, and around 10% of the SMN protein encoded by the *SMN2* gene is functional. In SMA patients, as the *SMN1* gene is not functional, the production of the SMN protein relies only on the *SMN2* gene. Consequently, it is not surprising that SMA severity is inversely correlated with the number of *SMN2* gene copies [6]. A phase I clinical study revealed the safety and efficacy of an *in vivo* gene therapy based on the delivery of a functional copy of the *SMN1* gene using AAV9 vectors [2].

Table 6.1 Clinical types of spinal muscular atrophy (SMA) and their main features

Type	Age of onset	Life expectancy	Maximal motor milestone	Number of SMN2 copies
0	Before birth	<1 month	None	1
1	0–6 months	<2 years	None	2
2	6–18 months	>2 years	Sitting	3–4
3[a]	18 months–3 years	Adult	Walking	3–4
4	10–31 years	Adult	Normal	4–8

[a]There is an additional categorization in SMA type 3a, 3b, and 3c, each having a different age of onset

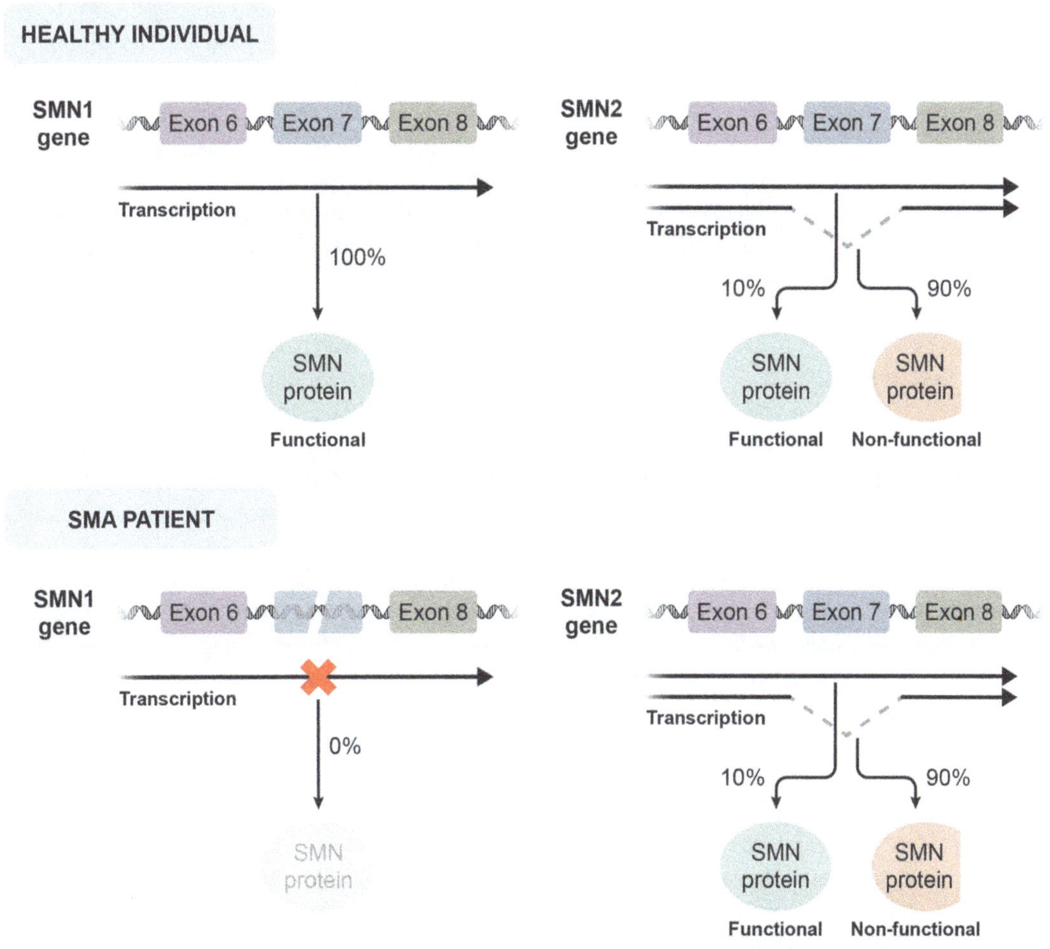

Fig. 6.2 Overview of the molecular mechanism underlying spinal muscular atrophy (SMA). In healthy individuals, the *SMN1* gene (*survival motor neuron 1*) encodes for a functional protein that is essential for motor neuron development. A second gene, *SMN2* (*survival motor neuron 2*), is also present, although it only encodes for a small percentage of the functional protein. In SMA patients, mutations in the *SMN1* gene produce a nonfunctional protein; therefore, the only functional protein is produced by the *SMN2* gene. That is why the number of copies of the *SMN2* gene modifies the severity of the condition.

The 15 children enrolled in the 2 cohorts of the study were homozygous for the deletion of exon 7 in the *SMN1* gene and had 2 copies of the *SMN2* gene. A 20-month follow-up after the gene therapy administration showed that all the children were alive, compared with an 8% survival rate of the natural history cohort. Importantly, no major adverse effects of the therapy were described. From the 12 children that received the higher dose (2.0×10^{14} vg per kilogram), 11 sat unassisted, 11 were able to feed orally and could speak, and 2 walked independently. These spectacular results and a wide range of preclinical studies opened an important avenue for gene replacement therapies that are in the R&D pipeline. It also led to a historical acquisition, with pharmaceutical company Novartis paying 8.7 billion US dollars for AveXis Inc., which was the company that developed the above phase I study. Recently, in May 2019, the FDA approved this therapy under the name Zolgensma, promising a onetime cure for SMA and with a marketing price of 2.125 million dollars.

6.2 Gene Addition

Gene replacement strategies are not suitable for monogenic dominant diseases, as even with an additional copy of the normal gene, the disease phenotype will be maintained. Also, for more complex diseases resulting from the combination of multiple genes and environmental factors, this strategy may have none, or very limited, therapeutic effects. Nevertheless, gene therapy can still represent an important therapeutic option for these diseases, as it may be used to produce an improvement in cellular function and homeostasis, therefore contributing to mitigate the disease phenotype. In fact, several studies both at the preclinical and clinical levels have focused on using gene addition therapy as a therapeutic approach for several complex conditions affecting human health. Different protein-coding genes can be used in these strategies, aiming to modulate diverse cell mechanisms and pathways.

This type of strategy is particularly useful in cancer applications, using, for example, genes that arrest cell proliferation, like cyclin-dependent kinase inhibitors or cell cycle checkpoint regulators, such as p53. In fact, the first gene therapy product ever approved (in 2003) is based on the delivery of p53 as a treatment for different types of cancer [7]. However, Gendicine® (developed by Sibiono GeneTech company) was only approved in China, and its beneficial effects are not completely consensual among the scientific community. Gendicine® was approved for treating head and neck cancer and is based on the expression of wild-type p53, upon direct intratumor, intracavity or intravascular delivery by an adenoviral vector. By 2018, more than 30,000 patients were treated with Gendicine®, and more than 30 clinical trials were published [8]. It was used alone or in combination with chemotherapy and/or radiotherapy for at least five different types of cancer, including nasopharyngeal cancer and hepatocellular carcinoma. Most of the published studies reported no major adverse effects. Despite the number of studies and subjects treated, currently Gendicine® continues to be commercialized exclusively in China.

Gene addition-based delivery of protein-coding genes was also studied as a possible means to modulate other cellular functions and achieve a therapeutic effect, for example aiming to, (i) potentiate the activity of cellular degradation systems (ubiquitin-proteasome system - UPS) - and autophagy), (ii) increase cellular proteostasis, and (iii) increase the expression of growth factors and cytokines. The use of genes activating autophagy or leading to the expression of chaperones is a gene addition strategy particularly interesting in context of neurodegenerative diseases, where frequently there is an abnormal accumulation of aggregate-prone proteins.

Strategies based on the expression of growth factors and cytokines have different goals and target other molecular mechanisms, aiming a protective role by potentiating cell survival or contributing to cellular homeostasis improvement. This idea was the basis for the development of CERE-110, a gene therapy product for Alzheimer's disease (AD) that relies on the delivery of nerve growth factor (NGF). AD is a progressive neurodegenerative disease characterized by a decline of cognitive functions and memory defects. A neuropathological hallmark of AD is the presence of extracellular amyloid plaques and the intracellular accumulation of hyperphosphorylated Tau protein. A small percentage (~5%) of AD cases display a Mendelian inheritance pattern, although most of the cases are sporadic and associated with late disease onset (after 65 years) [9, 10]. Despite the huge research efforts towards a therapy for AD, most of the results of the clinical trials have been disappointing, and even the approved drugs (such as memantine or rivastigmine) have very limited efficacy in stopping or reverting the disease progression. Considering the unmet and urgent need for therapies for AD, several studies focused on gene therapy as a possible route. One of the ideas behind gene therapy studies was to increase neuronal protection, trying to prevent or delay the neuronal loss observed in AD. The first strategies focused on the delivery of brain-derived neurotrophic factor (BDNF). A study in rodent and nonhuman primate models of AD showed a protective effect of BNDF expression, preventing neuronal loss and leading to cognitive improvement [11]. Later, the studies' attention moved to NGF, as a way to prevent the degeneration of cholinergic neurons, which is an early event and an important contributor to AD cognitive decline. A very early study using aged nonhuman primates showed a robust sprouting of cholinergic fibers upon NGF expression [12]. The animals received an intraventricular implant of polymer-encapsulated fibroblasts genetically modified to express human NGF (*ex vivo* gene therapy). Later, a phase I clinical trial used a similar procedure, transplanting autologous fibroblasts expressing NGF (transduced with a Moloney leukemia viral vector) into the basal forebrain of eight individuals with mild AD. The results showed an improvement in the rate of cognitive decline [13]. In a posterior phase I clinical trial, ten patients received an AAV2-NGF (CERE-110) dose directly injected through stereotaxic surgery into the nucleus basalis of Meynert. A 2-year follow-up demonstrated the safety profile of CERE-110 and showed a reduced rate of cognitive decline [14]. These encouraging results led to the development of a phase II multicenter clinical trial of CERE-110, following the acquisition of Ceregene (that initially developed the product) by Sangamo Therapeutics in 2013. The new clinical trial enrolled 49 patients with mild to moderate AD, from which 26 received a bilateral injection of AAV2-NGF and 23 a sham procedure. The first results of the study were somehow mixed, and 2 years after the procedure both groups showed a similar decline in cognitive function. In 2015, Sangamo Therapeutics ended the development of CERE-110. Despite the failure in CERE-110, the development of gene therapy products based on neurotrophic factors continues. For example, NurOwn, a cell therapy product, uses mesenchymal stem cells (MSCs) expressing NGF as a therapy for amyotrophic lateral sclerosis (ALS), and 3 clinical trials were already completed enrolling 74 patients. The first results showed no adverse effects and indicated mild improvement in the disease symptoms (assessed by specific scales) [15]. A future phase III clinical trial is planned.

Other gene addition therapy approaches aim at inducing cellular apoptosis, by delivering genes that trigger this event. This type of strategy, which is commonly referred to as suicide gene therapy, is specifically used for cancer applications and is discussed in further detail in Chap. 9.

6.3 Gene Addition to Modulate Autophagy

Autophagy is a very important and highly regulated cellular process, through which damaged organelles and proteins are degraded by the lysosome. It facilitates nutrient recycling in cells, and alterations in this pathway underlie the pathogenesis of several diseases. Autophagy was first described by Professor Yoshinori Ohsumi, who received the Nobel Prize in Medicine or Physiology in 2016 for the discovery. Currently, autophagy is one of the most consensual therapeutic targets for neurodegenerative diseases and cancer, both through gene therapy and pharmacological approaches. For this reason, the development of gene addition therapies to modulate autophagy is an important focus of research and interest in the gene therapy field.

6.3.1 The Autophagy Pathway

There are three known types of autophagy (Fig. 6.3): microautophagy, chaperone-mediated autophagy and macroautophagy. **Microautophagy** is a nonselective degradation pathway that involves the direct engulfment of cytosolic material through lysosomal invagination and its degradation in the lysosome [16]. **Chaperone-mediated autophagy** (CMA) is a selective degradation pathway in which proteins are targeted for degradation through a recognition motif in their amino acid sequences by chaperones. These selected proteins directly cross the lysosome membrane and enter its lumen for degradation [17]. Finally, **macroautophagy**, commonly simply referred to as autophagy, is characterized by the formation of double-membrane vesicles named autophagosomes that engulf cytoplasmic material and fuse with lysosomes, ultimately degrading that material [18]. Initially described as an apoptosis mechanism, autophagy is now recognized as a cellular survival mechanism, playing an essential role in cellular and energy homeostasis. Its cellular importance is highlighted by the evolutionary conservation found in the autophagy pathway, from yeast to mammals.

Autophagy is highly regulated by different proteins that control the initiation, elongation, maturation and fusion steps of the process, through mTOR-dependent and mTOR-independent pathways [19]. The initiation step of autophagy starts with the phosphorylation of the Unc51-like kinase (ULK) complex and the formation of the phagophore. Additional proteins including several Atg autophagy-related (Atg) proteins and beclin-1 are also essential for phagophore formation and autophagy initiation. The current view postulates that several cellular compartments contribute to the phagophore growth, including the cellular membrane, the Golgi complex and the endoplasmic reticulum [20]. In the elongation step, two ubiquitin-like systems are needed, the Atg12-Atg5-Atg16L1 complex, which dissociates after the autophagosome formation, and a second system resulting from the conjugation of microtubule-associated protein 1 light chain 3 (LC3) and phosphatidylethanolamine (PE). LC3 is cleaved by Atg4B, yielding LC3-I that conjugates with PE (involving Atg7 and Atg3), generating the LC3-II form. The end of this process completes the maturation step of autophagy, resulting in a mature autophagosome. The fusion of the autophagosome with the lysosome forms the autolysosome, where the engulfed substrates are degraded through acidification and lysosomal hydrolase activity, concluding the autophagy process.

As a very dynamic, complex and important cellular pathway, several guidelines have been established to study and monitor the autophagy process [21, 22], wherein two important autophagy markers emerge: LC3-II and SQSTM1/p62 (Sequestosome 1). During autophagy, the LC3-II residing in the inner membrane of the autolysosome is degraded, whereas the LC3-II in the outer membrane is recycled. Consequently, a decrease in LC3-II levels is probably related to an increase in degradation and in autophagy levels. SQSTM1/p62 is an essential cargo receptor involved in selective autophagy, delivering polyubiquitinated cargoes to the autophagy pathway. It interacts with LC3 and is specifically degraded by autophagy; thus, its levels are also correlated with activation or inhibition of this autophagy.

6.3.2 Autophagy Terms Glossary

- Phagophore – double-membrane structure that starts the sequestering of the cytoplasmic cargo for degradation. The continuous elongation of the phagophore leads to its complete closing, forming the autophagosome.
- Autophagosome – double-membrane compartment that contains the cytoplasmic material for degradation and that is formed by the expansion and closing of the phagophore.
- Lysosome – membrane-enclosed organelles containing a wide range of enzyme capable of degrading different biomolecules, such as proteins, lipids and carbohydrates.
- Autolysosome – autophagic compartment resulting from the direct fusion of an autophagosome with a lysosome.
- Autophagolysosome – specific situation that occurs during some types of xenophagy

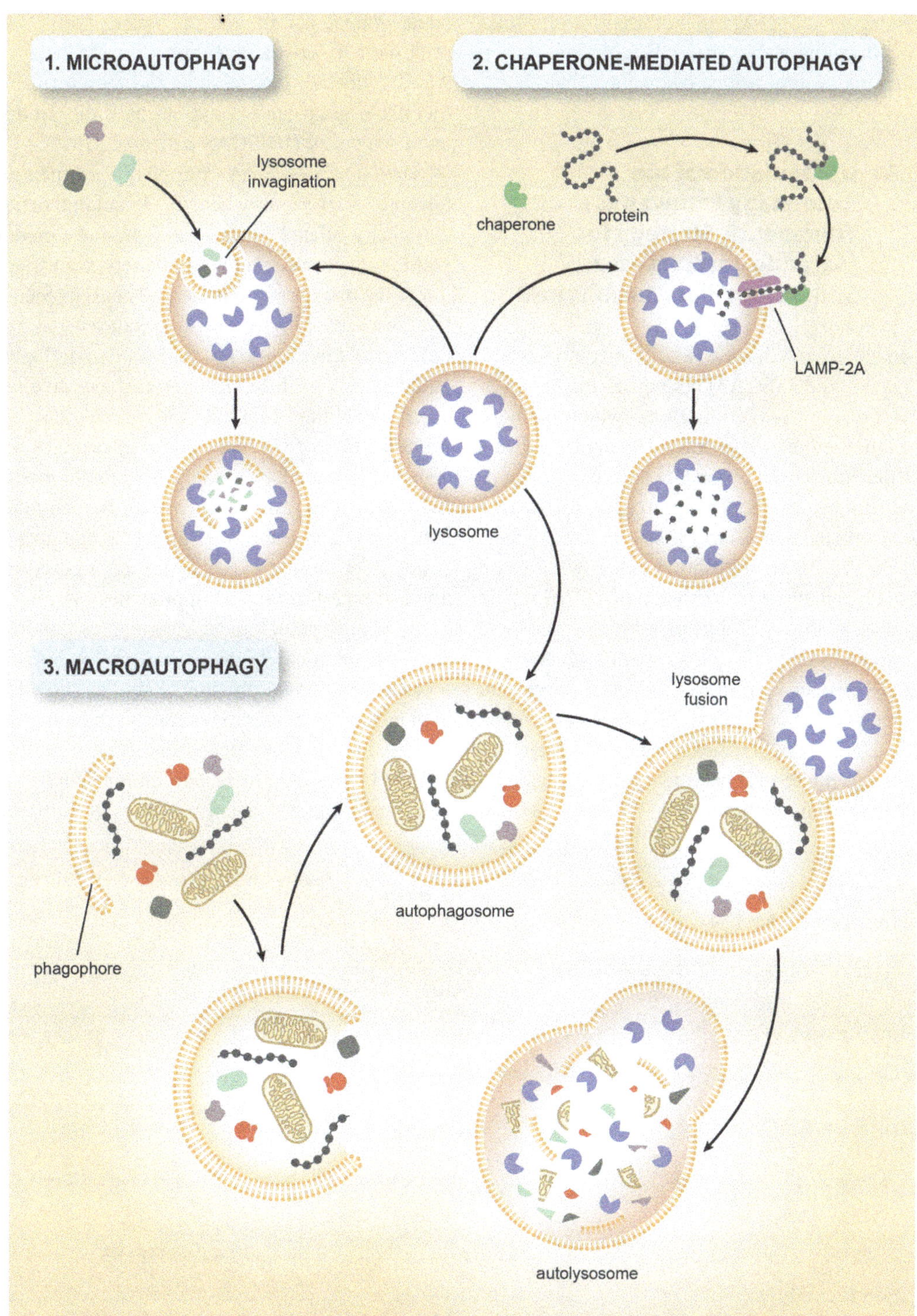

Fig. 6.3 The different autophagy mechanisms. *Microautophagy* (1) is a degradation pathway that involves the direct engulfment of cytosolic material through lysosomal invagination. *Chaperone-mediated autophagy* (2) is a selective degradation pathway in which proteins are targeted for degradation by chaperones, through a recognition motif in their amino acid sequences. *Macroautophagy* (3), commonly simply known as autophagy, starts with the formation of the phagophore that involves the cellular material to be degraded, which after closing forms double-membrane vesicles, named autophagosomes. These vesicles fuse with the lysosomes, forming the autolysosomes in which the material is degraded.

(selective autophagy that is used for eliminating invading pathogens), where there is fusion of a phagosome with a lysosome [23].

6.3.3 Upregulation of the Autophagy Pathway as a Therapeutic Strategy for Machado-Joseph Disease/ Spinocerebellar Ataxia Type 3

Machado-Joseph disease (MJD) or spinocerebellar ataxia type 3 (SCA3) belongs to the group of polyglutamine (polyQ) diseases, which includes nine inherited incurable neurodegenerative disorders that are caused by an abnormal expansion of CAG trinucleotide repeats present in the coding regions of single, and otherwise unrelated, genes. MJD/SCA3 is an autosomal inherited disease caused by an unstable expansion of a CAG repeat sequence in the *ATXN3* gene, which is translated into an abnormal polyQ tract within the ataxin-3 (atxn3) protein (Fig. 6.4) [24]. One important hallmark of MJD/SCA3 is the presence of intranuclear protein aggregates in neurons of selected regions of the central nervous system, especially the cerebellum, the brain stem, the basal ganglia, some cranial nerves and the spinal cord. Until now there is no therapeutic option to delay or stop the disease progression, and therefore gene therapy arises as an important possibility to treat and even cure MJD/SCA3 patients [25].

The importance of autophagy in neurons becomes very clear when the knockout of key autophagy-related genes in animal models yielded a neurodegeneration phenotype very similar to the one observed in several neurodegenerative diseases [26]. Now it is consensual that defects in the autophagy pathway underlie the pathogenesis of different neurodegenerative diseases, including MJD/SCA3, where abnormal accumulation of several autophagy proteins was detected in patients' samples and animal models [27, 28]. It is also consensual that targeting autophagy constitutes an effective therapeutic strategy for these diseases (using both gene therapy and pharmacological approaches), aiming to reduce the pathological aggregates and to improve neuronal homeostasis. Following this idea, two important preclinical studies show that a gene addition therapy strategy inducing the expression of the gene codifying beclin-1 (BECN1) was able to reduce motor deficits and neuropathological abnormalities in different

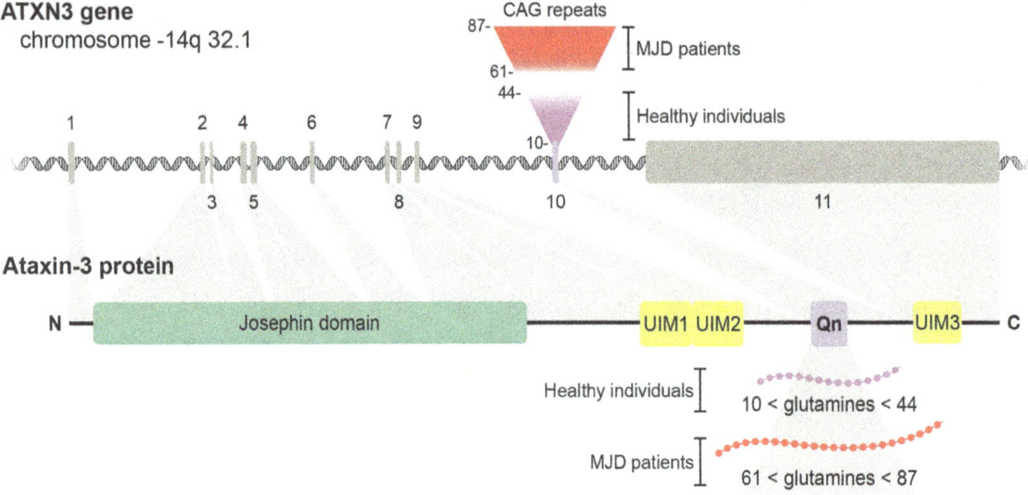

Fig. 6.4 ***ATXN3 gene*** **and ataxin-3 protein structures**, highlighting the CAG/polyglutamine repetition range within healthy individuals and the patients with Machado-Joseph disease (MJD). The *ATXN3* genes includes 11 exons (E1-11) and the CAG repeat region is localized in exon 10 (E10). The ataxin-3 protein is mainly composed of an N-terminal catalytic domain and a C-terminal tail with two or three ubiquitin-interacting motifs (UIMs) and the polyglutamine stretch.

MJD/SCA3 mouse models [27, 29]. The expression of beclin-1 was able to upregulate the autophagy pathway, which is dysregulated in the disease, thus clearing mutant ataxin-3 aggregates and improving neuronal homeostasis.

This Chapter in a Nutshell

- Gene augmentation consists on the delivery of a new protein-coding gene to treat a disease.
- Gene replacement therapy is directed at monogenic recessive diseases that are caused by nonfunctional genes, consisting in the delivery of a non-mutated form of the affected gene to treat the disease.
- Gene addition therapy aims to treat more complex diseases (including dominant monogenic diseases), by delivering a gene aiming to improve cellular function and therefore the disease phenotype.
- Several genes can be used in gene addition therapy, targeting, for example, the cellular degradation pathways or increasing the expression of growth factors and cytokines.
- Autophagy is an important cellular process through which damaged organelles and proteins are degraded by the lysosome.
- It is now known that defects in the autophagy pathway underlie the pathogenesis of several human conditions, including different neurodegenerative diseases.
- Gene therapy targeting the activation of autophagy was shown to be effective in counteracting the disease phenotype in several neurodegenerative diseases.

Review Questions

1. Classify as true or false each one of the following sentences:
   (a) Gene augmentation therapy can only be applied in monogenic diseases
   (b) Gene augmentation therapy involves the delivery of a protein-coding gene
   (c) Gendicine® is a gene therapy product based on gene replacement therapy
   (d) Autophagy is the only cellular degradation pathway
   (e) The activation of autophagy is a gene therapy strategy useful for neurodegenerative diseases
2. In a gene therapy study targeting a monogenic recessive disorder, which of the following is crucial to ensure the success of the wild-type inserted gene?
   (a) Remove the mutant gene
   (b) Produce more cytokines
   (c) Ensure that the correct amount of protein is produced
   (d) Insert the gene in the mitochondria
   (e) Use specifically a gene addition strategy
3. Which one(s) of the following are aims of gene addition therapy?
   (a) Focus on reducing mutant protein levels
   (b) Target monogenic dominant disorders
   (c) Deliver protein-coding genes
   (d) Improve cellular homeostasis
   (e) Cure the genetic cause of the disease
4. Why autophagy activation by gene therapy approaches is a valuable target for neurodegenerative diseases?
   (a) Because there is no other option
   (b) Because autophagy degrades misfolded aggregate proteins
   (c) Because the ubiquitin-proteasome system does not function in neurons
   (d) Because there are no pharmacological drugs to activate autophagy
   (e) Because autophagy is not a dynamic process

References and Further Reading

1. Wang D, Gao G (2014) State-of-the-art human gene therapy: part II. Gene therapy strategies and clinical applications. Discov Med 18:151–161
2. Mendell JR, Al-Zaidy S, Shell R, Arnold WD, Rodino-Klapac LR, Prior TW, Lowes L, Alfano L, Berry K, Church K et al (2017) Single-dose gene-replacement therapy for spinal muscular atrophy. N Engl J Med 377:1713–1722
3. Kolb SJ, Kissel JT (2015) Spinal muscular atrophy. Neurol Clin 33:831–846
4. Sugarman EA, Nagan N, Zhu H, Akmaev VR, Zhou Z, Rohlfs EM, Flynn K, Hendrickson BC, Scholl T, Sirko-Osadsa DA et al (2012) Pan-ethnic carrier screening and prenatal diagnosis for spinal muscular

atrophy: clinical laboratory analysis of >72,400 specimens. Eur J Hum Genet 20:27–32
5. Lefebvre S, Burglen L, Reboullet S, Clermont O, Burlet P, Viollet L, Benichou B, Cruaud C, Millasseau P, Zeviani M et al (1995) Identification and characterization of a spinal muscular atrophy-determining gene. Cell 80:155–165
6. Lefebvre S, Burlet P, Liu Q, Bertrandy S, Clermont O, Munnich A, Dreyfuss G, Melki J (1997) Correlation between severity and SMN protein level in spinal muscular atrophy. Nat Genet 16:265–269
7. Pearson S, Jia H, Kandachi K (2004) China approves first gene therapy. Nat Biotechnol 22:3–4
8. Zhang WW, Li L, Li D, Liu J, Li X, Li W, Xu X, Zhang MJ, Chandler LA, Lin H et al (2018) The first approved gene therapy product for cancer Ad-p53 (Gendicine): 12 years in the clinic. Hum Gene Ther 29:160–179
9. Bettens K, Sleegers K, Van Broeckhoven C (2013) Genetic insights in Alzheimer's disease. Lancet Neurol 12:92–104
10. Mangialasche F, Solomon A, Winblad B, Mecocci P, Kivipelto M (2010) Alzheimer's disease: clinical trials and drug development. Lancet Neurol 9:702–716
11. Nagahara AH, Merrill DA, Coppola G, Tsukada S, Schroeder BE, Shaked GM, Wang L, Blesch A, Kim A, Conner JM et al (2009) Neuroprotective effects of brain-derived neurotrophic factor in rodent and primate models of Alzheimer's disease. Nat Med 15:331–337
12. Kordower JH, Winn SR, Liu YT, Mufson EJ, Sladek JR Jr, Hammang JP, Baetge EE, Emerich DF (1994) The aged monkey basal forebrain: rescue and sprouting of axotomized basal forebrain neurons after grafts of encapsulated cells secreting human nerve growth factor. Proc Natl Acad Sci U S A 91:10898–10902
13. Tuszynski MH, Thal L, Pay M, Salmon DP, U HS, Bakay R, Patel P, Blesch A, Vahlsing HL, Ho G et al (2005) A phase 1 clinical trial of nerve growth factor gene therapy for Alzheimer disease. Nat Med 11:551–555
14. Rafii MS, Baumann TL, Bakay RA, Ostrove JM, Siffert J, Fleisher AS, Herzog CD, Barba D, Pay M, Salmon DP et al (2014) A phase 1 study of stereotactic gene delivery of AAV2-NGF for Alzheimer's disease. Alzheimers Dement 10:571–581
15. Petrou P, Gothelf Y, Argov Z, Gotkine M, Levy YS, Kassis I, Vaknin-Dembinsky A, Ben-Hur T, Offen D, Abramsky O et al (2016) Safety and clinical effects of mesenchymal stem cells secreting neurotrophic factor transplantation in patients with amyotrophic lateral sclerosis: results of phase 1/2 and 2a clinical trials. JAMA Neurol 73:337–344
16. Li WW, Li J, Bao JK (2012) Microautophagy: lesser-known self-eating. Cell Mol Life Sci 69:1125–1136
17. Kaushik S, Cuervo AM (2018) The coming of age of chaperone-mediated autophagy. Nat Rev Mol Cell Biol 19:365–381
18. Feng Y, He D, Yao Z, Klionsky DJ (2014) The machinery of macroautophagy. Cell Res 24:24–41
19. Sarkar S (2013) Regulation of autophagy by mTOR-dependent and mTOR-independent pathways: autophagy dysfunction in neurodegenerative diseases and therapeutic application of autophagy enhancers. Biochem Soc Trans 41:1103–1130
20. Yu L, Chen Y, Tooze SA (2018) Autophagy pathway: cellular and molecular mechanisms. Autophagy 14:207–215
21. Klionsky DJ, Abdalla FC, Abeliovich H, Abraham RT, Acevedo-Arozena A, Adeli K, Agholme L, Agnello M, Agostinis P, Aguirre-Ghiso JA et al (2012) Guidelines for the use and interpretation of assays for monitoring autophagy. Autophagy 8:445–544
22. Klionsky DJ, Abdelmohsen K, Abe A, Abedin MJ, Abeliovich H, Acevedo Arozena A, Adachi H, Adams CM, Adams PD, Adeli K et al (2016) Guidelines for the use and interpretation of assays for monitoring autophagy (3rd edition). Autophagy 12:1–222
23. Klionsky DJ, Eskelinen EL, Deretic V (2014) Autophagosomes, phagosomes, autolysosomes, phagolysosomes, autophagolysosomes... wait, I'm confused. Autophagy 10:549–551
24. Nobrega C, Simoes AT, Duarte-Neves J, Duarte S, Vasconcelos-Ferreira A, Cunha-Santos J, Pereira D, Santana M, Cavadas C, de Almeida LP (2018) Molecular mechanisms and cellular pathways implicated in Machado-Joseph disease pathogenesis. Adv Exp Med Biol 1049:349–367
25. Matos CA, de Almeida LP, Nobrega C (2019) Machado-Joseph disease/spinocerebellar ataxia type 3: lessons from disease pathogenesis and clues into therapy. J Neurochem 148:8–28
26. Hara T, Nakamura K, Matsui M, Yamamoto A, Nakahara Y, Suzuki-Migishima R, Yokoyama M, Mishima K, Saito I, Okano H et al (2006) Suppression of basal autophagy in neural cells causes neurodegenerative disease in mice. Nature 441:885–889
27. Nascimento-Ferreira I, Santos-Ferreira T, Sousa-Ferreira L, Auregan G, Onofre I, Alves S, Dufour N, Colomer Gould VF, Koeppen A, Deglon N et al (2011) Overexpression of the autophagic beclin-1 protein clears mutant ataxin-3 and alleviates Machado-Joseph disease. Brain 134:1400–1415
28. Onofre I, Mendonca N, Lopes S, Nobre R, de Melo JB, Carreira IM, Januario C, Goncalves AF, de Almeida LP (2016) Fibroblasts of Machado Joseph disease patients reveal autophagy impairment. Sci Rep 6:28220
29. Nascimento-Ferreira I, Nobrega C, Vasconcelos-Ferreira A, Onofre I, Albuquerque D, Aveleira C, Hirai H, Deglon N, Pereira de Almeida L (2013) Beclin 1 mitigates motor and neuropathological deficits in genetic mouse models of Machado-Joseph disease. Brain 136:2173–2188

7 Gene Therapy Strategies: Gene Silencing

It was already mentioned throughout this book that gene therapy was primarily developed as a strategy targeting recessive monogenic disorders caused by non-functional mutant genes, where in theory the simple addition of a functional (non-mutated) copy of the causative gene would be enough to counteract the disease phenotype and cure the disease. Nevertheless, for more complex diseases, the addition of a normal gene is not enough to revert the disease phenotype. For example, in dominantly inherited disorders, the presence of a single abnormal allele is sufficient to lead to disease manifestation, and thus a gene therapy should instead be able to shut down (or silence) the expression of that abnormal gene. Although this strategy is not so linear and easy, the possibility of silencing the expression of mutant genes using gene therapy approaches became possible with the development of antisense oligonucleotides (ASOs) and with the discovery of the RNA interference (RNAi) pathway. Despite several differences that will be explored further in this chapter, gene therapy strategies based on these two tools have been demonstrated to be efficient in treating dominant genetic diseases. More recently, gene editing tools have also been used to perform gene silencing, including at the DNA level, for example by introducing a premature stop codon or by removing the defective gene; however, this and other gene editing approaches will be discussed in Chap. 8.

7.1 Antisense Oligonucleotides

In a general way, antisense oligonucleoties (ASOs) are synthetic, unmodified or chemically modified, single-stranded nucleic acid molecules with a size ranging from 8 to 50 nucleotides, that hybridize to their messenger RNA (mRNA) target through Watson and Crick base pairing. ASOs were discovered in 1978 by Stephenson and Zamecnik [1], who observed that a tridecamer oligodeoxynucleotide complementary to Rous sarcoma virus 35S RNA was an efficient inhibitor of the translation of viral proteins and thus of the viral replication cycle. Two decades later, in 1998, the FDA approved the first antisense therapy product to treat cytomegalovirus-induced chorioretinitis, named Vitravene, which was developed by Isis Pharmaceuticals [2]. In fact, this company, now known as Ionis Pharmaceuticals, is the principal booster of ASOs technology, currently having two ASO-based therapies approved in the USA and one in Europe (see Chap. 1).

7.1.1 ASOs Generations

The first applications of natural DNA-mimicking, unmodified, ASOs demonstrated that the phosphoribose backbone undergoes rapid degradation by exonucleases and endonucleases, thus limiting their clinical potential. Moreover, these unmodified ASOs also demonstrated a weak affinity for their target. These important limita-

C. Nóbrega et al., *A Handbook of Gene and Cell Therapy*,
https://doi.org/10.1007/978-3-030-41333-0_7

tions led to the development of several chemical modifications in ASOs, aiming also at increasing their efficacy and bioavailability. The different modifications and improvements made to ASOs led to their classification into three generations (Fig. 7.1).

The **first generation** of ASOs was developed by replacing oxygen by a sulfur atom in the phosphate group to form phosphorothioates (PS) or a methyl group to form methylphosphonates [3]. These modifications facilitated the ASOs production, enhanced stability and increased the ASOs plasma half-life, by improving their resistance to nuclease degradation and reducing binding to serum proteins. One important feature distinguishes the two types of first-generation ASOs; the methyl-modified oligonucleotides do not allow RNAse H-mediated cleavage of the target mRNA, whereas the PS-modified oligonucleotides maintain this feature, allowing their use for RNA silencing purposes [4]. Nevertheless, these modifications also resulted in some disadvantages. For example, methyl-modified oligonucleotides showed reduced solubility and cellular uptake (due to the lack of charge), while the PS-modified oligonucleotides had reduced affinity for the target and displayed some toxic effects.

The **second generation** of ASOs aimed to overcome the main limitations of the previous generation, namely by improving the affinity for the target mRNA, increasing ASOs resistance against nuclease degradation and reducing their immunostimulatory activity [3]. The second generation of ASOs was developed through the introduction of alkyl modifications at the 2'-position of the ribose sugar. The most relevant second generation ASOs examples are the 2'-*O*-methyl (2'-OME) and 2'-*O*-methoxyethyl (2'-MOE) ASOs. Despite bringing some improvements and advantages, these modifications prevented the recruitment of RNAse H, not being able to induce mRNA degradation and thus exerting their activity by steric blocking [5].

Finally, the **third generation** of ASOs includes locked nucleic acids (LNA), peptide nucleic acids (PNA) and morpholino phosphoroamidates (MF) and was developed through the chemical modification of the furanose ring of the ASOs, along with changes to phosphate linkages [6]. These modifications enhanced stability, strengthened affinity to the target mRNA and improved their pharmacokinetic profiles. However, similarly to the second generation ASOs, third generation ASOs cannot induce RNAse H degradation. As their backbone has no charge, these ASOs do not have affinity for serum proteins, which reduces the unspecific interactions, but, on the other hand, leads to their rapid elimination from the body. Therefore, third generation ASOs require delivery systems that improve their cellular uptake.

From this description of each generation of ASOs, it is clear that there are advantages and limitations to each one of them, and therefore their choice as therapy agents should be weighted carefully (Table 7.1). Moreover, despite the fact that the introduced modifications aimed at improving different features of the ASOs, their application as a therapeutic strategy in preclinical or clinical studies is very dependent on the particular molecular targets, cells, disease and several other important aspects that are involved.

7.1.2 Important Considerations for the Use of ASOs in Gene Therapy

From the different features of each ASOs generation, several issues emerge that are important to appreciate when considering ASOs use in gene therapy.

First, depending on their chemical modification, the functional mechanism of ASOs can lead to target mRNA degradation or to a translational arrest, among others. These different functional mechanisms are important in terms of gene therapy application and outcomes and thus should be taken into consideration.

Second, the ASOs pharmacokinetic profile depends on their chemical modifications. For example, PS-ASOs exhibit high affinity for plasma proteins, preventing their elimination from the organism, although the increased unspecific interactions reduce their efficiency. On the contrary, neutrally charged ASOs (e.g., MF) have less tendency to bind plasma proteins, reducing

Basic nucleotide structure

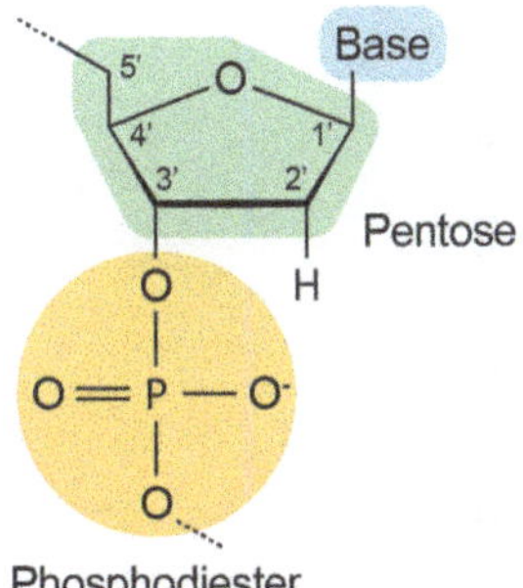

First generation ASOs

Base

O

O H

O=P—S-

O

Phosphorothioate (PS)

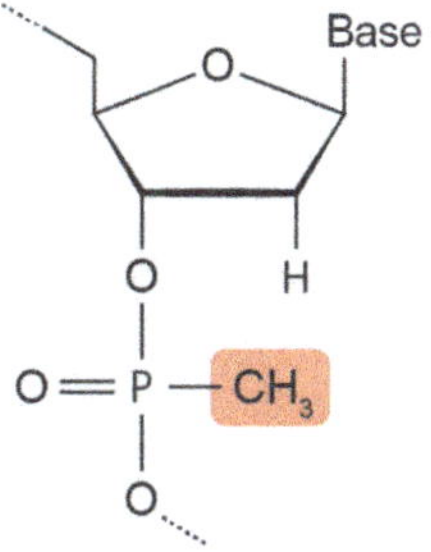

Methylphosphonate

Second generation ASOs

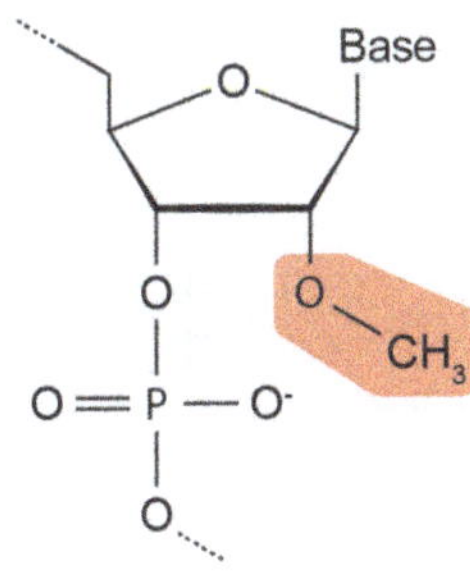

2'-O-methyl (2'-OME)

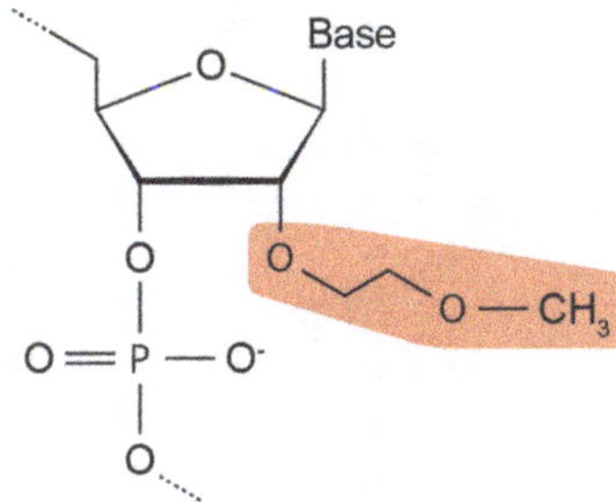

2'-O-methoxyethyl (2'-MOE)

Third generation ASOs

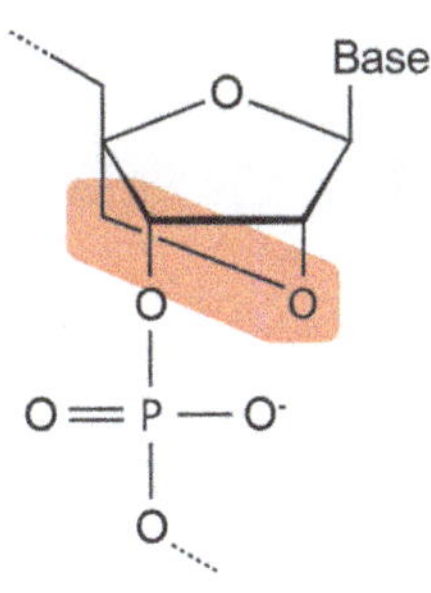

Locked nucleic acid (LNA)

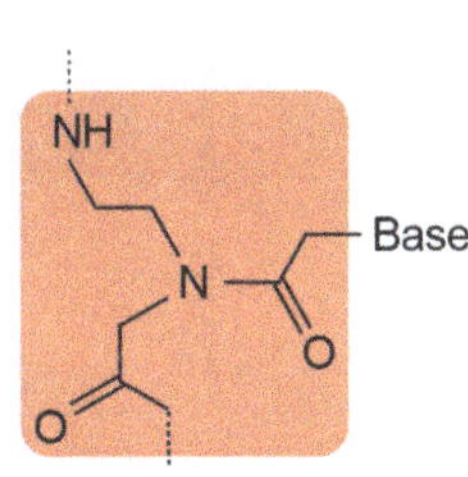

Peptide nucleic acid (PNA)

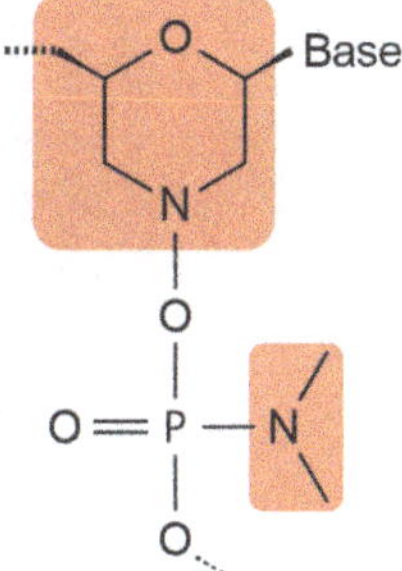

Morpholino phosphoroamidate (MF)

Fig. 7.1 **Chemical modifications made to the antisense oligonucleotides (ASOs) backbone** that are the basis of the different ASOs generations. These modifications were engineered to improve ASOs efficacy, tolerability profile and bioavailability. In the *first generation* of modified ASOs, the non-bridging oxygen atom of the ASOs backbone was replaced with a sulfur atom (or a methyl group), improving the resistance to nucleases and the cellular uptake. In the *second generation*, the changes were introduced in the ribose sugar, improving ASOs safety and efficacy profiles. Finally, the *third generation* of ASOs included the chemical modification of the furanose ring, along with changes to the phosphate linkages. These modifications resulted in enhanced stability, a stronger affinity for the target mRNA and better pharmacokinetic profiles.

Table 7.1 Main features of the different generations of antisense oligonucleotides.

Generation	Examples	Modifications	Functional mechanism	Nuclease resistance	Stability	Target affinity	Cellular uptake	Advantages	Limitations
1st	Phosphorothioate (PS) and methylphosphonates	Sulfur or methyl introduction	RNAse H activity[a]	+	+	–	+	Improved resistance to nucleases, good cellular uptake for PS-ASOs	Immune stimulation at high concentrations, reduced target mRNA affinity for the PS-ASOs, low cellular uptake for methyl-modified ASOs
2nd	2'-OME and 2'-MOE	Alkyl introduction	Steric hindrance	++	++	++	++	Higher nuclease resistance, decreased toxicity, improved uptake, higher affinity for the target mRNA	Cannot recruit RNAse H
3rd	PNA, MF and LNA	Chemical modification of the furanose ring along with modification of phosphate linkages	Steric hindrance	+++	+++	+++	–	Higher nuclease resistance, higher affinity to the target RNA, decreased unspecific interactions with serum proteins	Cannot recruit RNAse H, difficult cellular uptake (due to neutral backbone), rapid clearance from the body

[a]Only for phosphorothioate

unspecific interactions but increasing their clearance. Systemic delivery of ASOs results in a broad distribution to most tissues, with the exception of the central nervous system (CNS), as they do not cross the blood-brain barrier. To overcome this important limitation, several studies focused on the delivery of ASOs by intrathecal injection, leading to a rapid and broad distribution into the brain and spinal cord [7]. In fact, the ASOs therapy (Spinraza®) currently approved for spinal muscular atrophy (SMA) is based on intrathecal administration [8].

Third, an important limitation for the success of ASOs in clinical applications is their poor cellular uptake due to their negatively charged nature (at least for some types of ASOs). For this reason, delivery of ASOs has also employed different systems that increase cellular entry efficiency, for example electroporation and microinjection [9]. Finally, intracellular trafficking is another important issue for ASOs. Even with the poor cellular uptake, ASOs can enter the cell through endocytosis and eventually need to escape from the endosomes, which is now recognized as the main limiting step in oligonucleotide therapy. Several strategies can be used to overcome this limitation, such as altering the endosomal barrier or modulating intra-endosomal pH [7].

Despite all these considerations, ASO-mediated gene therapy is again in the center of the stage, with several therapies being tested for different conditions and many of them in advanced stages of clinical trials.

7.1.3 Functional Mechanisms

ASOs can exert their action through different functional ways, which can be categorized into two main types of mechanisms: an **RNAse H-dependent** mechanism, that leads to mRNA degradation, and **RNAse H-independent** mechanisms, that act through nucleic acid occupancy only (Fig. 7.2) [10].

In the first type of mechanism, upon base pairing, the ASOs form an RNA-DNA hybrid with the target mRNA, becoming a substrate for RNAse H, which then cleaves the mRNA and leaves the ASOs intact. However, to mediate this mechanism, ASOs need to have at least a portion of 2' unmodified nucleotides and, for this reason, only the first generation of ASOs displays this functionality.

On the contrary, the second type of mechanisms is based on the occupancy of target mRNA or even of the DNA. Several effects are possible, such as blocking interaction with RNA binding proteins, thus inhibiting translation, or altering RNA processing through the modulation of splicing. Recently, ASOs that inhibit microRNA (miRNA) function were also developed. These ASOs reduce the silencing activity of miRNAs and consequently increase the levels of the mRNAs they target.

7.1.4 ASOs Applications in Gene Therapy

Polyglutamine (PolyQ) Diseases

Polyglutamine (polyQ) diseases are a group of nine rare neurodegenerative disorders, each caused by an abnormal expansion of the trinucleotide CAG in the respective causative gene, giving rise to an abnormal polyQ tract in the protein therein encoded (Fig. 7.3). These abnormal proteins tend to aggregate, forming insoluble protein aggregates, which are a key hallmark of polyQ diseases. The toxic nature of the aggregate species remains unclear, since it is not completely understood whether they are the cause or the consequence of the progressive neurodegeneration observed in these diseases [11]. The group includes Huntington's disease (HD), dentatorubral-pallidoluysian atrophy (DRPLA), spinal and bulbar muscular atrophy (SBMA), and six different spinocerebellar ataxia types (SCA1, 2, 3, 6, 7, and 17). These disorders mostly affect the CNS and, so far, there is no available therapy that is capable of delaying or stopping disease progression [12].

Due to their mostly dominant genetics (SBMA is an exception) and the fact that disease is regarded as resulting from a toxic gain-of-fuction by the expanded proteins, the use of gene silencing strategies is a viable therapy option for polyQ diseases, aiming to reduce the expression of the mutant protein. In line with this idea, several preclinical stud-

1. RNAse H-DEPENDENT MECHANISMS

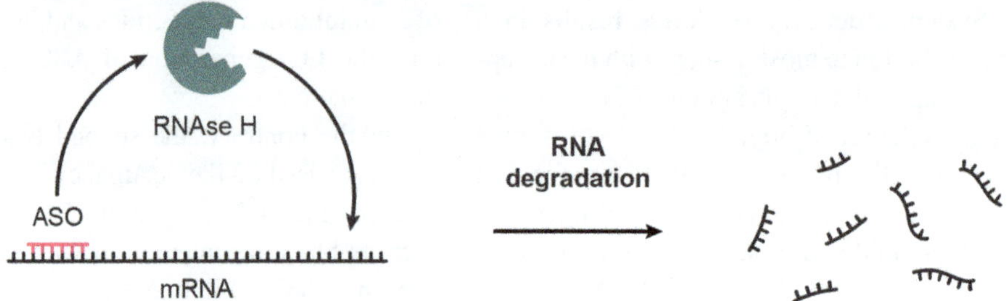

2. RNAse H-INDEPENDENT MECHANISMS

Splicing modulation

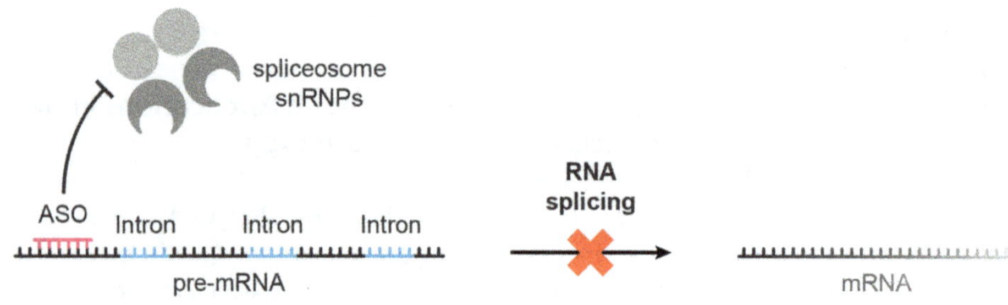

Translation inhibition

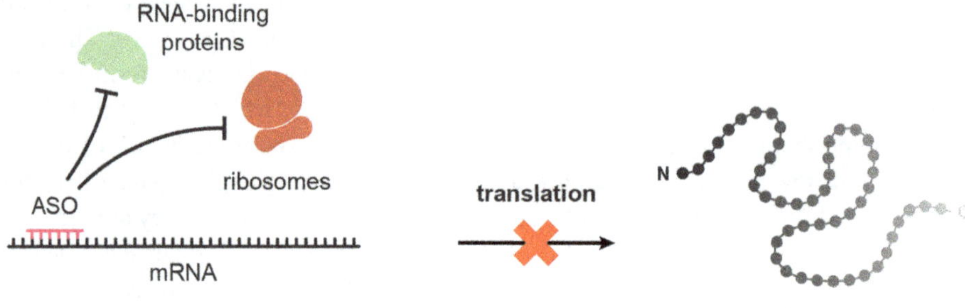

MicroRNA blocking

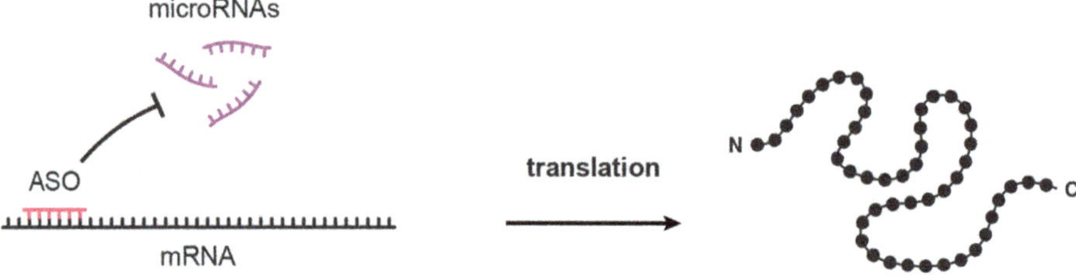

Fig. 7.2 Functional mechanisms of ASOs. ASOs can exert their action in different functional ways, which can be categorized into two main types of mechanisms: an *RNAse H-dependent* mechanism, which leads to mRNA degradation, and several *RNAse H-independent* mechanisms, based on the occupancy of the target nucleic acid. In this latter category, several effects can be produced by the ASOs, including the modulation of the splicing process (by interfering with the activity of small nuclear ribonucleoproteins - snRNPs), the inhibition of translation or blocking of microRNAs.

ies showed the potential of ASOs in specifically reducing mutant polyQ-expanded protein levels and disease-associated abnormalities [13]. Different generations of ASOs have been tested in animal models, thus exploring different functional mechanisms, such as mRNA degradation or translation inhibition (Table 7.2). Overall, these preclinical studies report results that are promising for future clinical studies. In fact, there is an ongoing clinical trial for HD using ASOs from Ionis Pharmaceuticals. The first results showed a safe profile for the therapy and a dose-dependent reduction of mutant huntingtin (the causative protein) in the cerebrospinal fluid [14].

Spinal Muscular Atrophy (SMA)

Spinal muscular atrophy (SMA) comprises a group of genetic disorders characterized by degeneration of anterior horn cells and resultant muscle atrophy and weakness. Around 95% of the SMA cases are caused by an autosomal recessive deletion or mutation in the survival of motor neuron (*SMN1*) gene [15]. *SMN2* is another gene form present in humans, responsible for around 10% of the physiological levels of the SMN protein (see Chap. 6 and Fig. 6.2). In SMA patients with the mutation in the *SMN1* gene, protein production relies on the *SMN2* gene alone, which, despite being low, ensures enough SMN protein for survival. Thus, it is not surprising that the clinical severity of SMA inversely correlates with the *SMN2* gene copy number.

The presence and features of the *SMN2* gene are the basis for the approved ASOs therapy for SMA patients. Spinraza® is a intrathecally delivered 2-MOE-PS oligonucleotide from Ionis Pharmaceuticals, whose functional mechanism is based on modulation of *SMN2* gene splicing [8]. Studies using this therapy yielded encouraging results, as the ASOs administration showed a good safety profile and tolerability, leading to its approval by the FDA and EMA.

Duchenne Muscular Dystrophy (DMD)

Duchenne muscular dystrophy (DMD) is an X-linked recessive disease characterized by progressive muscular degeneration and weakness. It is caused by different mutations (deletions, insertions and point mutations) in the *DMD* gene, resulting in a prematurely truncated, unstable dystrophin protein [16]. The protein is expressed at the muscle sarcolemma, in a protein complex that links the cytoskeleton to the basal lamina. However, the exact mechanism by which dystrophin deficiency leads to muscle fiber degeneration is not yet completely elucidated [17].

Although DMD is a recessive disease, the large size of the *DMD* gene, with more than 70 exons (more than 220 kb), makes it unsuitable, or at least challenging, to use a therapeutic strategy that would supply a normal copy of the gene. Interestingly, the functional versatility of ASOs, namely the possibility of modulating splicing, provides an important opportunity for DMD therapy, through the generation of a functional protein by skipping DMD-causing mutations. Based on this versatility, Sarepta Therapeutics developed an ASO-based therapy for DMD, which was approved by the FDA in 2016 under the name Exondys 51™ (eteplirsen). The ASOs in question is a phosphorodiamidate morpholino oligomers that promote specific skipping of *DMD* exon 51 (in defective gene variants) leading to the restoration of the normal *DMD* reading frame and the production of a functional dystrophin protein (Fig. 7.4) [18]. A pooled analysis of different studies using this ASOs-based drug revealed an increase in dystrophin-positive fibers in muscles and an improvement of walking distance in treated patients, although the clinical significance of these results is still unclear [19].

Other Diseases

The versatility of ASOs functional mechanisms, as well as their easy design and production, makes them attractive therapeutic agents for several diseases. Recently, an ASOs-based product named Tegsedi™ (inotersen) was approved in the USA as a treatment for hereditary transthyretin-mediated amyloidosis [20].

Apart from the approved products based on ASOs technology that were described in the above sections, several clinical trials are also currently ongoing for other diseases, including amyotrophic lateral sclerosis, Alzheimer's disease and frontotemporal dementia. These and other clinical trials highlight the interest and potential of ASOs technology for gene therapy, and new

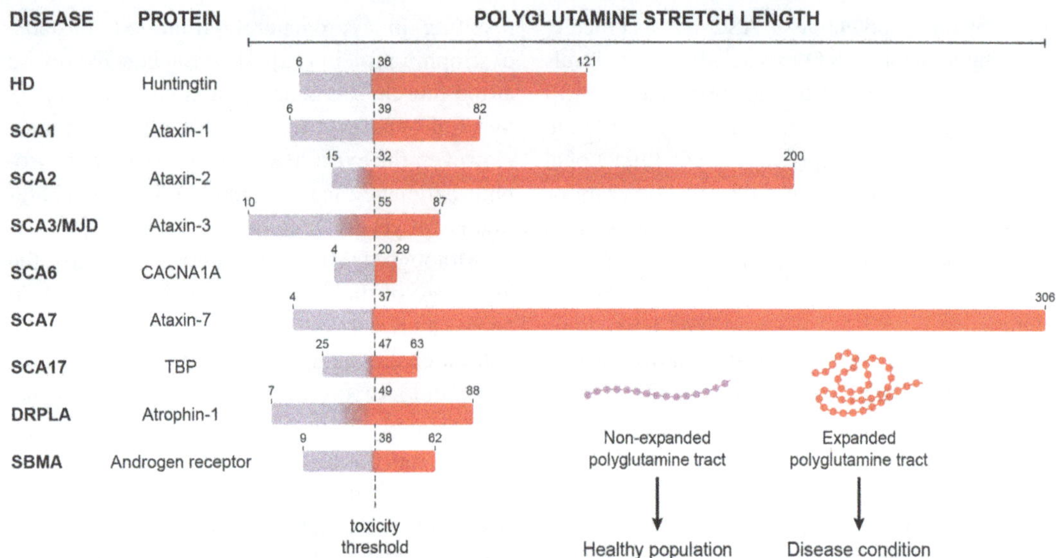

Fig. 7.3 Genes and proteins involved in polyglutamine diseases, highlighting the number of glutamine repetitions associated with the diseases phenotypes.

Table 7.2 Studies using antisense oligonucleotides to treat polyglutamine diseases.

Disease	Target	Chemistry	Mechanism	Administration route	Reference
Huntington's disease	mutHTT	cEt gapmer	RNAse H-mediated degradation	Intraparenchymal	Carroll *et al.* (2011)
	mutHTT	PS-2'OME gapmer	RNAse H-mediated degradation	ICV	Kordasiewicz *et al.* (2012)
	mutHTT	PS-2'OME cET gapmer	RNAse H-mediated degradation	ICV	Ostergaard *et al.* (2013)
	mutHTT	2'MOE gapmer	RNAse H-mediated degradation	ICV	Stanek *et al.* (2013)
	mutHTT	cET	RNAse H-mediated degradation	ICV	Southwell *et al.* (2014)
	mutHTT	PMO	Translation blockade	ICV	Sun *et al.* (2014)
	HTT	PS-2'OME	Exon skipping	Intraparenchyma	Evers *et al.* (2014)
SCA2	Ataxin-2	2'MOE gapmer	RNAse H-mediated degradation	ICV	Pulst (2016)
SCA3/MJD	Ataxin-3	PS-2'OME	Exon skipping	ICV	Toonen *et al.* (2017)
SBMA	Androgen receptor	cEt gapmer	RNAse H-mediated degradation	Subcutaneous	Lieberman *et al.* (2014)
	Androgen receptor	cEt/2'MOE gapmer	RNAse H-mediated degradation	ICV	Sahashi *et al.* (2015)

HTT Huntingtin
ICV intracerebroventricular

ASOs-based therapies may be approved in the next years. Nevertheless, the failure of some clinical trials using ASOs or the toxicity effects observed in some studies also recommend proceeding with caution, in order to build solid preclinical and clinical data ensuring the safety and efficacy of therapies.

7.2 RNA Interference

The discovery of the RNA interference (RNAi) pathway [21] revolutionized the molecular biology research field, providing important tools for studying genes and importantly constituting a new opportunity for gene therapy to treat genetic dominant disorders, where silencing of the mutant gene is essential to counteract the disease phenotype.

Several observations made in plant experiments, based on the introduction of antisense, exogenous genes, led researchers to postulate the existence of a posttranscriptional gene silencing mechanism [22–25]. In those studies, researchers observed a reduction in the expression of endogenous and exogenous genes, in a phenomenon named *co-suppression*. However, it was only in 1998 that Andrew Fire and Craig C. Mello demonstrated an efficient interference of gene expression by double-stranded RNA molecules in an animal species - *C. elegans* [21]. In this breakthrough study, the Nobel-awarded researchers (2006 Nobel Prize in Medicine or Physiology) observed that the introduction of exogenous double-stranded RNAs (dsRNAs) led to the silencing of a particular target gene. Now, it is known that RNAi is a cellular pathway conserved among several organisms, including in eukaryotes, being crucial in the regulation of gene expression and in the innate defense against invading viruses.

RNAi-based tools proved very successful in biomedical research, allowing researchers to apply gene silencing strategies for target identification and validation. Their use as therapeutics is also quite promising, and several clinical trials have been performed or are ongoing using small regulatory RNAi molecules. Importantly, their use is widespread in different biotechnology areas, including crop improvement [26].

7.2.1 Gene Expression Regulation in Eukaryotes

The central dogma of molecular biology establishes a unique direction of information flow from the DNA to RNA and finally to protein. It is now known that there are exceptions to this dogma, although it remains actual and accurate for most of the eukaryotic protein-coding genes. The human genome is estimated to have around 20,000 genes that codify for proteins (Table 7.3), whereas the number of genes that codify for different RNA species (e.g., noncoding RNAs, small regulatory RNA) is much higher. However, in a given cell and at a particular timepoint, only approximatively half of these genes are expressed, i.e., transcribed into RNA [27, 28]. This happens because gene expression in each cell is tightly, and differently, regulated. Differences in gene expression provide different features and properties to each cell type of a particular organism, despite all of them having the same DNA in their nuclei. Some important genes are expressed in all cells (constitutive genes), whereas the expression of others is dependent on internal cell factors and external factors and stimuli.

In order for cells to maintain their functions and homeostasis, gene expression is regulated at different points (Fig. 7.5). The first point of regulation is the **chromatin** control. Inside the cell nucleus, DNA binds to histone proteins, constituting the chromatin. When the DNA is tightly wrapped around histones, excess packaging prevents gene transcription, by preventing the coupling of transcription factors to the genes. Moreover, epigenetic events and the involvement of small regulatory RNAs contribute to gene silencing at the chromatin level, thus regulating gene expression in cells [29].

The next control step concerns the regulation of **transcription**. Even if the DNA is unwound and the genes are exposed, for transcription to occur several proteins need to be recruited. The binding of transcription factors can promote or repress transcription of a given gene, thus regulating its expression.

Next, there is the regulation of gene expression at the **RNA processing** level, in which the transcribed RNA undergoes different modifica-

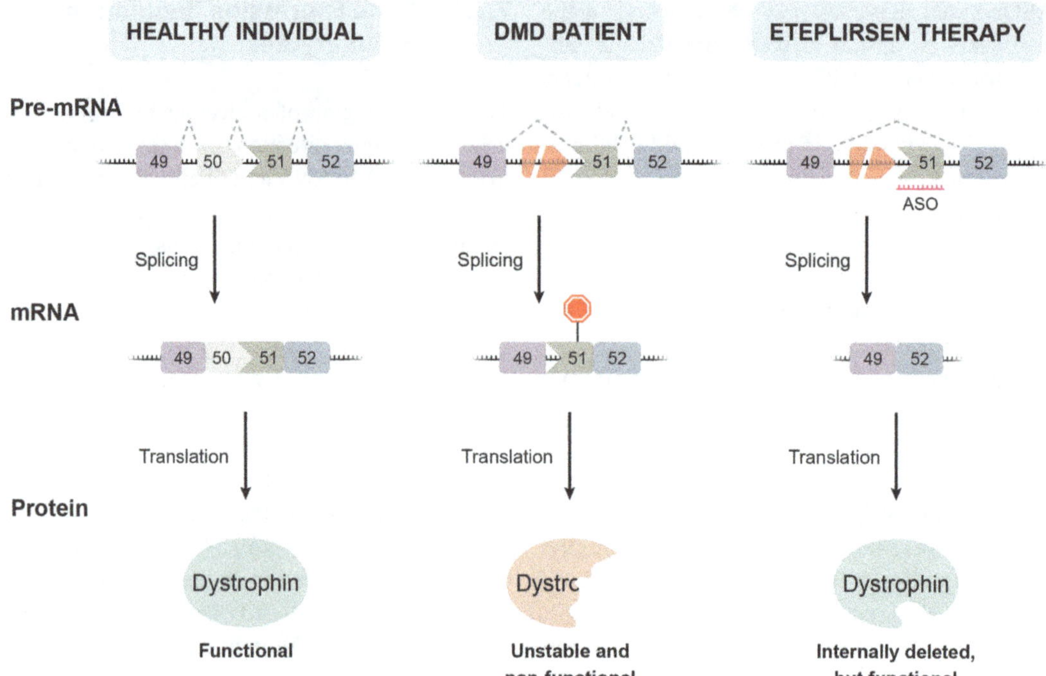

Fig. 7.4 Antisense oligonucleotides-based therapy for Duchenne muscular dystrophy (DMD). DMD is caused by different mutations in the *DMD* gene, leading to abnormal mRNA splicing and resulting in the production of a prematurely truncated, unstable, and nonfunctional dystrophin protein. A gene therapy for DMD has been approved, based on ASOs that target the splicing process. It leads to the skipping of exon 51, generating an internally deleted dystrophin protein that is nonetheless functional.

tions, for example splicing, capping and addition of a poly-A tail, in order to yield functional mature RNA molecules to be exported from the nucleus. The control of these types of modifications also constitutes a level of gene expression regulation [30]. At this point, **alternative splicing** of primary transcripts is yet another way for cells to regulate gene expression, by altering the mRNA products of the same gene.

The following regulation point concerns **RNA stability**. The lifetime of a mRNA molecule is an important factor for their translation into proteins. Binding of mRNA to RNA-binding proteins or to small regulatory RNAs such as RNAi molecules can prevent translation and target mRNA for degradation or, on the contrary, increase their stability promoting translation. Thus, RNA stability is now recognized as an important player in the regulation of gene expression [31].

The **translation** process is yet another important point, being divided into three phases: initiation, elongation and termination. Initiation is probably the most important step to be regulated, and players involved in the, process other than the mRNA molecules, such as initiation factors or ribosomes, can also be involved in the regulation of mRNA translation and, ultimately, of gene expression [32].

Finally, the **post-translational modifications** are the ultimate point of regulation of gene expression. Translated proteins can undergo a variety of modifications: some are permanent (e.g., proteolytic cleavage), whereas others are reversible (e.g., phosphorylation), but both types can affect protein activity, function and turnover, ultimately constituting a form of gene expression regulation.

The RNAi pathway interferes with gene expression through the action of particular small RNA

Table 7.3 Overview of the human genome genes and their function (according to GENCODE, Human Release 30).

Gene class	Function	Estimated number
Messenger RNA	Protein coding	19,986
Long noncoding RNA	Gene regulation	16,193
Small noncoding RNA	Gene regulation	7,576
MicroRNA	Translational inhibition and mRNA degradation	1,881
Small nuclear RNA	Processing of pre-mRNA	1,901
Small nucleolar RNA	Processing of rRNA, tRNA, and snRNA	942
Antisense RNA	Gene regulation	5,611

species that bind to mRNA targets and lead to their cleavage or degradation, or block their translation. The following sections describe these molecules and their functions, in the context of the two sub-pathways into which RNAi can be divided.

7.2.2 The Small Interfering RNA Pathway

Small or short interfering RNAs (siRNA) are ~22-nucleotide-long, double-stranded molecules with two-nucleotide 3'overhangs. The siRNA pathway starts with the cleavage of double-stranded RNAs (dsRNAs) by **Dicer** to form the siRNAs. The dsRNAs' can originate from viruses or transposons. The resulting siRNAs are then assembled into the minimal **RNA-induced silencing complex** (RISC) with an **Argonaute** protein. In this complex, the siRNA double strand is separated, with the **passenger strand** being discarded and the **guide strand** remaining linked to Argonaute. Then, RISC binds to a complementary mRNA sequence and silences it via the cleavage activity of Argonaute (Fig. 7.6). In humans, there are four Argonaute proteins (1–4), although only Ago2 presents slicer activity [33]. Nevertheless, other human Ago proteins can still participate in gene regulation by binding the target mRNA molecules and inhibiting their translation.

7.2.3 The MicroRNA Pathway

MicroRNAs (miRNAs) are ~22-nucleotide-long, double-stranded molecules encoded by specific genes in the nucleus that were first discovered in 1993, independently from the RNAi pathway [34]. The miRNA pathway starts in the nucleus, where miRNA genes are transcribed by RNA polymerase II, forming the **primary-miRNA** (pri-miRNA), which is at least 1000-nucleotide long, with single or clustered double-stranded hairpins (Fig. 7.7). The pri-miRNA is then cleaved by **Drosha** in the nucleus, resulting in the formation of a **precursor miRNA** (pre-miRNA) with around 70 nucleotides. The pre-miRNA associates with Exportin-5 to be exported to the cytoplasm. There, the miRNA pathway converges with the siRNA pathway, and the pre-miRNA is also processed by Dicer and assembled into RISC (Fig. 7.6).

7.2.4 Small Interfering RNAs Versus MicroRNAs

Despite converging in the final steps of the RNAi pathway, siRNAs and miRNAs share important differences that provide some specific features to each pathway, including their origin, number of mRNA targets or the molecular mechanism of gene regulation (Table 7.4).

One important difference between siRNAs and miRNAs is that siRNAs have a complete complementarity to the target mRNA (in the coding region), whereas miRNAs are only partially complementary, typically in the 3'UTR region of target mRNA (Fig. 7.8) (nevertheless, siRNAs designed with partially complementary binding sites have been implemented to study translational repression) [35]. From this difference emerge other important dissimilarities between the two pathways. First, if there is complete complemen-

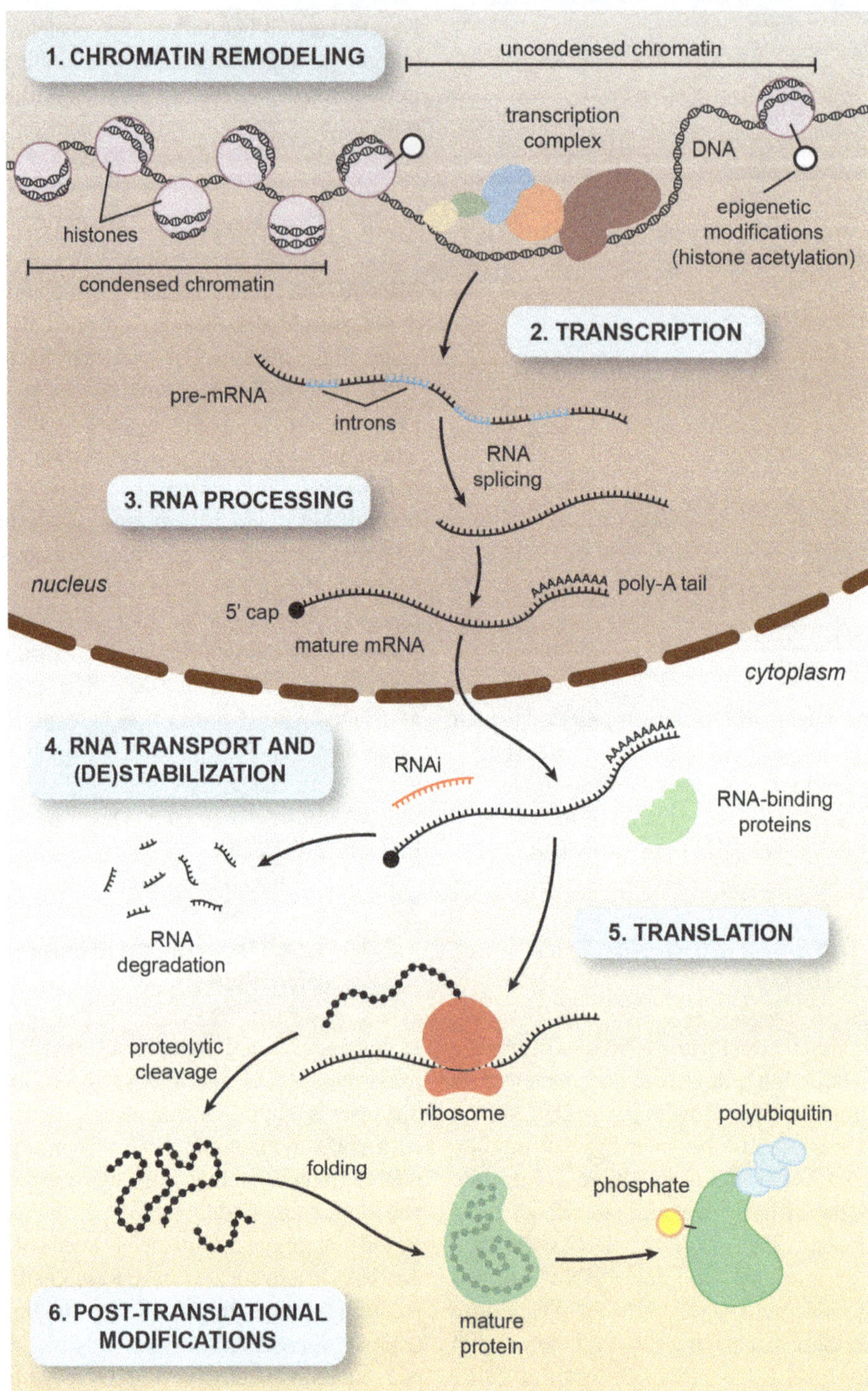

Fig. 7.5 Possible points of gene expression regulation in eukaryotes. The starting point correspondes to *chromatin remodeling* (1), as condensed chromatin prevents gene transcription. The second point of regulation involves the control of the *transcription* process (2), through the binding of transcription factors. Next, *RNA processing* in the nucleus (3) also constitutes an opportunity for regulation of gene expression. In the cytoplasm, the *transport* of mRNA and, importantly, control of its *stability* are other points of gene expression regulation (4). The next level of control is protein *translation* (5), since different mechanisms and players can promote or repress the translation of a particular mRNA. Finally, several *post-translational modifications* (6) of the proteins can alter their functionality and turnover, thus constituting a last level of control of gene expression.

tarity, which is the case of the siRNA pathway, then Argonaute cleaves the mRNA through its catalytic activity. On the other hand, in the miRNA pathway, while the incomplete complementarity may also lead to the degradation of the target mRNA, this does not occur through the catalytic activity of Argonaute, and the pathway may also lead to translational repression of the mRNA instead. Second, due to its complete complementarity, siRNAs target one single mRNA, whereas miRNAs can target multiple mRNAs. This particularity of miRNAs may be an important advantage in the treatment of complex multigenic diseases using gene therapy, which require the targeting of multiple genes.

Another important difference between the two RNAi pathways concerns the biogenesis of each regulatory molecule. MiRNAs are constitutively expressed by specific genes, whereas siRNAs have their origin in transposons and viruses. In the context of gene therapy applications, both siRNAs and miRNAs have advantages and disadvantages, although the former was far more tested in clinical studies. Nevertheless, they are both quite attractive as therapeutic tools, as they can target virtually any gene, which is difficult or almost impossible to achieve using small molecules or protein-based products.

7.2.5 Small Interfering RNAs Versus Short Hairpin RNAs

The use of siRNAs as a molecular tool for gene knockdown experiments or as a therapeutic agent has some important disadvantages. For example, siRNAs do not allow a long-term expression, limiting their use in cells to a 72 hours maximum and demanding continuous applications in their use as therapy. The transient effect also requires that siRNAs be administered in high doses, to allow a therapeutic effect. This important fact contributes to innate immune response and toxicity events, detected upon synthetic siRNAs administration, which strongly limits their clinical application as therapeutic agents. Moreover, in systemic delivery, siRNAs are rapidly degraded by blood nucleases, and their uptake is low in most organs and cells (except for the liver).

Trying to overcome these limitations, an alternative to siRNAs can be the use of short hairpin RNAs (shRNAs), which are artificial noncoding RNAs with a stem-loop structure, which can be expressed in the nucleus and use part of the miRNA pathway. They mimic a pri-miRNA and are processed by Drosha and exported to the cytoplasm by Exportin-5. The shRNAs are usually delivered through viral vectors and thus can achieve a permanent expression when using integrative virus. Like siRNAs, they are fully complementary to the target mRNA, although, in the last years, several studies showed less off-target effects of shRNAs comparing to siRNAs (Table 7.4) [36].

7.2.6 Gene Therapy Applications of RNAi

The discovery of the RNAi mechanism led to a huge development in gene therapy preclinical trial studies, which ultimately resulted in several clinical studies involving siRNAs, shRNAs and miRNAs. Besides the efficacy issue, clinical studies involving RNAi molecules also focus on toxicity and delivery aspects.

To date, more than 30 clinical trials were carried out using RNAi molecules (mainly siRNAs) for different therapeutic indications, especially cancer and ophthalmic conditions [37]. Cancer is a preferential target for RNA silencing molecules, aiming to inhibit genes related to uncontrolled cell growth, angiogenesis, metastasis and drug resistance. On the other hand, ocular conditions are also a good target as eyes are immune-privileged sites where local delivery can be achieved. Nevertheless, siRNA, shRNA or miRNA molecules were also studied as therapeutic effectors for several other conditions, including cardiovascular or infectious diseases.

Recently the first RNAi-based gene therapy product was approved in Europe and the USA for hereditary transthyretin-mediated amyloidosis. Onpattro™ (patisiran) is a siRNA encapsulated within a liposome nanoparticle, administrated intravenously once every 3 weeks, which was demonstrated to elicit an average knockdown of 87% (maximum of 96%) of the protein causing the disease (transthyretin - TTR) [20].

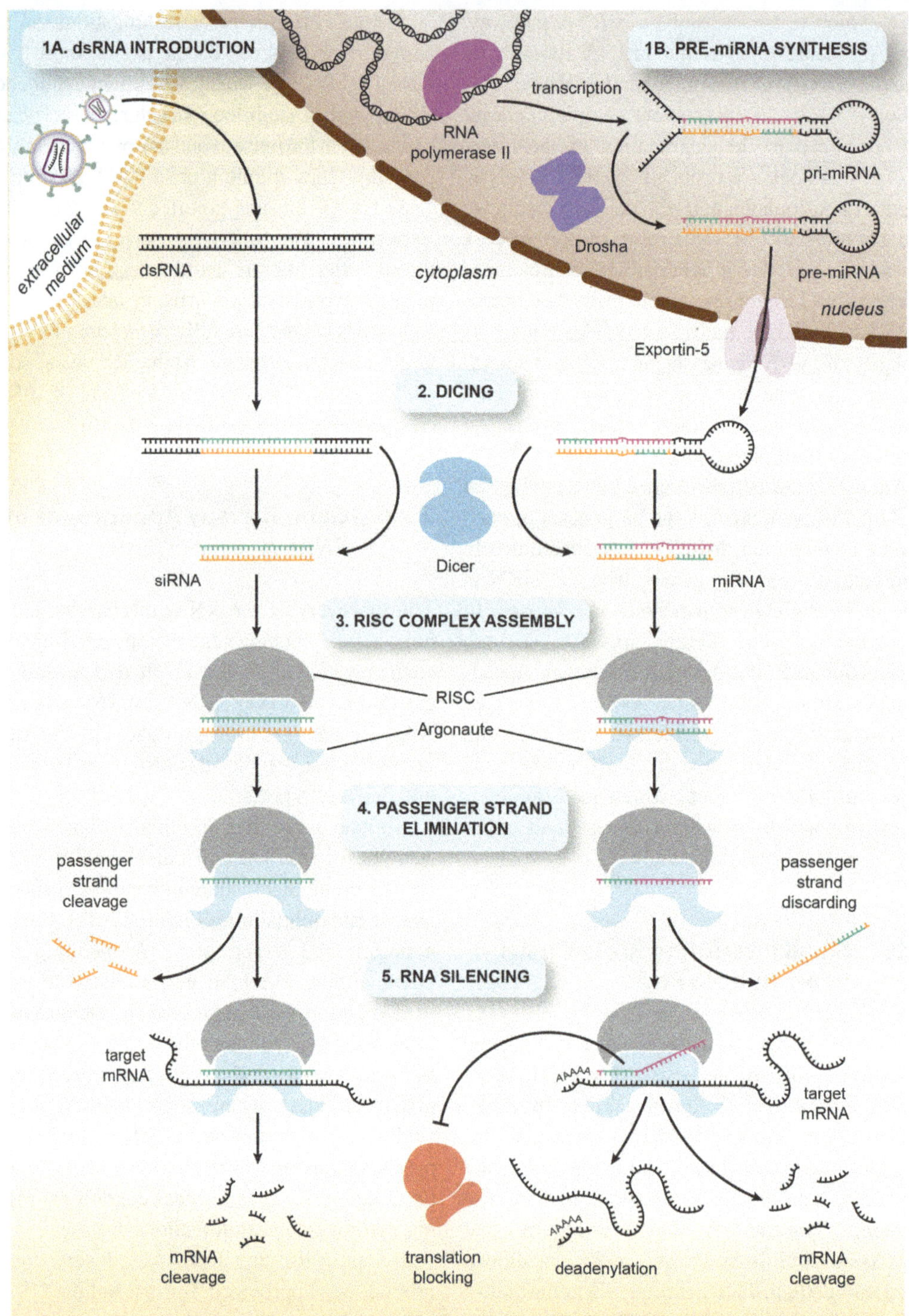

Fig. 7.6 **The RNA interference (RNAi) pathway** is a cellular pathway conserved among several organisms, including eukaryotes, being crucial in the regulation of gene expression and in the innate defense against invading viruses. RNAi has two main subpathways: the *small interfering RNA* (siRNA) and the **microRNA** (miRNA) pathways. The siRNA pathway starts with the introduction of a dsRNA molecule into the cell (1A), while the miRNA pathway is initiated with the transcription of pri-miRNA and its processing in the nucleus (1B). In the cytoplasm of the cell, the two pathways converge, as both molecules are processed by Dicer (2) and assembled into the RISC complex (3). This is followed by the elimination of the passenger strand (4) and the target mRNA silencing (5) through mRNA cleavage or by translational repression.

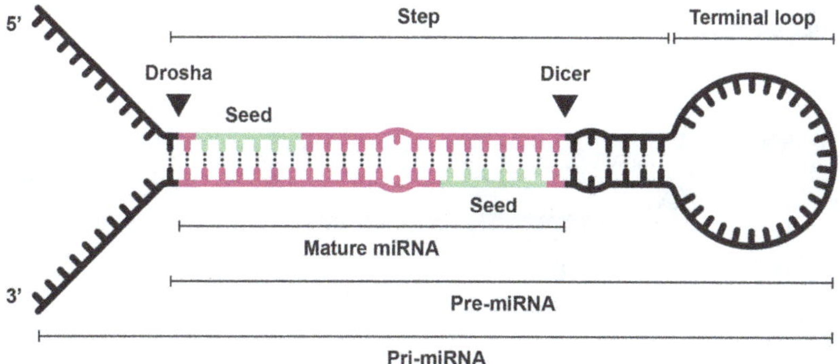

Fig. 7.7 MicroRNA secondary structure, highlighting the processing sites for Drosha and Dicer enzymes, as well as the seed region, which is the key binding location responsible for translation silencing. In the nucleus, the pri-miRNA is cleaved by Drosha into pre-miRNA, which is 60–70-nucleotide long. In the cytoplasm, the pre-miRNA, with its characteristic loop, is cleaved by Dicer, yielding the mature miRNA molecule.

Table 7.4 Comparison between the main characteristics of the different molecules of regulatory RNAs from the RNAi pathway that can be used in gene therapy.

Properties	siRNA	miRNA	shRNA
Origin	Transposons, viruses, exogenous (synthetic)	Encoded by their own genes, exogenous (synthetic)	Exogenous but encoded by genes (synthetic)
Structure (prior to Dicer processing)	Double-stranded RNA that contains 30 to over 100 nucleotides	Pre-miRNA contains 70–100 nucleotides with mismatches and hairpin structure	Double-stranded with loop
Structure	21–23 nucleotide RNA duplex	19–25 nucleotide RNA duplex	21–23 nucleotide RNA duplex
Complementarity	Fully complementary to target mRNA (coding region)	Partially complementary to mRNA, typically targeting the 3'UTR region of the target mRNA	Fully complementary to target mRNA (coding region)
mRNA target	One	Multiple	One
Mechanism of gene expression regulation	Cleavage of mRNA	Cleavage of mRNA / translational repression / degradation of mRNA	Cleavage of mRNA
Delivery to the cell	Non-viral systems	Viral and non-viral systems	Viral and non-viral systems
Persistence	Degraded after 48 hours	Expressed for up to 3 years	Expressed for up to 3 years
Dosage required	High	Low	Low
Likelihood of "off-target" effects	Low	High	Low

7.2.7 RNAi Terms Glossary

dsRNA – a long double-stranded noncoding RNA with more than 100 nucleotides, from different sources (e.g., transposons, viruses or artificially introduced in cells), which is processed by Dicer generating siRNAs.

siRNA – a noncoding regulatory small double-stranded RNA molecule with 21–23 nucleotides, with two-nucleotide 3' overhangs, originated from Dicer cleavage of dsRNAs.

microRNA – a noncoding regulatory double-stranded RNA molecule (~22 nucleotides) encoded by specific genes in the nucleus and processed by Drosha and Dicer.

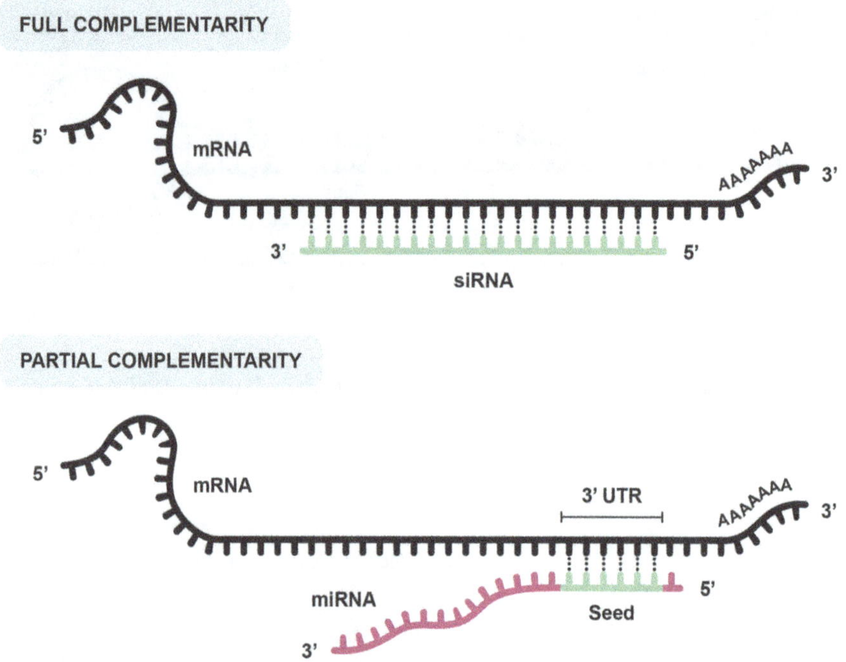

Fig. 7.8 Complementarity of siRNA and miRNA to target mRNA. siRNAs have a complete complementarity to target mRNA (in the coding region), whereas miRNAs are only partially complementary, through the seed region, typically to the 3'UTR region of the target mRNA.

shRNA – a noncoding small regulatory double-stranded RNA molecule that contains a loop structure, which is artificially introduced in cells, expressed in the nucleus, and processed to siRNA by Dicer in the cytoplasm.

Drosha – a ribonuclease (RNase) III enzyme that processes pri-miRNAs and shRNAs in the nucleus.

Dicer – an RNase III enzyme vital for the siRNA and miRNA pathways, generating small double-stranded RNA molecules suitable to be loaded into Argonaute protein.

RISC – a ribonucleoprotein complex, consisting of an Argonaute protein bound to a guide strand from a siRNA or miRNA molecule, that fragments the target mRNA.

Argonaute – an essential protein for RISC assembly, functioning in the recognition of the guide strand, target cleavage and recruitment of other important proteins involved in the silencing process.

dsRBP – a protein important for the Dicer processing of dsRNAs and subsequent passage to the RISC (e.g., TAR RNA-binding protein, TRBP).

Guide strand – a single-stranded RNA molecule resulting from the separation of the siRNA or miRNA, that has complementarity the target mRNA, providing specificity to the RNAi process.

Seed sequence – a region of 2–8 nucleotides in the guide strand whose complementarity to the target mRNA is critical to the silencing success.

7.2.8 Gene Silencing as Therapy for Machado-Joseph Disease/ Spinocerebellar Ataxia Type 3

Machado-Joseph disease is a fatal dominant inherited neurodegenerative disease, caused by an abnormal CAG expansion in the coding region of the *ATXN3* gene. Along with the other eight polyglutamine diseases, it constitutes a larger group of monogenic inherited neurodegenerative disorders, for which there is currently no cure or treatment able to delay or stop disease progression. Due to its monogenic cause, gene therapy strategies are very promising, and its dominant character puts the strategy of gene silencing in the frontline of the possible approaches studied. In fact, several gene silencing preclinical studies were performed in the last years, showing both the efficiency and safety of the RNAi-based silencing of the *ATXN3* gene.

In 2008, a first study showed that allele-specific silencing of mutant *ATXN3* led to a significant decrease in the neuropathological abnormalities associated with mutant ataxin-3 expression in the rat striatum [38]. This specific silencing was possible due to the presence of a nucleotide polymorphism that is present in 70% of MJD/SCA3 patients, and the strategy was mediated by the delivery of an shRNA by lentiviral vectors. Later, the strategy also proved successful in reducing the neuropathological and behavior deficits when the shRNA was delivered into the cerebellum of two different mouse models [39, 40]. More recently, it was shown that long-term expression of this shRNA is safe, as no toxicity or RNAi pathway saturation was observed at 6 months after the lentiviral injection of the shRNA [41]. A non-viral delivery system was also studied, using lipid particles (SNALPs) encapsulating siRNA specifically targeting mutant ataxin-3. This strategy also reduced the neuropathological and behavior deficits in two MJD/SCA3 mouse models [42].

Other studies explored the potential of the miRNA pathway as a form of silencing the expression of mutant ataxin-3 and thus mitigate the disease phenotype. For example, an artificial miRNA mimic targeting human ataxin-3 mRNA led to a reduction of ataxin-3 expression in a mouse model; however, it failed to rescue its motor deficits [43]. Later, the overexpression of several miRNAs dysregulated in MJD/SCA3 was able to reduce mutant ataxin-3 levels, aggregates and neuronal dysfunction in a lentiviral MJD/SCA3 mouse model [44].

Altogether, these preclinical studies clearly showed the potential of RNAi-based gene therapy strategies to mitigate MJD/SCA3-associated abnormalities. However, until now, there are no studies testing these RNAi-based gene therapy strategies in clinical trials. Several factors contribute to this lack of translation, including the rare condition of MJD/SCA3 and the recent implication of the expanded RNA species in the toxicity observed in polyglutamine diseases. Moreover, the advent of gene editing strategies, where the DNA is the target, greatly contributed to a loss of interest in gene silencing strategies based on RNAi.

7.3 Future Prospects on Gene Silencing

The potential of gene silencing strategies for research and gene therapy is clear and well established for different diseases. The possibility to target particular mRNAs, thus controlling protein translation and ultimately gene expression, is a major advantage over conventional drugs that target proteins, such as enzymes and receptors. For the gene therapy field, it promises to cure genetic diseases with a dominant character, by reducing the expression of the mutant gene and protein. Both ASOs and RNAi strategies yielded very promising and encouraging preclinical results, which could provide newly approved therapies in the near future. The choice of one tool or the other is much dependent, among other aspects, on the target disease, type of administration, delivery method and also the advantages and disadvantages of each system (Table 7.5). For example, the diversity of functional mechanisms and chemical modifications of ASOs provide a strong advantage for this system. On the other hand, shRNAs and microRNAs are more easily packaged into different delivery vectors and therefore can provide a more robust and persistent gene silencing than ASOs.

Despite the promising perspectives on the use of gene silencing techniques in human gene therapy, several challenges need to be addressed and overcome. Further studies should focus on possible off-target effects, the improvement of delivery techniques, a better understanding of the disease physiopathology in order to comprehend

Table 7.5 Advantages and disadvantages of the antisense oligonucleotide technology and of the RNA interference pathway.

Technology	Advantages	Disadvantages
ASOs	Can lead to a considerable downregulation of gene expression	May lead to off-target effects
	Several chemical modifications can be made	Limited uptake *in vivo*
	Different functional mechanisms can be used	Need for repeated administrations
	Easy production	Limited choice of delivery vectors
		Potential toxicity events
RNAi	Robust downregulation of gene expression	May lead to off-target effects
	Multiple small RNAs to choose from (siRNA, shRNA, miRNA)	Saturation of the endogenous RNAi pathway
	Multiple delivery vectors can be used	Potential toxicity events
	Can achieve a long-term effect	

if gene silencing is enough to revert the phenotype, the long-term safety profile and toxicity of the strategies, among other important subjects.

The approval of gene therapy products based on ASOs technology and on the RNAi pathway exploitation, as well as the existence of several others in the pipeline for marketing approval, suggests that gene silencing can be an efficient and safe gene therapy strategy for several human diseases.

This Chapter in a Nutshell

- Silencing the expression of mutant genes using gene therapy approaches became possible with the development of antisense oligonucleotides (ASOs) and with the discovery of the RNA interference (RNAi) pathway, thus allowing the treatment of genetically dominant diseases.
- ASOs are synthetic, unmodified or chemically modified, single-stranded DNA molecules with 8 to 50 nucleotides that hybridize to their target mRNA through Watson and Crick base pairing.
- ASOs can exert their action in many functional ways, which can be categorized into two main mechanisms: RNAse H-dependent (mRNA degradation) and RNAse H-independent (nucleic acid occupancy only) mechanisms.
- RNAi is a cellular pathway conserved among several organisms, including eukaryotes, that is crucial in the regulation of gene expression and in the innate defense against invading viruses.
- In eukaryotes, gene expression is regulated at different points and levels, including through chromatin control, transcription regulation, RNA processing, RNA stability modulation, translation regulation, and posttranslational modifications.
- Small/short interfering RNAs (siRNA) are ~22-nucleotide long molecules, formed through the cleavage of double-stranded RNAs (dsRNAs). They present full complementarity with a single target mRNA molecule, leading to its cleavage.
- MicroRNAs (miRNAs) are ~22-nucleotide long molecules encoded by their own genes, thus having a processing step in the nucleus. They exhibit partial complementarity with several target mRNAs. Besides leading to target mRNA cleavage, they can also lead to its degradation or to its translational repression.

Review Questions

1. Which of the following features is common between ASOs (antisense oligonucleotides) and siRNAs (small interfering RNAs)?
   (a) Composed of more than 100 nucleotides
   (b) Only of synthetic origin
   (c) Degradation involves the RISC complex
   (d) Degradation involves RNase H
   (e) High specificity
2. Which of the following features is not improved in third-generation ASOs compared with the second generation?

   (a) Nuclease resistance
   (b) Stability
   (c) Target affinity
   (d) Cellular uptake
3. Which of the following is not a functional action mechanism of ASOs?
   (a) RNAse H-mediated degradation
   (b) Translational arrest
   (c) Splicing modulation
   (d) Inhibition of transcription
   (e) Inhibition of RNA-binding proteins
4. Which of the following is not a feature of microRNAs?
   (a) Encoded by their own genes
   (b) Fully complementary to mRNA
   (c) Multiple mRNA targets
   (d) Different mechanisms of gene expression regulation
   (e) 19–25 nucleotide RNA duplex
5. Which of the following has/have one single mRNA target?
   (a) siRNA
   (b) siRNA and shRNA
   (c) shRNA and microRNA
   (d) microRNA
   (e) dsRNA and microRNA
6. Which of the following elements does not belong to the siRNA pathway of the RNA interference mechanism?
   (a) Argonaute
   (b) dsRNAs
   (c) RISC
   (d) Exportin
   (e) Dicer

References and Further Reading

1. Stephenson ML, Zamecnik PC (1978) Inhibition of Rous sarcoma viral RNA translation by a specific oligodeoxyribonucleotide. Proc Natl Acad Sci U S A 75:285–288
2. Roehr B (1998) Fomivirsen approved for CMV retinitis. J Int Assoc Phys AIDS Care 4:14–16
3. Rinaldi C, Wood MJA (2018) Antisense oligonucleotides: the next frontier for treatment of neurological disorders. Nat Rev Neurol 14:9–21
4. Zhao Q, Matson S, Herrera CJ, Fisher E, Yu H, Krieg AM (1993) Comparison of cellular binding and uptake of antisense phosphodiester, phosphorothioate, and mixed phosphorothioate and methylphosphonate oligonucleotides. Antisense Res Dev 3:53–66
5. Mansoor M, Melendez AJ (2008) Advances in antisense oligonucleotide development for target identification, validation, and as novel therapeutics. Gene Regul Syst Bio 2:275–295
6. Juliano RL (2016) The delivery of therapeutic oligonucleotides. Nucleic Acids Res 44:6518–6548
7. Southwell AL, Skotte NH, Bennett CF, Hayden MR (2012) Antisense oligonucleotide therapeutics for inherited neurodegenerative diseases. Trends Mol Med 18:634–643
8. Finkel RS, Chiriboga CA, Vajsar J, Day JW, Montes J, De Vivo DC, Yamashita M, Rigo F, Hung G, Schneider E et al (2016) Treatment of infantile-onset spinal muscular atrophy with nusinersen: a phase 2, open-label, dose-escalation study. Lancet 388:3017–3026
9. Zhao X, Pan F, Holt CM, Lewis AL, Lu JR (2009) Controlled delivery of antisense oligonucleotides: a brief review of current strategies. Expert Opin Drug Deliv 6:673–686
10. Crooke ST, Witztum JL, Bennett CF, Baker BF (2018) RNA-targeted therapeutics. Cell Metab 27:714–739
11. Matos CA, Almeida LP, Nobrega C (2017) Proteolytic cleavage of Polyglutamine disease-causing proteins: revisiting the toxic fragment hypothesis. Curr Pharm Des 23:753–775
12. Nóbrega C, Almeida LP (eds) (2018) Polyglutamine diseases. Springer: Gewerbestrasse, Cham, Switzerland
13. Matos CA, Carmona V, Vijayakumar UG, Lopes S, Albuquerque P, Conceicao M, Nobre RJ, Nobrega C, de Almeida LP (2018) Gene therapies for Polyglutamine diseases. Polyglutamine Disord 1049:395–438
14. Rodrigues FB, Wild EJ (2018) Huntington's disease clinical trials corner: february 2018. J Huntingtons Dis 7:88–97
15. Kolb SJ, Kissel JT (2015) Spinal muscular atrophy. Neurol Clin 33:831
16. Graf H, Wurster CD (2018) Antisense oligonucleotides in psychiatric disorders. J Clin Psychopharmacol 38:651–652
17. Yiu EM, Kornberg AJ (2015) Duchenne muscular dystrophy. J Paediatr Child Health 51:759–764
18. Lim KRQ, Maruyama R, Yokota T (2017) Eteplirsen in the treatment of Duchenne muscular dystrophy. Drug Des Devel Ther 11:533–545
19. Randeree L, Eslick GD (2018) Eteplirsen for paediatric patients with Duchenne muscular dystrophy: a pooled-analysis. J Clin Neurosci 49:1–6
20. Shaer DA, Musaimi OA, Albericio F, de la Torre BG (2019) 2018 FDA tides harvest. Pharmaceuticas 12:52
21. Fire A, Xu SQ, Montgomery MK, Kostas SA, Driver SE, Mello CC (1998) Potent and specific genetic interference by double-stranded RNA in Caenorhabditis elegans. Nature 391:806–811
22. Vanderkrol AR, Lenting PE, Veenstra J, Vandermeer IM, Koes RE, Gerats AGM, Mol JNM, Stuitje AR (1988) An anti-sense Chalcone Synthase Gene in

transgenic plants inhibits flower pigmentatioN. Nature 333:866–869
23. Vanderkrol AR, Mur LA, Beld M, Mol JNM, Stuitje AR (1990) Flavonoid genes in petunia - addition of a limited number of gene copies may lead to a suppression of gene-expression. Plant Cell 2:291–299
24. Napoli C, Lemieux C, Jorgensen R (1990) Introduction of a Chimeric Chalcone Synthase gene into Petunia results in reversible co-suppression of homologous genes in trans. Plant Cell 2:279–289
25. Jorgensen RA, Cluster PD, English J, Que QD, Napoli CA (1996) Chalcone synthase cosuppression phenotypes in petunia flowers: comparison of sense vs antisense constructs and single-copy vs complex T-DNA sequences. Plant Mol Biol 31:957–973
26. Kamthan A, Chaudhuri A, Kamthan M, Datta A (2015) Small RNAs in plants: recent development and application for crop improvement. Front Plant Sci 6:208
27. Carlberg C, Molnár F (2014) Mechanisms of gene regulation. Springer: Dordrecht
28. Lee TI, Young RA (2013) Transcriptional regulation and its misregulation in disease. Cell 152:1237–1251
29. Hoopes L (2008) Introduction to the gene expression and regulation topic room. Nat Educ 1(1):160
30. Lodish H, Berk A, Zipursky SL et al (2000) Molecular cell biology. W.H. Freeman and Company, New York
31. Hollams EM, Giles KM, Thomson AM, Leedman PJ (2002) MRNA stability and the control of gene expression: implications for human disease. Neurochem Res 27:957–980
32. Lackner DH, Bahler J (2008) Translational control of gene expression from transcripts to transcriptomes. Int Rev Cell Mol Biol 271:199–251
33. Wilson RC, Doudna JA (2013) Molecular mechanisms of RNA interference. Annu Rev Biophys 42:217–239
34. Lee RC, Feinbaum RL, Ambros V (1993) The C. elegans heterochronic gene lin-4 encodes small RNAs with antisense complementarity to lin-14. Cell 75:843–854
35. Doench JG, Petersen CP, Sharp PA (2003) siRNAs can function as miRNAs. Genes Dev 17(4):438–442. doi:10.1101/gad.1064703
36. Singh S, Narang AS, Mahato RI (2011) Subcellular fate and off-target effects of siRNA, shRNA, and miRNA. Pharm Res 28:2996–3015
37. Bobbin ML, Rossi JJ (2016) RNA interference (RNAi)-based therapeutics: delivering on the promise? Annu Rev Pharmacol Toxicol 56:103–122
38. Alves S, Regulier E, Nascimento-Ferreira I, Hassig R, Dufour N, Koeppen A, Carvalho AL, Simoes S, de Lima MC, Brouillet E et al (2008) Striatal and nigral pathology in a lentiviral rat model of Machado-Joseph disease. Hum Mol Genet 17:2071–2083
39. Nobrega C, Nascimento-Ferreira I, Onofre I, Albuquerque D, Hirai H, Deglon N, de Almeida LP (2013) Silencing mutant ataxin-3 rescues motor deficits and neuropathology in Machado-Joseph disease transgenic mice. PLoS One 8:e52396
40. Nobrega C, Nascimento-Ferreira I, Onofre I, Albuquerque D, Deglon N, de Almeida LP (2014) RNA interference mitigates motor and neuropathological deficits in a cerebellar mouse model of Machado-Joseph disease. PLoS One 9:e100086
41. Nobrega C, Codesso JM, Mendonca L, Pereira de Almeida L (2019) RNAi therapy for Machado-Joseph disease: long-term safety profile of lentiviral vectors encoding shRNAs targeting mutant ataxin-3. Hum Gene Ther 30:841
42. Conceicao M, Mendonca L, Nobrega C, Gomes C, Costa P, Hirai H, Moreira JN, Lima MC, Manjunath N, Pereira de Almeida L (2016) Intravenous administration of brain-targeted stable nucleic acid lipid particles alleviates Machado-Joseph disease neurological phenotype. Biomaterials 82:124–137
43. Costa Mdo C, Luna-Cancalon K, Fischer S, Ashraf NS, Ouyang M, Dharia RM, Martin-Fishman L, Yang Y, Shakkottai VG, Davidson BL et al (2013) Toward RNAi therapy for the polyglutamine disease Machado-Joseph disease. Mol Ther 21:1898–1908
44. Carmona V, Cunha-Santos J, Onofre I, Simoes AT, Vijayakumar U, Davidson BL, Pereira de Almeida L (2017) Unravelling endogenous microRNA system dysfunction as a new pathophysiological mechanism in Machado-Joseph disease. Mol Ther 25:1038–1055

Gene Editing

8

8.1 Rewriting the Typewriter

For many decades, scientists have tried to develop methods to modify the fundamental code that underlies every single organism – the genome. The reasons behind this intent have been diverse: altering genes and proteins in order to study their roles and functions; modifying cells and organisms with scientific, or commercial interest; or developing strategies and approaches to tackle human diseases, among many others. Beginning in the 1920s, scientists have made use of electromagnetic radiation, mutagenic chemical compounds, recombinant DNA and molecular cloning technology, transfection methods, viruses and transposons in order to alter the nucleic acids inside a cell, generating random mutations, inserting or removing genes, or modifying existing ones [1]. Each of these techniques has revolutionized the fields of molecular biology and biotechnology and has contributed to the development of many other areas. However, the degree of precision and fidelity these techniques allow is below the ideal, perhaps utopic, scenario of being downright able to "freely rewrite" an existing genome.

In eukaryotic cells, perhaps the approach that has come consistently closest to this objective relies on the exploitation of homologous recombination mechanisms as a means to target and alter specific genetic loci [2, 3]. When a DNA sequence is inserted into a cell, there is the chance that it will be randomly inserted into one of its chromosomes. However, if that DNA fragment is additionally flanked by two regions that are homologous to a particular DNA sequence of the target cell genome, it is possible that the endogenous homologous recombination mechanisms will lead to the substitution of the homologous site with the exogenous sequence, thereby inserting or substituting the region in between the homology arms. This gene targeting strategy has been recurrently and reliably applied to the generation of knockout mouse models, as well as knock-in animals in which particular genetic sequences have been altered or inserted. These types of models have come to be some of the most valuable tools in biomedical research, providing the bases for studies that have unveiled crucial aspects of human pathologies and contributing to the development and preclinical testing of therapeutic approaches.

Nonetheless, this proven method of modifying genomes may not be practical, feasible, or at all possible in every type of setting. The dominion scientists have acquired over mouse genetics is not always easily translated to other organisms, which may have their own particularities regarding life cycle, reproductive behavior, development, or the very molecular mechanisms that take place inside their cells. Moreover, generation of knock-in and knockout animals relies on a series of procedures, using, for example, stem cell cultures and animal crossings, which are not

C. Nóbrega et al., *A Handbook of Gene and Cell Therapy*,
https://doi.org/10.1007/978-3-030-41333-0_8

conceivable in several other contexts, such as when aiming at more straightforward forms of cell manipulation.

Genome manipulation as a possible pathway to the treatment of human diseases should ideally target a patient's cells directly and allow a high degree of control over the changes that are introduced. It is tempting to conjecture that techniques may be developed that, by allowing the direct rewriting of the four-letter blueprints of the human organism, will be able to delete disease-associated genes and correct pathogenic mutations.

Over the last few years, the promise of more direct, precise, and versatile methods for "rewriting" the eukaryotic genome has increasingly drawn the attention of the scientific community. Researchers are now employing a variety of new ways to manipulate genes, collectively termed as gene, or genome, editing. These techniques offer the promise of being able to target precise regions of the genome of any cell and produce a variety of customizable modifications with an unprecedented degree of consistency.

8.2 The Basis of Gene Editing

Present-day approaches to gene editing rely on two conditions: (a) the ability to define the region of the genome that is to be altered and (b) the capacity to effect the actual changes or, more precisely, to create the conditions for the desired alterations to occur [4, 5]. These two abilities combine to generate the desired modification(s), at the intended locus (or loci).

Definition of the target site is accomplished by molecules that specifically bind to a particular nucleotide sequence and subsequently cleave both chains of the DNA, producing a **double-strand break** (DBS) [6]. These molecules are **endonucleases** – enzymes that are able to separate nucleotides adjacently localized in the middle of a polynucleotide chain, by cutting the phosphodiester bond existing between them. In order to target a particular sequence with as much specificity as possible, endonucleases used in gene editing must be as selective as possible in regard to the DNA sequence that they bind to and cut. The region to be altered can coincide with the sequence targeted by the endonucleases or, alternatively, that region can be in the close vicinity of the targeted sequece.

Because of the central role endonucleases have in current approaches to gene editing, the term gene editing could, and perhaps should, be more appropriately substituted by "nuclease-based gene editing" [6]. However, DNA DSBs in isolation would be insufficient to edit genes and genomes. The actual modifications that then take place in the nucleotide sequence are in fact enacted by the cell, namely, through its endogenous **DNA repair mechanisms** [4, 5]. As a direct consequence of a DSB, DNA repair systems are recruited to the vicinity of the DSB site [7]. Changes to the nucleotide sequence may be introduced upon DNA repair, and, if appropriate conditions are established, those changes will result in the desired modification of the target locus.

The following section briefly outlines the two main mechanisms through which DNA DSBs are repaired in a cell and the ways they can be exploited in order to generate a particular desired alteration in the genome. The section after that will describe the four main classes of endonucleases that can be used to introduce the DSBs responsible for triggering modifications at the intended sites.

8.3 DNA Double-Strand Break Repair Mechanisms

Maintaining the integrity of the genome is of crucial importance for the preservation of cellular homeostasis and for the overall health of the organisms the cells compose. For this reason, Life has evolved diverse systems and molecular pathways that ensure that the DNA is appropriately repaired in case an insult threatens its integrity. DNA DSBs in particular are mainly repaired through one of two mechanisms: nonhomologous end-joining (NHEJ) and homology-directed repair (HDR; Fig. 8.1). A vast array of proteins participates in both pathways, performing intricate biochemical and

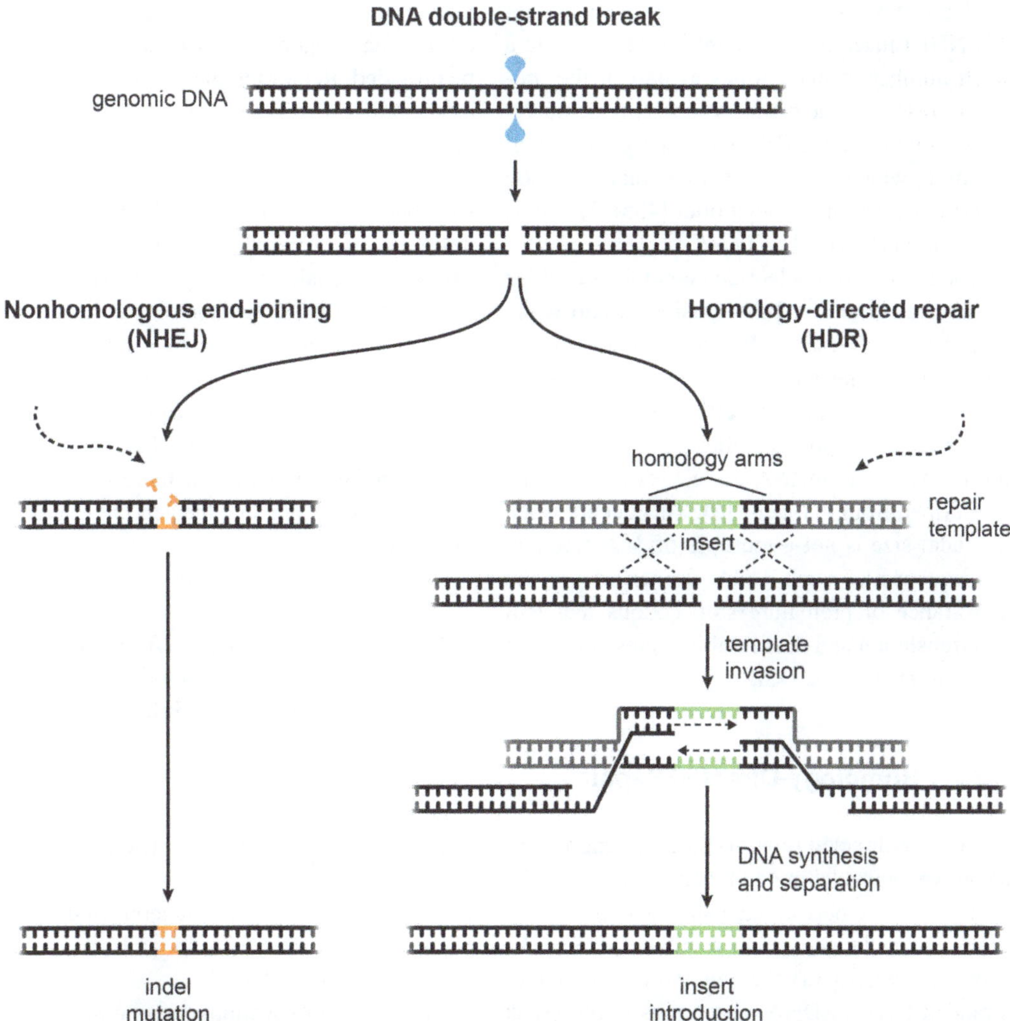

Fig. 8.1 Main mechanisms of DNA double-strand break (DSB) repair. *Nonhomologous end-joining* (NHEJ) involves resealing of the DSB site by simple linking of the two free ends of the DNA double strand. However, this process is prone to errors and often leads to the introduction or deletion of nucleotides. *Homology-directed repair* (HDR) is a more complex mechanism, in which the DSB is repaired using a homologous DNA molecule as template. While NHEJ may be exploited in order to generate gene knockouts or delete genetic elements, HDR may be used to introduce or substitute specific nucleotide sequences in a genome.

structural operations that are beyond the scope of this chapter. Nonetheless, regarding their role in gene editing, it is important to understand that, perhaps ironically, these DNA repair mechanisms are not error-proof and can thus be manipulated in order to produce intentional changes in the genome.

8.3.1 Nonhomologous End-Joining

Between the two DSB repair pathways, nonhomologous end-joining (NHEJ) is the simpler mechanism, leading to the straightforward resealing of the DSB by "regluing" the free ends left at each side of the break site [7, 8]. The proteins mediating this pathway "mend" the "wound" that was introduced in the DNA, but, importantly, a nucleotidic "scar" may be left behind. In fact, this mechanism is somewhat error-prone, considering that NHEJ machinery may introduce or delete a small number of nucleotides as part of the process of resealing the break. As a result, the nucleotide sequence at the DSB site undergoes a small mutation, which may consist of a small *in*sertion or a small *del*etion of nucleotides [4, 5]. This type of mutation is termed an *indel*.

The number of nucleotide pairs that are added or removed from the DNA chains as part of an indel varies. If the DSB occurs at an exonic region of a gene and an indel is subsequently introduced, these small insertions or deletions of nucleotides may produce alterations in the reading frame of the mRNA molecules that will be transcribed from that gene [6]. This occurs when the indel size is not a multiple of 3. A frequent consequence of such DNA frameshifts is the appearance of premature stop codons that will halt translation and thus inhibit expression of the gene targeted by the DSB.

8.3.2 Homology-Directed Repair

Homology-directed repair (HDR) is a more conservative, and elaborate, mechanism, whereby DSB repair is performed using a homologous DNA molecule as a repair template [7, 9]. The process involves (a) the generation of single-stranded DNA (ssDNA) overhangs at the break site; (b) homology-directed invasion of the DNA template by the ssDNA; and (c) synthesis of DNA primed by the invading DNA strand and using the homologous DNA duplex as a template. Both DNA molecules are then separated through one of several different possible mechanisms that may, or may not, involve DNA crossover. Whatever the case, at the end of process, the DNA molecule that underwent DSB and HDR is seamlessly repaired, in the large majority of cases [7].

Though HDR is generally not as error-prone as NHEJ, it may also be directed to produce desirable alterations of the DNA sequence at the vicinity of the DSB site [4]. In a normal biological context, the repair template for HDR corresponds to the sister chromatid of the one that underwent the DSB [9]. In the context of gene editing, an exogenous DNA repair template can be provided. Repair templates can be designed so as to induce precise alterations to the genome; they must include sequences bearing complete homology to regions in the vicinity of the DSB site, but they can also include an intentionally designed, and altered, sequence. Usually, these exogenous repair templates consist of a DNA sequence including two homology arms, flanking the region that is to be inserted in the vicinity of the break site or that will substitute a portion of the genome at that vicinity. Upon DSB, the HDR machinery will repair the break using the exogenous template as a model, thereby introducing the altered sequence into the genome that is being repaired.

8.3.3 Manipulating DNA Double-Strand Break Repair Mechanisms to Edit Genomes

NHEJ and HDR define, and limit, what kind of genome alterations can be achieved through nuclease-based gene editing. The action of both mechanisms can be directed to produce different changes that may be advantageous in the context of biomedical investigation and gene therapy development [6].

NHEJ is an exogenous template-independent mechanism and relies only on the ability to precisely define the site at which DSBs will be introduced. As explained above, simply producing a DSB in the codifying region of a gene can be sufficient to knock out that gene: the DSB can produce an indel mutation that will generate a premature stop codon. In the same way, an indel can be enough to restore the reading frame of a gene bearing a frameshift-inducing mutation. Moreover, NHEJ can also be used to "excise" a particular genetic region. Upon producing two DSBs, one upstream and another downstream of a region that is to be deleted, the NHEJ pathway may reseal the DNA by uniting the end upstream of the first break site and the end downstream of the second, thus excluding the intervening region.

Taking advantage of HDR requires not only the ability to direct the repair machinery to the target site by inducing a DNA DSB but also the provision of a homologous repair template that will bear the particular alterations to be introduced. Broadly speaking, HDR allows for both substitutions and insertions of particular nucleotide sequences. In principle, HDR can be used to introduce a mutation of one or more nucleotides, to correct a particular mutation, or to eliminate a particular gene sequence, by providing a template in which that sequence was removed. Through HDR, particular genes can be inserted in the genome: a particular therapeutic gene may be introduced at a designated site, or a particular tag or fluorescent protein can be introduced in frame with another existing gene.

It must be noted, however, that this type of outline assumes that scientists would have complete control over the DSB repair mechanisms employed by the cell. This is not the current reality. The factors that determine whether DSBs are repaired through one path or the other are still being elucidated [7, 9]. Overall, NHEJ is favored over HDR, making its applications more reliable. What is more, the absence of a repair template would completely preclude HDR in favor of NHEJ, making it more reliable still. HDR-dependent strategies are more challenging. Given its endogenous dependence on homologous sister chromatids, HDR occurs only in dividing cells, excluding any HDR-based strategy from use on postmitotic cells. Additionally, the principles governing exogenous repair template design are still not completely clear. Although small insertions or substitutions can be reliably enacted, introduction of longer sequences is still fraught with several experimental limitations [10].

8.4 Programmable Nucleases Used in Gene Editing

Gene editing relies on endonucleases to precisely define the region of the genome that will be altered. In order for a particular class of endonucleases to be suitable for this end, they have to possess a series of characteristics that are not transversal to all nucleases that can be found in Nature. For these reasons, while some nucleases used in gene editing are, in fact, more or less similar to their natural cognates, others are artificial chimeric proteins, engineered from naturally occurring proteins and protein domains.

Nucleases used in gene editing must be able to cut both chains of a DNA duplex and be highly specific, in regard to the nucleotidic sequences they target. If that was not the case, DSBs could be inserted in several different sites of the genome at the same time, producing unintended changes. A high specificity minimizes putative off-target effects. Overall, the longer a particular base sequence that a nuclease recognizes, the greater the specificity of the nuclease, since the probability of that sequence being repeated in the genome is lower.

Additionally, nucleases used in gene editing are preferably programmable, i.e., they are amenable to being redesigned and reengineered in order to target them to different gene loci, with high specificity, according to the aims of the gene editing approach at hand.

Since the 1980s, four classes of endonucleases have been selected and engineered for use in gene editing approaches (Fig. 8.2; Table 8.1).

8.4.1 Meganucleases

Meganucleases, also named homing endonucleases, are naturally occurring restriction enzymes that are found in diverse organisms, including bacteria, archaea, fungi, algae and plants [11, 12]. Contrary to the restriction enzymes that are routinely employed in molecular cloning, such as EcoRI or HindIII, meganucleases recognize extended base pair sequences: from 12 to 40 base pairs, contrasting with the 6 base pairs of those, and many other, traditional restriction enzymes. This long recognition sequences are responsible for the *mega*nuclease designation and for the high degree of target discrimination these enzymes possess. Among the meganucleases most used in genome engineering are I-SceI from

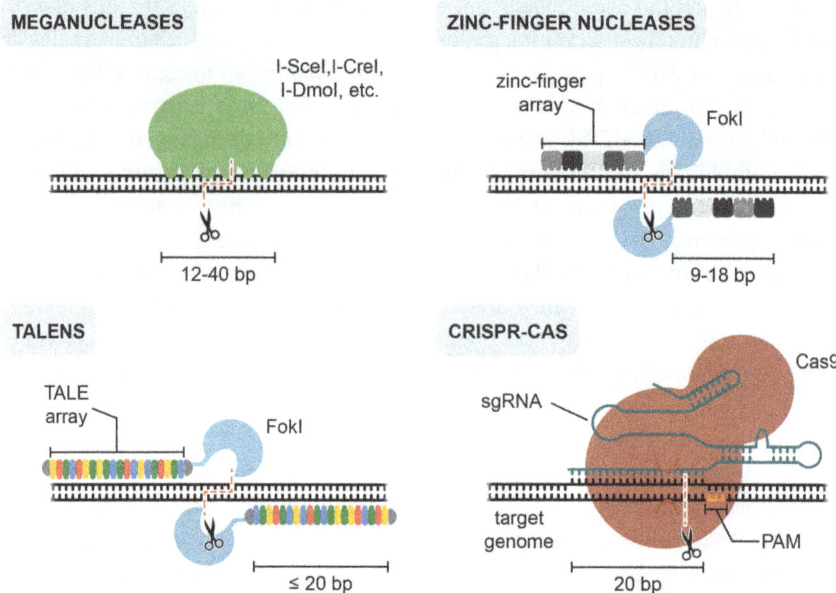

Fig. 8.2 Overview of the main programmable nucleases systems. *Meganucleases* are naturally occurring restriction enzymes, recognizing sequences of 12 to 40 base pairs. *Zinc-finger nucleases* (ZFNs) are chimeric proteins formed by two domains: the DNA recognition motif, composed of a tandem sequence of zinc-finger units, and the DNA-cleaving domain, harboring the nuclease activity – the bacterial FokI enzyme. *Transcription activator-like effector nucleases* (TALENs) are a similar set of engineered nucleases, with the DNA recognition motif derived from bacterial transcription activator-like effectors (TALEs). Both ZFNs and TALENs function in pairs. The main *CRISPR-Cas* system is based on the *Streptococcus pyogenes* Cas9 ribonucleoprotein. This protein is an endonuclease with two DNA-cutting active sites and binds to guide RNA sequences that are complementary to the target DNA loci.

Table 8.1 Main features of the four programmable nuclease platforms used in gene editing.

	Meganucleases	Zinc-finger nucleases	TALENs	CRISPR-Cas (Cas9)
Origin	Prokaryotes and eukaryotes	Eukaryotes	Bacteria of the genus *Xanthomonas*	Prokaryotes (*Streptococcus pyogenes*)
Agents required for DSB	Single protein (may dimerize)	Pair of proteins	Pair of proteins	Protein + guide RNA (gRNA or crRNA-tracrRNA)
DNA-binding mechanism	Protein-DNA interaction	Protein-DNA interaction	Protein-DNA interaction	RNA-DNA base pairing
Target DNA site size	12–40 bp	9–18 (single); 18–36 bp (pair)	Up to 20 (single); up to 40 bp (pair)	20 bp
Permissivity to mismatches	Mildly tolerated	Mildly tolerated	Mildly tolerated	Tolerated
Ease of reprogramming	Very difficult	Possible, but time-consuming	Possible, but time-consuming	Easy

the yeast *Saccharomyces cerevisiae* and I-CreI from the green algae *Chlamydomonas reinhardtii* [13].

Although successfully employed for diverse approaches, meganucleases present some significant limitations. Since the DNA-binding and DNA-cleaving domain of these proteins are one and the same, it is hard to engineer meganucleases in order to target different base sequences without affecting their cutting performance. Additionally, there is no clear correspondence between the amino acid sequence of the DNA recognition site of the proteins and the DNA base pair sequence the domain recognizes, turning rational reengineering difficult, if not impossible.

The above disadvantages have precluded meganuclease-based gene editing from ever achieving widespread use. In fact, meganucleases have been largely substituted by other programmable nuclease platforms that have been developed in recent years.

8.4.2 Zinc-Finger Nucleases

Zinc-fingers are protein motifs that can be found in several eukaryotic transcription factors, mediating interaction of these proteins with their specific DNA targets [14]. Because of the high degree of specificity with which the zinc-finger motifs bind to particular base pair sequences, they have been used to generate a class of programmable nucleases that was first described in 1996 [15].

Zinc-finger nucleases (ZFNs) are chimeric proteins mainly composed of two distinct domains: a DNA-binding domain and a DNA-cleaving domain, harboring the actual nuclease activity [16]. The DNA-binding domain is composed of a tandem sequence of zinc-finger units, forming a zinc-finger array. Since each of those units recognizes a particular sequence of three base pairs, combining different zinc-finger units yields a zinc-finger array that is able to recognize a longer base pair sequence. For example, an array of 3 zinc-finger units is capable of recognizing a sequence of 9 base pairs, while an array of 6 zinc-finger units will recognize an 18-base pair sequence.

The DNA-cleaving domain linked to the zinc-finger array is a bacterial restriction enzyme – usually FokI, derived from *Flavobacterium okeanokoites*. Since FokI requires dimerization to elicit DNA cleavage, a pair of ZFNs is necessary for DNA DSB to occur; one ZFN will bind to one DNA strand, and the other will bind to the complementary strand. If the target sites of each ZFN are properly spaced, FokI dimerizes and cuts both DNA strands, generating a DSB. Since two ZFNs are required for DSB to occur and each one of them recognizes 9–18 base pairs, a ZFN pair can recognize 18–36 base pairs, providing the system a high degree of specificity.

8.4.3 Transcription Activator-Like Effector Nucleases

Transcription activator-like effector nucleases (TALENs), first described in 2010 [17], are a set of engineered nucleases that are very similar to ZFNs in terms of the rationale behind their design. TALENs are also made up of two domains – a DNA-binding domain and a DNA-cleavage domain – and they also function in pairs. The DNA-cleavage domain is once again the bacterial FokI enzyme, but they differ from ZFNs in their DNA recognition domain.

In the case of TALENs, recognition and binding of DNA is mediated by an array derived from bacterial transcription activator-like effectors (TALEs). TALEs were first described as an integral part of DNA-binding proteins used by plant pathogens of the genus *Xanthomonas* [18]. Each TALE array is composed of a collection of 34-amino acid-long modules arranged in tandem, and each of those units is capable of binding one particular DNA base. Binding specificity is determined by only two residues that vary between modules (repeat-variable diresidue – RVD) [19]. Importantly, TALE arrays are costumizable and can be assembled with a particular repeat order that will define their ability to bind a designated base pair sequence [20, 21]. Since each unit recognizes one nucleotide, an array of 20 units, for example, will distinguish a particular 20 base pair sequence; considering a pair of TALENs is needed for DSB to occur, the system is overall able to recognize sequences of up to 40 base pairs.

8.4.4 CRISPR-Cas Systems

In 2013, a new programmable nuclease system was introduced, and the advantageous features it presents over the previously existing gene editing platforms have since gathered unprecedented interest by the scientific community – the CRISPR-Cas system.

"CRISPR" is an acronym that was first used to describe particular genetic loci found in prokaryotes – clustered regularly interspaced short palindromic repeats [4, 22–24]. Puzzling at first, the extensive research work that then followed to understand the biological importance of these loci led to the recognition that CRISPR loci function as a prokaryotic adaptive immune mechanism, moved against viruses and other external sources of nucleic acids. Briefly put, CRISPR loci function as data banks that allow prokaryotes to rapidly counteract the action of invading pathogens through the action of proteins – CRISPR-associated (Cas) proteins – that are guided by RNA molecules, complementary to DNA sequences of the pathogen. The invading DNA is destroyed by the action of Cas proteins with endonuclease activity.

Several classes of CRISPR systems have since been described, varying in the Cas proteins associated with the CRISPR loci, the RNA molecules that participate in the immune mechanisms, and the processes responsible for the maturation of those RNA molecules [25]. New CRISPR variants are continuously being identified, but there is one in particular that is being widely used as a gene editing platform – the one based on the Cas9 protein from a type II CRISPR system of *Streptococcus pyogenes*, sometimes designated as SpCas9 [26, 27]. Contrary to other CRISPR system types, type II systems require only one protein – Cas9 – to elicit cleavage of DNA. This, combined with the DNA recognition features of SpCas9 (described below) and the very history of its implementation, has led SpCas9 to be the CRISPR system that is currently more well established as a gene editing platform.

SpCas9, henceforth designated simply as **Cas9**, is a ribonucleoprotein with endonuclease activity that is able to generate DNA DSBs through the concerted action of its two nuclease domains, each responsible for the cleavage of one DNA strand: RuvC and HNR. In the biological context of *S. pyogenes*, Cas9 binding to the target DNA sequence is mediated by an RNA molecule that is complementary to 20 bases of the target DNA – the **crRNA**. Binding occurs through complementary base pairing. Maturation of the crRNA and binding of crRNA to Cas9 are mediated by another RNA molecule – the **trans-activating crRNA** (tracrRNA) [28, 29].

In an experimental, or biotechnological, context, Cas9 and the **guide RNA** (gRNA) molecules can be artificially introduced in a eukaryotic cell, leading to a DNA DSB in the region that is complementary to the crRNA. In order to further simplify the system, the two RNA molecules have been rationally fused to create a **single guide RNA** (sgRNA) sequence that is sufficient to guide the nuclease activity of Cas9 [27].

Cas9 cannot be freely targeted to every 20-nucleotide sequences of the genome, and one particular requirement must be met in order for it to recognize and bind a designated genetic locus. A specific protospacer-associated motif (PAM) must be localized immediately downstream of the 20-nucleotide sequence that is to be targeted. Among other things, the PAM dictates and defines the search for putative molecular targets and is responsible for initiating binding to the target DNA sequence [28, 29]. In the biological context, the PAM requirement also prevents self-recognition and cleavage of the bacterial DNA, since crRNA-codifying sequences (spacers) lack the PAM [30]. The PAM of *S. pyogenes* Cas9 corresponds to an NGG motif, where N can be any nucleotide and GG are two sequential guanine nucleotides; however, it should be noted that different Cas nucleases have different PAM requirements. In a gene editing setting, the PAM actually circumscribes the range of possible targets the Cas9 protein can have, but since the motif is very short, this requirement does not constitute a great limitation. In other words, two sequential guanine nucleotides – or two sequential cytosines, in the complementary strand – are sufficiently common for Cas9 to be directed to almost anywhere in the genome.

8.4.5 Comparing the Four Classes of Programmable Nucleases Used in Gene Editing

As with any other biotechnological tools, each class of designer nucleases employed in gene editing presents advantages and disadvantages. Although the CRISPR-Cas system is regarded as having several unprecedented advantages over the other systems, the main shortcoming of this platform is its relative proneness to elicit off-target effects. In fact, Cas9 is known to tolerate mismatches in the 20-nucleotide sequence complementarity and is thus capable of producing DSBs at unintended sites [24]. Meganucleases, ZFNs and TALENs, while also capable of tolerating mismatches, are more restringing [31].

However, the use of CRISPR-Cas presents diverse benefits over the other platforms, which were arguably responsible for its noteworthy increase in popularity since its inception [32]. Perhaps the principal advantage of CRISPR-Cas is its simple design, in what concerns to the engineering that is necessary to reprogram it to target a particular desired locus. While meganucleases are near-impossible to reprogram and both ZFNs and TALENs reengineering entail expensive and time-consuming rounds of carefully planned molecular cloning [33], directing Cas9 to a new site requires only the definition of a new 20-nucleotide sequence upstream of a PAM, which should be in the vicinity of the target site. That gRNA containing the 20-nucleotide sequence can then be introduced in cells along with Cas9, which does not require reengineering.

Additionally, editing using the CRISPR-Cas system requires only one protein, instead of a pair as was the case of ZFNs and TALENs. The system is highly efficient, meaning that the probability of Cas9 inducinhg DSBs in a large percentage of the cells in a population is high [4]. CRISPR-Cas is also amenable to multiplexing, i.e., targeting several sites simultaneously [24]; using just one protein and several different gRNAs, it is possible to induce DSB at various locations at once. Finally, and as will be described later in this chapter, CRISPR-Cas is a very versatile system and can be easily employed to perform operations that go beyond the introduction of DSBs in the DNA.

The process of selecting a particular nuclease-based gene editing platform must contemplate the experimental objectives at hand, as well as the technical requirements, time constraints and costs that a particular approach may entail. The relevance of the diverse existing nuclease platforms notwithstanding, taking into account the relative simplicity of the CRISPR-Cas system and the growing interest it congregates, the following sections will focus on some practical considerations concerning gene editing using the CRISPR-Cas system.

8.5 Editing Genes Using CRISPR-Cas

As mentioned above, nuclease-based gene editing relies on the action of a molecular scissor – an endonuclease – that cuts a DNA molecule in the vicinity of the site that is to be edited, and on endogenous DNA repair mechanisms that will elicit changes in the nucleotide sequence.

In order to knock out a gene using CRISPR-Cas, Cas9 can be directed by a sgRNA (or a crRNA-tracrRNA pair) to an early exon of that target gene. The subsequent DSB and the ensuing NHEJ may produce a small indel mutation, which in turn can lead to a shift in the DNA reading frame, generating a premature stop codon. Alternatively, Cas9 can be directed to regions both upstream and downstream of the initiation codon (ATG), utterly removing it and preventing translation from taking place. These excision approaches can also be used to remove particular genetic regions from the genome.

If a particular insertion or substitution is intended, gRNAs should direct Cas9 to the vicinity of the region to be altered. If a homology template is provided, either in the form of a single-stranded DNA molecule (for short, <200 nucleotides insertions or substitutions) or a donor plasmid (for longer insertions or substitutions), there is a possibility that the region suffering DSB will be repaired by HDR using the exogenous template as a model, leading to the introduction of the desired sequence [10].

Design of the gRNA sequences can be performed using several bioinformatic tools that allow searching a particular sequence for PAMs and the corresponding 20 nucleotide sequences. Many of these platforms also curate every possible gRNA sequence in a designated genome and have algorithms that allow predicting the probability of off-target effects [34]. Sequences should be selected using criteria that minimize putative off-target effects, by decreasing the number of overall off-target effects and/or minimizing the number of putative off-targets in coding regions.

The CRISPR-Cas tools can then be introduced in cells through a variety of methods (Fig. 8.3). Cas9 and the sgRNA (or the crRNA-tracrRNA pair) can be directly introduced in the cells in their "natural" form: as a protein and as RNA molecules, respectively. *In vitro*-synthetized or recombinant Cas9, along with *in vitro*-synthetized gRNAs, are assembled as a ribonucleoprotein complex, also *in vitro*, and then introduced in cells through transfection, electroporation or microinjection. Both agents can also be delivered in the form of RNA, through the same methods. Cas9-codifying mRNA will then be translated inside the cytoplasm, and the ribonucleoprotein complexes will assemble intercellularly. Finally, Cas9 and the gRNAs can be administered in the form of DNA plasmids, which will be transcribed in the cell. These plasmids may be amenable to viral production, allowing viral delivery of the CRISPR-Cas system.

8.6 Expanding the Possibilities of Gene Editing with CRISPR-Cas

Even 1 year after its introduction, CRISPR-Cas had already been successfully used to modify the genome of diverse organisms, from bacteria to yeast and from agriculturally interesting plants to human cells [5]. CRISPR-Cas-based gene editing is regarded as a time- and cost-efficient method for generating new animal models, in particular of nontraditional animal species, for which previous attempts at genetic manipulation have been unfruitful [35]. Numerous publications in recent years underline the potential of CRISPR-Cas as a very versatile tool for manipulating genomes, in ways that go beyond those dependent on the introduction of DNA DSBs.

8.6.1 Gene Editing Beyond DNA Double-Strand Breaks

The above sections probably demonstrate the potential of Cas9 endonuclease activity in easily generating the conditions for gene editing to occur. The applicability of the CRISPR-Cas system goes, however, way beyond what can be achieved through DNA DSBs, and, in fact, gene editing as a whole can be regarded as a means to alter not only the genome but also its context and its products [4]. Cas9-derived tools developed so far offer the promise of altering epigenetic markers in the DNA, the architecture of the chromatin and the levels of transcribed RNA molecules, amid many other possibilities.

Mutating a single amino acid in one or both nuclease activity sites of Cas9 renders the protein partially, or completely, inactive, respectively [5, 22]. When only one site is mutated, the resulting protein is termed a nickase (nCas9), since it is still able to make a nick in the DNA duplex, by cutting one of the DNA strands. The completely inactivated Cas9 is termed dead Cas9 (dCas9), and, although no longer able to produce DNA breaks, it retains its ability to specifically bind to the DNA, through complementary base pairing between the gRNAs and their targets. dCas9 can be fused to other proteins or effector domains, and several such fusion variants have been developed so far. In common they all have the fact that, by taking advantage of the homing capacity of dCas9, they are able to direct the activity of the particular effectors they harbor to specific genetic loci [32] (Fig. 8.4). Notably, zinc-fingers and TALEs can also be used as homing devices of proteins other than the FokI endonuclease, but the simplicity of CRISPR-Cas has allowed a quick and diverse expansion of its potential in this regard.

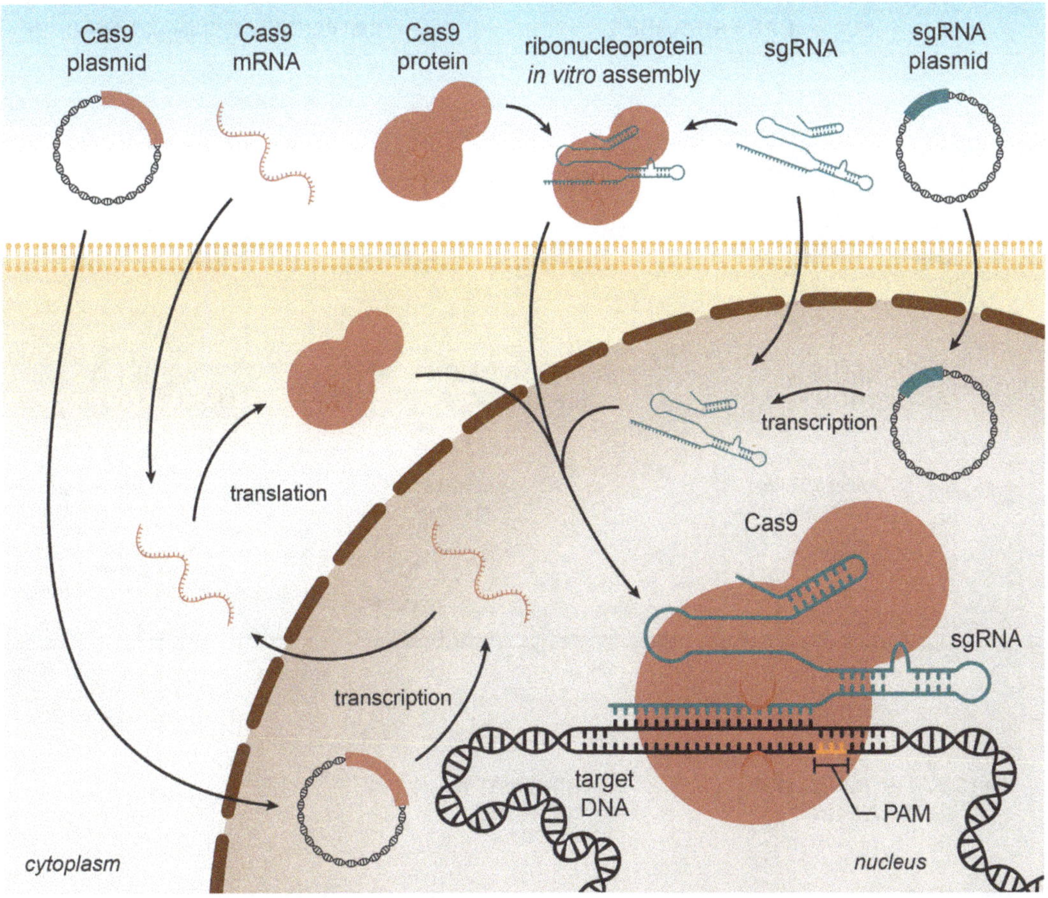

Fig. 8.3 Possible delivery strategies for the CRISPR-Cas9 system. The different components of the CRISPR-Cas9 system can be provided in multiple forms. Cas9 can be delivered as a Cas9 cDNA-containing plasmid, as an mRNA molecule or as a recombinant protein, already complexed with the guide RNA molecule(s). Apart from this route, the guide RNA can also be delivered in an unconjugated form to cells, as well as in the form of DNA plasmids.

dCas9 fused to a transcriptional activator such as VP64 can be directed to a particular gene, recruiting transcriptional machinery that will lead to an increase of the expression levels of that gene. Conversely, dCas9 conjugated with a transcriptional repressor such as the Krüppel-associated box (KRAB) domain can be used to decrease the expression of a target gene, without introducing a DNA DSB and thus preventing the possibility of an undesirable indel mutation. Transcription regulation can also be achieved using dCas9 fused to epigenetic modifiers that alter the acetylation levels of histones, or the methylation levels of the DNA, or pairs of dCas9 fused with proteins that are able to interact and thereby "bend" the chromatin, generating loops in its topology [32]. Fusing fluorescent proteins such as the green fluorescent protein (GFP) can be used to pinpoint the localization of a particular genetic sequence in a chromosome.

What is more, mutation of the target genome has also been shown to be achievable without recurring to DNA DSBs. Fusing deaminase proteins such as APOBEC1 along with excision repair inhibitor UGI to dCas9 or nCas9 has been shown to produce direct conversion between nucleotides. This type of strategy, termed base

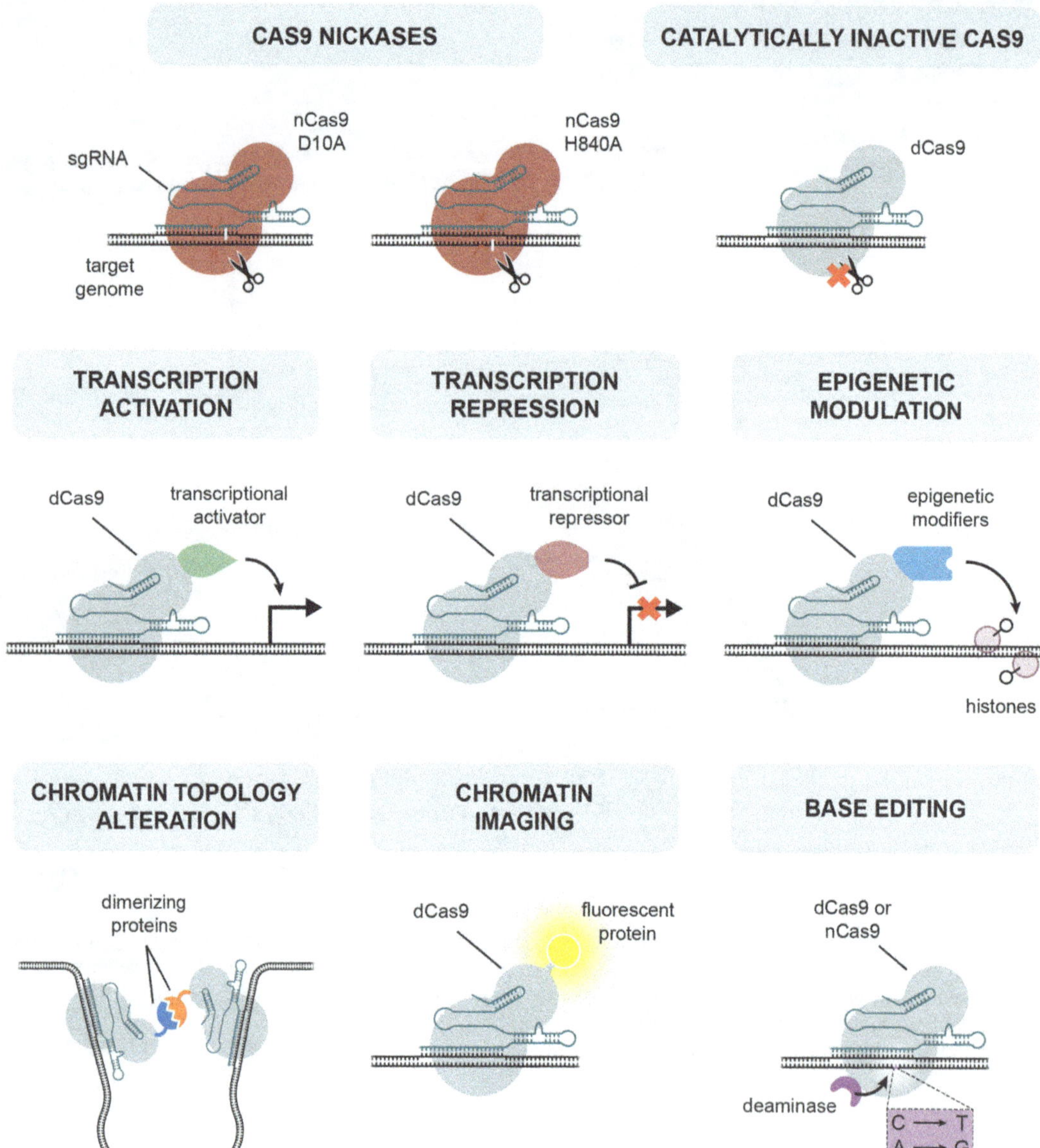

Fig. 8.4 Versatility of the CRISPR-Cas9 system. The figure highlights possible applications of the CRISPR-Cas9 system that go beyond the introduction of DNA double-strand breaks. These application are based on the use of partially (nCas9) or completely (dCas9) catalytically inactivated Cas9 and include transcription activation or repression, epigenetic modulation and base editing, among others.

editing, has been successfully employed in converting cytosines to thymines and adenines to guanines [36, 37].

In addition to this search to expand *what* Cas9 can do, scientists have also looked for ways to reliably control *when* and *where* its action takes place. In fact, several variants of inducible Cas9 have also been developed. Among them, some respond to chemical compounds and others to ligand-binding, and still others are activated by optical light [32].

In recent years, reengineering of Cas9 and even of the gRNA sequences has yielded a vast array of tools that is continuously being expanded on and that is thus optimizing the applications of the CRISPR-Cas9 system.

8.6.2 Cas Variants

The Cas9 protein from *S. pyogenes* is not the only Cas that has drawn the attention of scientists looking at improving the existing gene editing platforms. Many more Cas proteins have been identified and more or less extensively characterized, leading to the creation of an ever-expanding Cas protein library. Its remaining members, although not having such a widespread use as SpCas9, certainly contribute to the ever-growing possibilities offered by the CRISPR-Cas system.

Different Cas proteins, including different Cas9 proteins from species other than *S. pyogenes*, have been described to be diversely prone to off-target effects and are known to display distinctive PAM requirements [32]. Some of these PAMs are longer than three nucleotides, and although this restricts the range of targets the respective Cas proteins can have, the increase can be beneficial in further decreasing the number of off-targets. Different PAMs may also allow targeting regions that the NGG PAM requirement of SpCas9 does not allow.

Different Cas proteins also display different molecular sizes, and smaller Cas variants may be advantageous regarding their delivery to cells, for example. SpCas9 is too large for its codifying DNA to be included in AAV particles along with the gRNA; as such, currently each agent has to be provided in a different AAV vector. Smaller Cas variants may be more amenable to AVV-mediated delivery.

Finally, some Cas proteins have been described to target nucleic acids differently from SpCas9 [38]. Cas12a, for example, has been shown to produce staggered DNA breaks, leading to the generation of sticky ends at the cut site. Cas13a has been shown to target RNA.

New Cas variants are continuously coming to light, in a constant search for a particular Cas protein, or set of Cas proteins, that may display advantages over all others. A smaller molecular size, a low incidence of off-target effects, and a high versatility are important factors in this ongoing search for the "ultimate" Cas protein.

8.7 Limitations and Challenges to Current Gene Editing Strategies

As promising as it may be, implementation of gene editing in an experimental setting, let alone in the development of a gene therapy approach, is subject to several limitations and challenges that must be accounted for. Some of them can be avoided or controlled for, but others will require continuous investigation and development before they can be tackled with unquestionable success. Overall, challenges to the application of gene editing include (a) the ability to properly deliver the molecular tools, (b) problems with specificity, (c) guarantees of fidelity, and (d) control over DSB repair and HDR [4–6].

As explained in Chap. 4, delivery of gene therapy agents faces several physiological barriers that limit their efficacy. What is more, it is important to ensure that the molecular tools reach the proper organ or tissue where their therapeutic effect will take place. All the while, the delivery mechanisms should pose no threat to the health and safety of the organism. Selection and design of methods for the delivery of gene editing tools should abide by the same principles as other gene therapy approaches, possibly preferring *ex vivo* strategies over local or systemic administration as a means to improve safety and minimize delivery to unintended tissues or cells [31].

As mentioned above, none of the gene editing nuclease platforms is devoid of possible off-target effects, by virtue of the permissiveness every system has, on a higher or lesser degree, to mismatch tolerance. Ideally, the nuclease systems would have no off-targets, but since it is currently impossible to ensure this, it is crucial that gene editing strategies minimize the occurrence of unintended modifications. This can be done by selecting variants with a proven lower degree of off-target activity or, in the case of CRISPR-Cas, selecting gRNA sequences with low levels of predicted off-targets. After editing, it is important to be able to search the targeted genome for putative off-target mutations. This can be done by whole-genome sequencing and through other, recently developed, targeted approaches, but such techniques may not be available to every lab.

Fidelity, in the context of gene editing, describes the degree to which the intended alteration was inserted in the genome and chiefly concerns mutations introduced by HDR [6]. As a result of a DSB and in the presence of an HDR repair template, the intended insert may be introduced at a site other than the one expected, more than one copy of the insert may be introduced, and translocations may occur, among many other unintended changes with disastrous effects. Techniques that allow confirmation that only the intended changes took place are necessary.

Finally, the infrequent occurrence of HDR is a limiting factor for strategies that rely on this pathway of DSB repair for the intended genome alteration to occur. HDR is limited to dividing cells, and its rate is generally low, compared to NHEJ [9]. Several lines of research have invested in increasing the rate of HDR, but, as with any other manipulation that may interfere with DNA repair mechanisms, the possibility of unexpected outcomes may outbalance the benefits of the intervention. Nonetheless, in the case of the CRISPR-Cas system, diverse approaches have been described to increase success of HDR-mediated editing, including employing nickases instead of wild-type Cas9, since a DNA nick is less prone to induce NHEJ-derived indels while still potentiating HDR; enriching cell cultures with cells in the G2/M phase of the cell cycle; inducing "cold shocks"; overexpressing the Rad51 protein; employing small molecules and other chemical compounds (RS-1, Brefeldin A, L755507, Nocodazole); and fusing Cas9 with proteins involved in HDR, among many others [39–43]. Studies aiming to define the best parameters for donor repair template design are also ongoing.

8.8 Gene Editing as a Tool for Human Disease Therapy

The possibilities offered by existing gene editing strategies can be translated into diverse approaches to tackle human health conditions and disease. Importantly, gene editing may be a significant tool not only for direct therapeutic intervention but also in other, no less relevant, steps of the therapy development pipeline [44].

Nuclease-based gene editing can be employed in basic research, assisting in the investigation of gene functions. The CRISPR-Cas system, in particular, can be used to perform high-throughput screening of disease modifiers, which may yield important clues into disease pathogenesis and possibly constitute relevant therapeutic targets [45]. Gene editing is also a potent method for generating disease models and can be used to develop isogenic cell lines, i.e., cell cultures derived from individuals, where disease-causing genes can be introduced or removed, thus producing cultures with the same genetic background, in which the only modifying factor is the designated genetic factor [46]. Additionally, gene editing allows for the rapid and inexpensive generation of animal models, compared with traditional methods used for generating transgenic knockout, and knock-in animals [35]. CRISPR-Cas9 is particularly well-suited for modeling complexed diseases, by allowing the alteration of several genes simultaneously through its multiplexing capability [47].

Concerning the use of gene editing as a therapeutic approach, the interest this field is drawing has led research teams to develop and test diverse gene therapy strategies that are based on the growing capabilities of the described systems to operate changes in the DNA. Current literature offers diverse examples, overall focusing on one of several routes: (a) correcting pathogenic mutations; (b) inactivating disease-causing genes; (c) re-establishing gene functions; (d) eliminating disease-causing elements; (e) introducing protective mutations; and (f) generating cells with therapeutic activities. Examples of the potential of these approaches abound in reports using both cell cultures and animal models, and some strategies have already transitioned to clinical testing. Two examples of gene editing-based approaches to therapy follow: one focusing on a genetic disorder and another on an infectious disease.

As described in Chap. 7, Duchenne muscular atrophy (DMD) is a hereditary disorder that arises as a consequence of mutations in the *DMD* gene, which in healthy individuals codifies dystrophin, a protein that plays a crucial role in muscular structure and physiology (Fig. 7.4). Amoasii and collaborators systemically administered AVVs codifying CRISPR-Cas tools to dogs harboring a deletion of exon 50 of the DMD gene, which is a hot spot for mutations linked to DMD, in humans. CRISPR-Cas activity was directed at an early region of exon 51. Overall, treated animals displayed an improvement in muscular histology and dystrophin levels. Indels were detected at the targeted genomic site, and the amelioration observed was related to reestablishment of the reading frame of the gene or exon 51 skipping [48].

Human immunodeficiency virus (HIV) infection of T-cells relies on interaction of the virus with cell receptors that mediate its internalization (Fig. 3.4). As explained in Chap. 3, one of these receptors is CCR5, and it has been known for some time that individuals bearing mutated forms of this receptor are refractory to HIV infection. It has been thus hypothesized that knocking out *CCR5* gene function through the action of programmable nucleases may be beneficial in preventing HIV entry into lymphocytes [49, 50]. In 2014, in the very first clinical trial using gene editing, HIV patient cells were edited *ex vivo* with ZFNs and then autologously reinfused, with promising results in what regards to viral loads in circulation [51–53]. Other clinical trials using similar rationales were also underway at the time, and others followed, with a clinical trial for HIV using CRISPR-Cas-edited stem cells being currently underway [54].

This Chapter in a Nutshell

- Current gene editing strategies rely on two conditions, (i) the ability to define the specific region of the genome to be altered and (ii) the ability to create conditions for the desired alterations to occur.
- The definition of the target site is accomplished by molecules that specifically bind to a nucleotide sequence and then cleave the DNA producing a double-strand break (DSB). These are then repaired through different endogenous mechanisms of DNA repair.
- There are two main mechanisms of DNA DSB repair, nonhomologous end-joining (NHEJ) and homology-direcgted repair (HDR), with the former being simpler and the latter involving a DNA molecule as template for the repair.
- Four main systems of programmable nucleases have been used in gene editing: meganucleases, zinc-finger nucleases, TALENs, and the CRISPR-Cas system.
- The CRISPR-Cas system presents advantageous features over the other programmable nuclease systems. Several variations of the system have been developed, or example in order to alter epigenetic markers, the architecture of the chromatin and the levels of transcribed RNA molecules.
- Despite being an important promise for gene therapy, gene editing faces important limitations and challenges. Some are also found in other gene therapy strategies, but other are especific to this approach.

Review Questions

1. In the context of gene therapy, gene editing systems can be used to:
   (a) Disrupt a mutated gene
   (b) Substitute a malfunctioning gene
   (c) Regulate the expression of a mutated gene
   (d) None of the above
   (e) All of the above

2. From the following, select the ones that apply to the CRISPR-Cas system of gene editing:
   (a) Direct binding to the target DNA through a protein domain
   (b) DNA recognition motif composed of several repeated units
   (c) Ligation of target DNA to RNA through base complementarity
   (d) Protein with only one site with nuclease activity
   (e) Derived from a prokaryote "immune" system
   (f) Requires a guide RNA sequence
   (g) Consists necessarily of a fusion protein
   (h) Has the ability to induce DNA double-strand breaks
3. In the CRISPR-Cas system, what is the role of the gRNA?
   (a) It ensures that the cell recognizes that the DNA is damaged and tries to repair it
   (b) It determines the site at which the Cas9 enzyme cuts the genome
   (c) It uses the DNA repair machinery to introduce changes to one or more genes
   (d) It makes a cut across both strands of the DNA
4. From the following, select the ones that apply to the zinc-finger nucleases and TALENs systems of gene editing:
   (a) Direct binding to the target DNA through a protein domain
   (b) DNA recognition motif composed of several repeated units
   (c) Function in pairs
   (d) Use fusion proteins
   (e) Can lead to the introduction of indels
   (f) Require a guide RNA sequence

References and Further Reading

1. Stanford WL, Cohn JB, Cordes SP (2001) Gene-trap mutagenesis: past, present and beyond. Nat Rev Genet 2:756–768. https://doi.org/10.1038/35093548
2. Capecchi MR (1989) Altering the genome by homologous recombination. Science 244:1288–1292. https://doi.org/10.1126/science.2660260
3. Capecchi MR (2005) Gene targeting in mice: functional analysis of the mammalian genome for the twenty-first century. Nat Rev Genet 6:507–512. https://doi.org/10.1038/nrg1619
4. Hsu PD, Lander ES, Zhang F (2014) Development and applications of CRISPR-Cas9 for genome engineering. Cell 157:1262–1278. https://doi.org/10.1016/j.cell.2014.05.010
5. Sander JD, Joung JK (2014) CRISPR-Cas systems for editing, regulating and targeting genomes. Nat Biotechnol 32:347–355. https://doi.org/10.1038/nbt.2842
6. Maggio I, Goncalves MA (2015) Genome editing at the crossroads of delivery, specificity, and fidelity. Trends Biotechnol 33:280–291. https://doi.org/10.1016/j.tibtech.2015.02.011
7. Chapman JR, Taylor MR, Boulton SJ (2012) Playing the end game: DNA double-strand break repair pathway choice. Mol Cell 47:497–510. https://doi.org/10.1016/j.molcel.2012.07.029
8. Lieber MR (2010) The mechanism of double-strand DNA break repair by the nonhomologous DNA end-joining pathway. Annu Rev Biochem 79:181–211. https://doi.org/10.1146/annurev.biochem.052308.093131
9. Heyer WD, Ehmsen KT, Liu J (2010) Regulation of homologous recombination in eukaryotes. Annu Rev Genet 44:113–139. https://doi.org/10.1146/annurev-genet-051710-150955
10. Song F, Stieger K (2017) Optimizing the DNA donor template for homology-directed repair of double-strand breaks. Mol Ther Nucleic Acids 7:53–60. https://doi.org/10.1016/j.omtn.2017.02.006
11. Silva G et al (2011) Meganucleases and other tools for targeted genome engineering: perspectives and challenges for gene therapy. Curr Gene Ther 11:11–27. https://doi.org/10.2174/156652311794520111
12. Stoddard BL (2005) Homing endonuclease structure and function. Q Rev Biophys 38:49–95. https://doi.org/10.1017/S0033583505004063
13. Paques F, Duchateau P (2007) Meganucleases and DNA double-strand break-induced recombination: perspectives for gene therapy. Curr Gene Ther 7:49–66. https://doi.org/10.2174/156652307779940216
14. Klug A (2010) The discovery of zinc fingers and their applications in gene regulation and genome manipulation. Annu Rev Biochem 79:213–231. https://doi.org/10.1146/annurev-biochem-010909-095056
15. Kim YG, Cha J, Chandrasegaran S (1996) Hybrid restriction enzymes: zinc finger fusions to Fok I cleavage domain. Proc Natl Acad Sci U S A 93:1156–1160. https://doi.org/10.1073/pnas.93.3.1156
16. Durai S et al (2005) Zinc finger nucleases: custom-designed molecular scissors for genome engineering of plant and mammalian cells. Nucleic Acids Res 33:5978–5990. https://doi.org/10.1093/nar/gki912
17. Christian M et al (2010) Targeting DNA double-strand breaks with TAL effector nucleases. Genetics 186:757–761. https://doi.org/10.1534/genetics.110.120717
18. Kay S, Bonas U (2009) How Xanthomonas type III effectors manipulate the host plant. Curr Opin

Microbiol 12:37–43. https://doi.org/10.1016/j.mib.2008.12.006
19. Moscou MJ, Bogdanove AJ (2009) A simple cipher governs DNA recognition by TAL effectors. Science 326:1501. https://doi.org/10.1126/science.1178817
20. Miller JC et al (2011) A TALE nuclease architecture for efficient genome editing. Nat Biotechnol 29:143–148. https://doi.org/10.1038/nbt.1755
21. Zhang F et al (2011) Efficient construction of sequence-specific TAL effectors for modulating mammalian transcription. Nat Biotechnol 29:149–153. https://doi.org/10.1038/nbt.1775
22. Doudna JA, Charpentier E (2014) Genome editing. The new frontier of genome engineering with CRISPR-Cas9. Science 346:1258096. https://doi.org/10.1126/science.1258096
23. Ishino Y, Shinagawa H, Makino K, Amemura M, Nakata A (1987) Nucleotide sequence of the iap gene, responsible for alkaline phosphatase isozyme conversion in Escherichia coli, and identification of the gene product. J Bacteriol 169:5429–5433. https://doi.org/10.1128/jb.169.12.5429-5433.1987
24. Mali P, Esvelt KM, Church GM (2013) Cas9 as a versatile tool for engineering biology. Nat Methods 10:957–963. https://doi.org/10.1038/nmeth.2649
25. Koonin EV, Makarova KS, Zhang F (2017) Diversity, classification and evolution of CRISPR-Cas systems. Curr Opin Microbiol 37:67–78. https://doi.org/10.1016/j.mib.2017.05.008
26. Gasiunas G, Barrangou R, Horvath P, Siksnys V (2012) Cas9-crRNA ribonucleoprotein complex mediates specific DNA cleavage for adaptive immunity in bacteria. Proc Natl Acad Sci U S A 109:E2579–E2586. https://doi.org/10.1073/pnas.1208507109
27. Jinek M et al (2012) A programmable dual-RNA-guided DNA endonuclease in adaptive bacterial immunity. Science 337:816–821. https://doi.org/10.1126/science.1225829
28. Jinek M et al (2014) Structures of Cas9 endonucleases reveal RNA-mediated conformational activation. Science 343:1247997. https://doi.org/10.1126/science.1247997
29. Nishimasu H et al (2014) Crystal structure of Cas9 in complex with guide RNA and target DNA. Cell 156:935–949. https://doi.org/10.1016/j.cell.2014.02.001
30. Shah SA, Erdmann S, Mojica FJ, Garrett RA (2013) Protospacer recognition motifs: mixed identities and functional diversity. RNA Biol 10:891–899. https://doi.org/10.4161/rna.23764
31. Cox DB, Platt RJ, Zhang F (2015) Therapeutic genome editing: prospects and challenges. Nat Med 21:121–131. https://doi.org/10.1038/nm.3793
32. Adli M (2018) The CRISPR tool kit for genome editing and beyond. Nat Commun 9:1911. https://doi.org/10.1038/s41467-018-04252-2
33. Sanjana NE et al (2012) A transcription activator-like effector toolbox for genome engineering. Nat Protoc 7:171–192. https://doi.org/10.1038/nprot.2011.431
34. Doench JG et al (2016) Optimized sgRNA design to maximize activity and minimize off-target effects of CRISPR-Cas9. Nat Biotechnol 34:184–191. https://doi.org/10.1038/nbt.3437
35. Liu ET et al (2017) Of mice and CRISPR: the post-CRISPR future of the mouse as a model system for the human condition. EMBO Rep 18:187–193. https://doi.org/10.15252/embr.201643717
36. Gaudelli NM et al (2017) Programmable base editing of A∗T to G∗C in genomic DNA without DNA cleavage. Nature 551:464–471. https://doi.org/10.1038/nature24644
37. Komor AC, Kim YB, Packer MS, Zuris JA, Liu DR (2016) Programmable editing of a target base in genomic DNA without double-stranded DNA cleavage. Nature 533:420–424. https://doi.org/10.1038/nature17946
38. Knott GJ, Doudna JA (2018) CRISPR-Cas guides the future of genetic engineering. Science 361:866–869. https://doi.org/10.1126/science.aat5011
39. Charpentier M et al (2018) CtIP fusion to Cas9 enhances transgene integration by homology-dependent repair. Nat Commun 9:1133. https://doi.org/10.1038/s41467-018-03475-7
40. Danner E et al (2017) Control of gene editing by manipulation of DNA repair mechanisms. Mamm Genome 28:262–274. https://doi.org/10.1007/s00335-017-9688-5
41. Guo Q et al (2018) 'Cold shock' increases the frequency of homology directed repair gene editing in induced pluripotent stem cells. Sci Rep 8:2080. https://doi.org/10.1038/s41598-018-20358-5
42. Ran FA et al (2013) Double nicking by RNA-guided CRISPR Cas9 for enhanced genome editing specificity. Cell 154:1380–1389. https://doi.org/10.1016/j.cell.2013.08.021
43. Yang D et al (2016) Enrichment of G2/M cell cycle phase in human pluripotent stem cells enhances HDR-mediated gene repair with customizable endonucleases. Sci Rep 6:21264. https://doi.org/10.1038/srep21264
44. Fellmann C, Gowen BG, Lin PC, Doudna JA, Corn JE (2017) Cornerstones of CRISPR-Cas in drug discovery and therapy. Nat Rev Drug Discov 16:89–100. https://doi.org/10.1038/nrd.2016.238
45. Agrotis A, Ketteler R (2015) A new age in functional genomics using CRISPR/Cas9 in arrayed library screening. Front Genet 6:300. https://doi.org/10.3389/fgene.2015.00300
46. Bassett AR (2017) Editing the genome of hiPSC with CRISPR/Cas9: disease models. Mamm Genome 28:348–364. https://doi.org/10.1007/s00335-017-9684-9
47. Niu Y et al (2014) Generation of gene-modified cynomolgus monkey via Cas9/RNA-mediated gene targeting in one-cell embryos. Cell 156:836–843. https://doi.org/10.1016/j.cell.2014.01.027
48. Amoasii L et al (2018) Gene editing restores dystrophin expression in a canine model of Duchenne

muscular dystrophy. Science 362:86–91. https://doi.org/10.1126/science.aau1549

49. Lombardo A, Naldini L (2014) Genome editing: a tool for research and therapy: targeted genome editing hits the clinic. Nat Med 20:1101–1103. https://doi.org/10.1038/nm.3721
50. Xiao Q, Guo D, Chen S (2019) Application of CRISPR/Cas9-based gene editing in HIV-1/AIDS therapy. Front Cell Infect Microbiol 9:69. https://doi.org/10.3389/fcimb.2019.00069
51. Rittie L et al (2019) The landscape of early clinical gene therapies outside of oncology. Mol Ther 27:1706–1717. https://doi.org/10.1016/j.ymthe.2019.09.002
52. Schacker M, Seimetz D (2019) From fiction to science: clinical potentials and regulatory considerations of gene editing. Clin Transl Med 8:27. https://doi.org/10.1186/s40169-019-0244-7
53. Tebas P et al (2014) Gene editing of CCR5 in autologous CD4 T cells of persons infected with HIV. N Engl J Med 370:901–910. https://doi.org/10.1056/NEJMoa1300662
54. Xu L et al (2019) CRISPR-edited stem cells in a patient with HIV and acute lymphocytic leukemia. N Engl J Med 381:1240–1247. https://doi.org/10.1056/NEJMoa1817426

9 Gene Therapy Applications

9.1 Preclinical Studies and Clinical Trials

The development of a therapy for human use entails a complex and long process of studies and validations, first in cells and animal models and later in human subjects. The first step is commonly referred to as **preclinical research** or development, being important to access the safety and also the efficacy of the therapy proposed. In a gene therapy context, the *in vivo* preclinical studies, using relevant animal models, are particularly important, as most of the therapies are being studied for the first time. If preclinical studies yield relevant results, therapy can continue its process of development, being then evaluated in **clinical trials**, which refer to research studies performed in human subjects. In their simplest design, clinical trials aim to evaluate the outcomes of a particular therapy in human subjects under an experimental condition, being also referred to as interventional clinical studies.

The current foundations, regulations, and importance of clinical trials were established as a response to the unethical, and in some cases even criminal, experimentations performed in human subjects, until the mid of the twentieth century. To regulate human experimentation, in 1964 the World Medical Association adopted the Declaration of Helsinki, describing the basic principles of human research [1]. Further improvements and regulations were introduced with the Belmont Report, that was created in response to the Tuskegee syphilis experiment. In 1932, the US Public Health Services initiated an experiment to determine the natural course of untreated, latent syphilis in black males. Even when penicillin became available to treat the disease (in the 1950s), the 400 men involved did not receive the therapy. Only in 1972 was the study stopped, after more than 100 men died directly from syphilis lesions [2]. The Belmont Report introduced the informed consent, which is now a mandatory component of clinical trials. This consent must be signed by all the participants enrolled in a clinical trial and must clearly state (i) that the study is conducted for research purposes, and include the risks, the benefits and the possible alternatives; (ii) that participation is voluntary; (iii) the extension of the confidentiality of the study and results; and (iv) the contact for questions. These important rules are, of course, not infallible, and therefore a close supervision must be performed by national authorities, such as the EMA in Europe and the FDA in the USA.

The design of a clinical trial is complex, as several variables and components must be addressed to ensure the safety of the study and to potentiate the results of the intervention. For example, the criteria for participant selection and the number of patients should be carefully planned, as well as the doses of the therapy and the duration of the study. In general, before a therapy receives its marketing authorization and

C. Nóbrega et al., *A Handbook of Gene and Cell Therapy*,
https://doi.org/10.1007/978-3-030-41333-0_9

becomes available to human use, it has to pass through three phases of clinical trials (Fig. 9.1). The **phase I** studies aim to assess the safety of the therapy, the side effects and the maximum tolerated dose. Normally, these are open-label studies, in which both participants and researchers know which treatment is administrated, with a small number of volunteer participants (healthy or diseased). A **phase II** clinical trial is usually larger than the previous phase study and primarily focuses on the effectiveness of the therapy. The safety profile of the therapy is also evaluated in these studies, along with its side effects. The phase II studies are also important to plan the phase III trials, which are the final step before the marketing authorization. **Phase III** clinical trials are conducted in even larger samples, aiming to demonstrate or confirm the efficacy and importantly to monitor adverse reactions. Often, phase III trials are blinded both for participants and researchers, to reduce the bias of subjective outcomes.

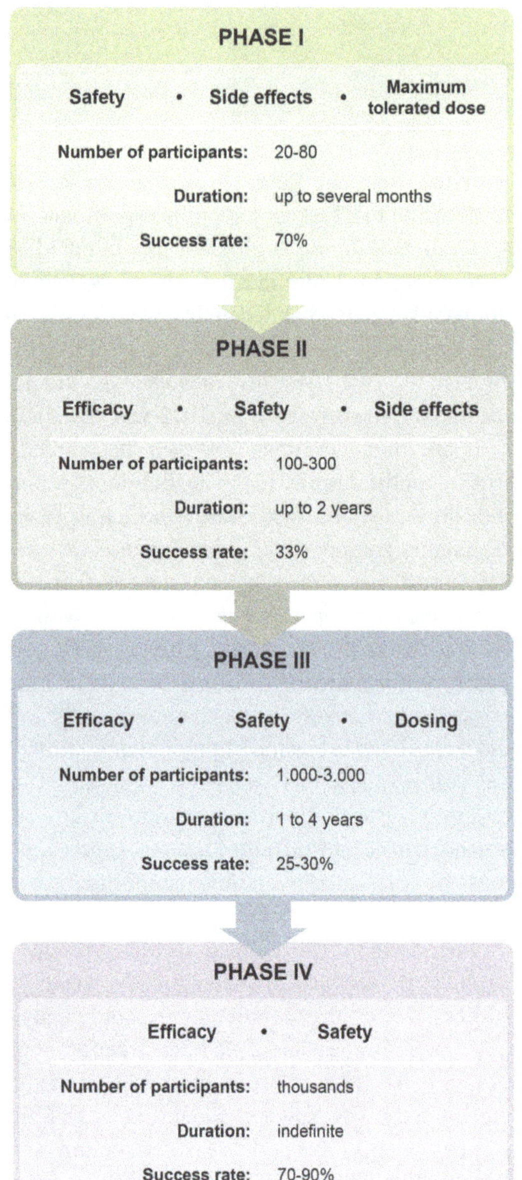

Fig. 9.1 Overview of the different phases of a clinical trial and their main features.

If the results of phase III trials prove the therapy's efficacy and the regulatory agency agrees with the positivity of the results, the experimental therapy is approved. After this approval, the therapy continues to be monitored, in the so-called **phase IV** studies, in which there is a continuous assessment of the safety and efficacy of the therapy in the target subjects.

This simple classification and description of clinical trials hides a very complex, long, and expensive process, which strongly limits the number of therapies that receive marketing authorization. However, they are needed to ensure the safety, quality, and efficacy of the therapies for human use. There are some exceptions to this process and phases, namely for rare diseases, for which it would be impossible to have the recommended number of subjects to perform the different phases of the trial and also because the economic value of a potential therapy is not comparable to the economic value predicted for common diseases. Thus, for these rare/orphan diseases, the rules of each clinical trial phase are more relaxed, and the duration of the process is normally shorter.

9.2 Gene Therapy for Cancer

Cancer is the second leading cause of death worldwide, despite the important advances made to its treatment, both in classical therapies and in novel, gene-based therapies. In fact, one important breakthrough in gene therapy was the development of suicide gene therapy (Fig. 9.2). The idea behind this strategy is to deliver into tumor cells a gene encoding a cytotoxic protein. However, currently, different other types of gene therapy targeting cancer are being investigated and tested. Overall, gene therapy for cancer employs one of four strategies [3]: (1) introduction of suicide genes, which induce the generation of compounds that are toxic to tumor cells; (2) induction of cell lysis using modified viruses; (3) introduction of immunomodulatory genes, aiming to induce or increase the immune system response; and (4) introduction of tumor-suppressor genes, which will block cell division. Of course, this simple classification of gene therapy strategies targeting cancer might be reductive, as cancer is a very complex disease. Furthermore, this general strategies hide the fact that different approaches are strongly interconnected, and one particular approach may, for example, cause cell lysis and induce the immune system response at the same time.

Cancer is one of the main targets for gene therapy, and advances in this field led to the approval of Imlygic®, which is a suicide gene therapy product for melanoma treatment. It is based on the intratumoral delivery of a weakened form of herpes simplex virus type 1 (e.g., the neurovirulence factor ICP34.5 was deleted) that can infect and multiply inside melanoma cells, killing them; moreover, it makes the infected cells produce the human cytokine granulocyte macrophage colony-stimulating factor gene (GM-CSF), which stimulates the patient's immune system to destroy the cancer cells. The therapy, now approved both in Europe and the USA, is able to selectively recognize and destroy malignant cells with a minimal effect on normal cells [4].

9.2.1 Suicide Gene Therapy

Suicide gene therapy can be achieved using at least two strategies [5]: an indirect form in which gene therapy is accomplished through an enzyme-activated prodrug that will be converted into a lethal drug inside the cells, or the direct delivery of proapoptotic genes (Fig. 9.3). Importantly, both forms aim to produce cell death without affecting normal cells.

The first strategy uses genes that codify for enzymes that convert a prodrug (which is afterward administrated) into a compound that is actively toxic to the cell, leading to its death. One of the most popular systems uses the herpes simplex virus thymidine kinase gene (HSV-tk), which converts the antiviral drug ganciclovir (GCV) to monophosphate, that is then metabolized by cell kinases into GCV-triphosphate, inhibiting DNA synthesis and leading to cell death [6]. This suicide gene therapy strategy also has the advantage of the bystander effect, which is a potentiation of the effect whereby the prodrug efficacy will be extended to neighboring cells. Other popular systems employ the enzyme cytosine deaminase and the prodrug 5-flurocytosine (5-FC) or the cytochrome P450 and the prodrugs cyclophosphamide (CPA) and ifosfamide (IFO).

The other type of suicide gene therapy is based on the introduction of proapoptotic genes. For example, the delivery of the gene codifying wild-type p53 (which is frequently mutated in several

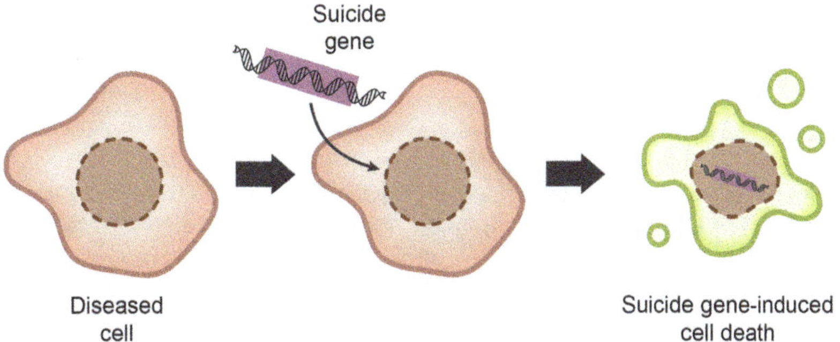

Fig. 9.2 Overview of suicide gene therapy targeting cancer, aiming to cause cell death through the introduction of a toxic gene.

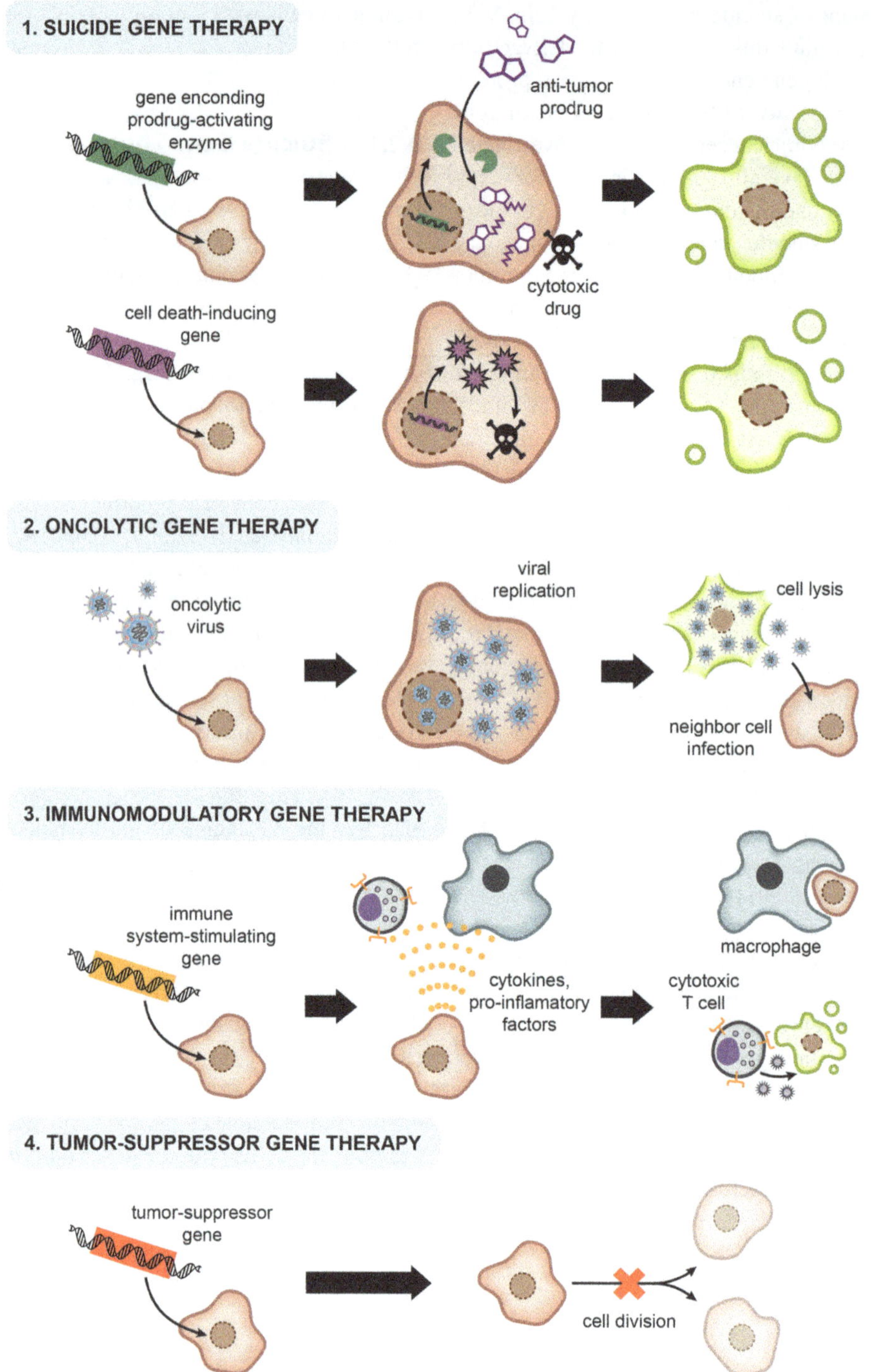

Fig. 9.3 The different types of gene therapy strategies targeting cancer. *Suicide gene therapy* can be achieved using at least two strategies, an indirect form in which gene therapy is accomplished through an enzyme-activated prodrug that will be converted into a lethal drug in cells, or a direct form consisting on the delivery of pro-apoptotic genes (1). The idea of *oncolytic gene therapy* is to apply viruses directly into the tumor, and the pathogenic agents will induce tumor cell lysis (2). In the *immunomodulatory gene therapy*, there is the introduction of a gene leading to the stimulation or enhancement of antitumor immunity (3). In the *tumor-suppressor gene therapy*, there is the introduction of tumor-suppressor genes as a way to fight cancer, preventing uncontrolled cellular growth (4).

human tumors) can restore the normal function of the protein and promote cellular apoptosis in tumors. In fact, Gendicine®, the first gene therapy product approved in China, is based on this strategy, delivering wild-type p53 gene through adenoviral vectors. The use of other genes aiming to promote cell death is also under investigation, including the genes codifying caspase 3 and Smac.

9.2.2 Oncolytic Gene Therapy

Besides the important use of viruses as carriers, their natural infectious features may also be used for cancer gene therapy applications. The idea of oncolytic gene therapy is to apply viruses directly into the tumor, where the pathogenic agents will induce tumor cell lysis (Fig. 9.3). These so-called oncolytic viruses can lead to cellular death through oncolysis, destruction of tumor blood vessels, antitumor immune activation, or even through the promotion of the expression of therapeutic genes in parallel with the lysis effect.

For example, attenuated forms of herpes simplex virus type 1 (HSV-1) only able to replicate in tumor cells can be introduced into these cells leading to lysis upon infection. This strategy presents two important features. First, there is a potentiation effect, since the new viruses produced (these are replicative viruses) are able to infect neighboring tumor cells. Second, the use of these oncolytic viruses and their replication will promote an immune response, thus enhancing the therapeutic effect. This strategy is the basis of Imlygic®, an HSV-1-based system that included several genome modifications, such as deletion of ICP34.5 and ICP47 genes and the insertion of a transgene encoding GM-CSF, as previously explained. Moreover, the insertion of cellular promoters for tumor-targeted replication was also used to increase tumor specificity [7].

In China, another oncolytic gene therapy named Oncorine (H101) was also approved. Oncorine is an oncolytic adenovirus encoding the p53 protein, to be used in combination with chemotherapy for the treatment of nasopharyngeal carcinoma [8]. Other viruses are also being investigated in clinical trials as oncolytic gene therapy strategy effectors, namely vaccinia virus, reovirus or measles virus [9].

9.2.3 Immunomodulatory Gene Therapy

The idea behind immunomodulatory gene therapy is the recognition that growing tumors actively evade the immune system and that stimulation or enhancement of antitumor immunity can be a therapeutic strategy to fight cancer. In fact, the 2018 Nobel Prize in Physiology or Medicine was awarded to James P. Allison and Tasuku Honjo for their discovery of cancer therapy by inhibition of negative immune regulation.

In a gene therapy setting, this strategy can take advantage of the enhancement of the immune response through the delivery of proinflammatory genes (Fig. 9.3). Furthermore, the use of modified cells (e.g., T-lymphocytes or dendritic cells) is another strategy for immunomodulatory therapy, thus combining gene and cell therapy [10]. For example, the combination of an immunological checkpoint blockade with CAVTAK™, which is an oncolytic coxsackievirus, led to a very good response and disease control rates in a phase Ib clinical trial (NCT02307149) for advanced melanoma [11]. Also, preclinical studies showed that the combination of oncolytic virus (expressing C-X-C motif of chemokine 11 precursor) and immunological checkpoint blockade reduced tumor immunosuppressive activity and boosted T-cell infiltration in the tumor [12].

9.2.4 Tumor-Suppressor Gene Therapy

The introduction of tumor-suppressor genes as a way to fight cancer is a particular case of suicide gene therapy, as most cancer patients present mutations or even deletions in tumor-suppressor genes. The main goal of this gene therapy strategy is the functional restoration of those mutated/deleted genes and prevention of the uncontrolled cellular growth (Fig. 9.3). Additionally, this group of strategies includes therapies targeting angiogenesis, aiming to cut off the nutrient supply to tumors, thus preventing their growth and contributing to their shrinkage. As well as for suicide gene therapy, the use of the p53-codifying gene, or several downstream mediators, can help to prevent tumor growth [13]. Additionally, preclinical studies show that the transfer of genes encoding cyclin inhibitors, and even microRNAs, could be used in this strategy. This type of strategies also includes the blocking of oncogenes expression through gene silencing approaches (based on RNAi or ASOs) and the expression of genes controlling checkpoints of the cell cycle, trying to stop cellular division.

9.2.5 Comparing Gene Therapy Strategies for Cancer

Along with rare neurodegenerative diseases and eye-affecting conditions, gene therapy strategies targeting cancer are probably the most advanced in terms of preclinical and clinical research. Like it was mentioned, several of the described strategies are used together and take advantage of different features both of the delivered gene and the delivery vector. As they present advantages and disadvantages, choosing between particular strategies should be based on the application purpose and the type of tumor, among other considerations (Table 9.1). Moreover, gene therapy strategies for cancer may also be used in combination with conventional treatments such as chemotherapy or radiotherapy, which is another important point that should be considered in the definition of the strategy to study.

9.2.6 Challenges to Gene Therapy Targeting Cancer

As mentioned throughout this book, several conditions must be addressed and considered to ensure the success of a gene therapy approach, regardless of the target disease. However, gene therapy targeting cancer has some particularities that should also be taken into consideration.

In the first line of obstacles or challenges to tackle is the heterogeneity of cancers, ranging from the anatomical differences to the individual differences among the same type of cancer and even to the heterogeneity of the cell types whithin a cancer. All these features make the application

Table 9.1 Main advantages and disadvantages of the different strategies of gene therapy targeting cancer.

Gene therapy strategy	Advantages	Disadvantages
Suicide gene therapy	Proved safe in clinical trials	Difficulty in targeting all tumor cells
	Action potentiation – bystander effect	Low efficacy of gene transfer *in vivo*
	May be used in combination with conventional therapies	
Oncolytic gene therapy	Replication only in tumor cells	Viruses may be cleared by the immune system
	Action potentiation – bystander effect	Low efficacy of gene transfer *in vivo*
	May be used in combination with conventional therapies	
	May enhance the action of the immune system	
Immunomodulatory gene therapy	Promotion of long-term immunity	Must overcome tumor-induced immunosuppression
	Good efficacy results *in vivo*	Autoimmune side effects
	May be used in combination with conventional therapies	
Tumor-suppressor gene therapy	Proved safe in clinical trials	Existence of redundant pathways in tumors
	May be used in combination with conventional therapies	Low efficacy of gene transfer *in vivo*

of gene therapies (and also of conventional therapies) very difficult, bringing with them a high degree of variation in their efficacy. Another crucial aspect is the ability to specifically target tumor cells, especially when using oncolytic suicide gene therapy. One way to try to ensure specificity is targeting the vectors to specific receptors of tumor cells, or using promoters specific to those cells, allowing the preservation of normal, non-diseased, cells. The safety profile of vectors is another aspect that must me guaranteed, especially in the gene therapy strategies that use replicative viral vectors. Safety can be improved also by engineering the viral particles so as to specifically infect the tumor cells. Another major concern of gene therapy applied to cancer is that efficacy of several strategies was shown to be reduced due to the presence of circulating antibodies, which is particularly relevant when using viruses as a delivery vector.

Despite all these and other challenges, the future of gene therapy for cancer seems bright, with the prediction that several products presently in advanced stages of development could be approved for marketing in the next years (Table 9.2).

9.3 Gene Therapy for Eye Conditions

Since the beginning of gene therapy, the eye and more specifically the retina was considered a privileged target organ for the development of new therapeutics. Briefly, the eye comprises two main regions: the anterior part (with the cornea, lens and conjunctiva) and the posterior part, with the retina as the prominent structure. The retina is considered part of the brain, consisting of different cellular layers such as ganglion cells, nerve fibers, light-sensing photoreceptors and the retinal pigment epithelium (RPE), among others. Despite being a highly specialized structure, compared to other organs or tissue, the retina has several advantages that make it very suitable for gene therapy. First, it can be easily and directly accessed, which allows direct and accurate delivery of the therapy. Moreover, as a closed system, the delivery of genes is limited to the eye structures, without propagation into peripheral organs. Second, the retina allows an easy visualization, which permits a noninvasive follow-up of the intervention throughout time. Third, the eye is an immunologically privileged site, which limits the immune response to the gene delivery. Finally, there are two eyes, allowing the possibility of maintaining an untreated contralateral eye, as a valuable control for the disease natural history, to assess the treatment efficacy and to evaluate its safety profile. Because of all these advantageous features, there is a wide range of preclinical and clinical studies evaluating different gene therapy approaches for several retinal diseases. It is thus also not surprising that one of the currently approved gene therapy products targets a retinal disease. Luxturna™ (Spark Therapeutics) is a gene therapy product consisting of AAV vectors that mediate delivery of a normal copy of the *RPE65* gene, administered through subretinal injection, to treat an inherited retinal disease caused by mutations in both copies of the aforementioned gene.

9.3.1 A Privileged Immunologic Organ

One of the important obstacles to the success of gene therapy is the immune response, which is particularly relevant when genes are delivered by viral vectors. Several features are present in the eye to maintain its immune privilege [14], circumventing the immune response and making them a very good target for gene therapy. First, several physical barriers (such as the blood-retina barrier and the lack of efferent lymphatic vessels) prevent the free exchange of cells and large molecules between the eye and the rest of the organism. Second, the ocular microenvironment inhibits the immune-competent cells, due to the production of several factors (e.g., TGF-β) and the direct action of ocular cells (e.g., retinal glial Muller cells). Third, the eye is able to regulate the immune response directly. All these properties of the eye and the retina constitute advantages that make them good targets for gene therapy strategies.

Table 9.2 Gene therapies targeting cancer that are in the final stages for marketing approval.

Name	Type of gene therapy	Vector	Indication	Therapeutic gene/cell	Therapy mechanism	Clinical study phase	Manufacturer
DNX-2401 (tasadenoturev)	*In vivo*	Adenovirus	Glioblastoma/gliosarcoma	Modified Ad genome	Oncolytic + immunomodulatory	II	DNAtrix
ONCOS-102	*In vivo*	Adenovirus	Mesothelioma	CPI (PD1 inhibitor)	Tumor-suppressor	I/II	Targovax
VB-111 (ofranergene obadenovec)	*In vivo*	Adenovirus	Glioblastoma	VBL's PPE-1-3x proprietary promoter	Oncolytic + immunomodulatory	III	VBL Therapeutics
Pexa-Vec (pexastimogene devacirepvec, JX-594)	*In vivo*	Vaccinia virus	Hepatocellular carcinoma	*GM-CSF*	Oncolytic + immunomodulatory	III	SillaJen
JCAR017 (lisocabtagene maraleucel)	*Ex vivo*	T-cells	Relapsed or refractory DLBCL	CAR T-cells	Immunomodulatory	I	Juno Therapeutics, Celgene
bb2121	*Ex vivo*	T-cells	Relapsed or refractory MM	CAR T-cells	Immunomodulatory	I/II	bluebird bio, Celgene
CG0070	*In vivo*	Adenovirus	NMIBC/MIBC	GM-CSF	Oncolytic + immunomodulatory	II	Cold Genesys
E10A	*In vivo*	Adenovirus	Squamous cell carcinoma of the head and neck	Endostatin	Tumor-suppressor	III	Marsala Biotech
Nadofaragene firadenovec/Syn3, rAd-IFN/Syn3 (Instiladrin)	*In vivo*	Adenovirus	BCG-unresponsive NMIBC	Interferon alfa-2b	Immunomodulatory	III	FKD Therapies, Ferring Pharmaceuticals
ProstAtak	*In vivo*	Adenovirus	Prostate cancer/hepatocellular carcinoma, pancreatic adenocarcinoma/NSCLC	*TK* + valacyclovir (drug)	Suicide + immunomodulatory	III	Advantagene
Toca 511 (vocimagene amiretrorepvec)	*In vivo*	Retroviral vector	Recurrent high-grade glioma	Cytosine deaminase +5-FU (drug)	Suicide + immunomodulatory	II/III	Tocagen

DLBCL – Diffuse large B-cell lymphoma
GM – CSF - granulocyte macrophage colony-stimulating factor
MM – Multiple myeloma
CAR – chimeric antigen receptor
NMIBC – Non-muscle invasive bladder cancer
TK – thymidine kinase
MIBC – Muscle invasive bladder cancer
BCG – Bacillus Calmette-Guérin
NSCLC – Non-small cell lung cancer

9.3.2 Gene Delivery Routes and Vectors

The delivery of genes to the retina takes advantage of the extensive research made concerning the pharmacokinetic profiles of conventional drugs and the multiplicity of routes and types of administration tested [15]. Nevertheless, despite the variety of delivery routes that could be used, due to efficacy issues, gene delivery is mainly performed by intravitreal or subretinal injection. The latter route leads to more rapid expression of the transgene compared to intravitreal injection. However, both routes also have limitations and barriers that need to be surpassed. For example, the genes delivered by subretinal injection need to overcome the external membrane of the photoreceptors. On the other hand, the intravitreal route dilutes the genes in the vitreous and has to overcome the inner limiting membrane [16].

Systemic administration, in the blood circulation, of recombinant AAV targeting the eye was also tested in different studies, as it is a less invasive and challenging pathway. However, this route needs high-titer injections of viral vectors and could potentially lead to off-target toxicity effects.

Concerning the delivery vehicle, several studies have used non-viral vectors to deliver genes to the eye (Table 9.3). However, these methods frequently lack efficiency, and some of them are clinically inviable due to ocular complications related to their application. Therefore, the use of recombinant viruses as vectors to deliver a gene directly to the eye rapidly increased in preclinical and clinical studies. Out of all the possibilities, AAV emerges as the preferred vector, due to their low immunogenic profile, the stable transgene expression they induce and the natural tropism of specific serotypes to eye cells. A huge effort is continuously being made to improve AAV efficacy, including optimization of the viral transduction or of the specificity of viral tropism. For example, to overcome the main limitation of AAV vectors, which is the limited cloning capacity, "overstuffed" and dual AAV vectors [17] were developed (see Chap. 3). However, despite being very useful for large transgenes, the applicability of dual AAV vectors is currently limited, because the system requires the co-expression of two vectors in the same cells, and the subsequent occurrence of homologous recombination. The simultaneous success of both steps is needed, which has been limiting the efficiency of dual vectors in gene therapy.

9.3.3 Gene Therapy Trials for Ocular Conditions

The heterogeneity of conditions affecting the eye and of the genes and cells involved is reflected by the variety of gene therapy trials developed so far, using a multiplicity of vectors, genes and administration routes. Nevertheless, currently, only 1.4% of gene therapy trials have targeted

Table 9.3 Examples of gene therapy studies targeting the retina using different delivery methods.

	Method	Gene	Model	Route of administration	References
Physical methods	Naked DNA	*VEGF*	CD-1 mice	Intrastromal corneal injection	Stechschulte *et al.* [34]
	Electroporation	*sFlt-1*	Albino Lewis rats	Injection in the suprachoroidal space	Touchard *et al.* [35]
	Gene gun	*GFP*	Rabbits	Cornea	Tanelian *et al.* [36]
	Ultrasounds	*GFP*	New Zealand albino rabbits	Intravitreal	Sonoda *et al.* [37]
Chemical methods	Liposomes	*Rpe65*	Blind mice	Subretinal injection	Rajala *et al.* [38]
	Polymers	Oligonucleotides	Lewis female rats	Intravitreal injection	Gomes dos Santos *et al.* [39]

ocular diseases. From these, most of the clinical studies have focused on the *RPE65* gene, especially for treating Leber's congenital amaurosis (LCA). Several strategies were also studied for other ocular conditions, such as choroideremia, achromatopsia or retinitis pigmentosa. In Table 9.4, selected gene therapy clinical trials targeting eye conditions are shown, providing an overview of the different target conditions, strategies and vectors used.

9.3.4 Challenges to Gene Therapy Targeting the Eye

The current gene therapy tools targeting the retina are well improved and developed. There is a wide range of delivery routes, several vectors, and diverse gene transfer reagents available, and there is data from animal studies on the safety profiles of gene therapy strategies, as well as extensive data on their immune response. Nevertheless, the increasing accuracy in diagnosing retinal diseases has led to the discovery of more than 250 different genetically distinct retinal diseases that are now known [18] and being curated in a public database (https://sph.uth.edu/RETNET/). This number highlights the high diversity of retinal diseases-causing mutations, and suggests the consequent, difficulty in developing gene therapy products for each one of them.

Another important challenge that gene therapy targeting the retina needs to overcome is the difficult transition of the preclinical animal results to human studies. Contrary to other targets, the human eye is far more specialized than that of current rodent models and, for this reason, several studies need to be performed in species with a more complex eye, like nonhuman primates. Of course, this raises ethical and economic issues that need to be carefully considered and discussed. Additionally, it is particularly important to develop vectors specifically directed to the human retina, as frequently the transduction pattern is different in rodents compared to humans or larger animals.

9.4 Gene Therapy for Cardiovascular Diseases (CVDs)

Cardiovascular diseases (CVDs) constitute a serious health problem, being the major cause of mortality and morbidity worldwide, and their prevalence is still increasing. CVDs comprise a group of disorders of the heart and blood vessels, for example, coronary heart disease, cerebrovascular disease and peripheral arterial disease, among others. Heart attacks and strokes, despite usually corresponding to acute events, are also classified as CVDs. There has been a huge advance in the pharmacological and surgical therapies for CVDs, which resulted in symptom mitigation and in the reduction of disease progression; however, there is still a lack of effective therapies to effectively cure and treat CVDs. Therefore, gene therapy appears as a promising strategy for the treatment of both inherited and acquired CVDs gene. The potential of gene therapy in the context of CVDs became very clear a long time ago when direct intra-arterial gene transfer was performed using endovascular catheter techniques [19]. In fact, the multiplicity of surgical techniques used for the different CVDs contributed to the extensive gene therapy trials performed for these diseases, as several administration routes have already been tested and established (Table 9.5).

In terms of vectors, the first attempts used non-viral methods; however, similarly to other applications, viral vectors have become the preferable system in CVD studies. In this group, there are studies using adenoviral vectors, AAV vectors and lentiviral vectors. In the case of AAV vectors, serotypes 1, 6, 8 and 9 have been identified to have a good tropism for cardiac tissue, after systemic delivery.

Another important issue regarding gene therapy studies for CDVs is the therapeutic genes to transfer. The most promising results were obtained with genes that induce angiogenesis or vasculogenesis or genes that encode for proteins involved in the cardiomyocytes Ca^{2+} pathway (Table 9.6). Nevertheless, several other genes were also studied [20], although until now no gene therapy product has been approved.

Table 9.4 Selected gene therapy clinical trials targeting eye diseases.

Condition	Therapeutic gene	Phase	Vector	Route of administration	Sponsor	Trial ID
Leber's congenital amaurosis	*RPE65*	III	AAV2	Subretinal	Spark Therapeutics	NCT00999609
Choroideremia	*CHM*	III	AAV2	Subretinal	Nightstar Therapeutics	NCT03496012
Achromatopsia	*CNGB3*	I/II	AAV8	Subretinal	MeiraGTx	NCT03001310
Retinitis pigmentosa	*PDE6B*	I/II	AAV2/5	Subretinal	Horama S.A.	NCT03328130
Stargardt disease	*ABCA4*	I/II	Lentivirus	Subretinal	Sanofi	NCT01367444
Neovascular age-related macular degeneration	Anti-VEGF protein	I/II	AAV8	Subretinal	REGENXBIO Inc.	NCT03066258

Therefore, more studies are needed both at the preclinical and clinical levels to deliver a safe and efficient gene therapy product for CVDs.

9.5 Gene Therapy for Neurodegenerative Diseases

The brain and its molecular networks and cellular circuits are quite complex, which complicates its study in normal and pathological conditions. Neurodegenerative diseases are defined by a progressive dysfunction of the nervous system, which is normally translated into severe symptoms and a debilitating phenotype. Normally, these are late-onset diseases, affecting a large part of the world population, which is a reflex of an increased elderly population worldwide. Therefore, they represent a huge burden for current societies, without any therapy available to stop or delay disease progression. In fact, current therapies are only symptomatic, with very limited effects and do not target the underlying molecular mechanisms of pathogenesis and degeneration.

The difficulty in treating neurodegenerative diseases has several causes; however, their complex nature and the limited accessibility of the central nervous system (CNS) are probably the two most important ones. First, neurodegenerative diseases such as Alzheimer's disease and Parkinson's diseases are mainly sporadic, without a clear cause, which strongly limits the ability to rationally develop therapies for them. Even monogenic neurodegenerative diseases, like the polyglutamine diseases, have a complex molecular pathogenesis that only now is starting to be completely elucidated. The other important point is the existence of the blood-brain barrier (BBB), preventing the access of many drugs to the CNS, which is the part of the nervous system that is most commonly affected in these diseases. Furthermore, there is a high number of different neurodegenerative diseases, and each one is characterized by a considerable degree of heterogeneity and by a variable incidence; combined, these aspects constitute additional challenges to the search for therapeutic solutions. Two other important factors contribute to the lack of efficient therapies: the CNS has a poor ability to repair neuronal damage, which strongly limits the success of a therapy in actually reverting neuronal loss, and the evaluation of the clinical outcomes after a therapy administration is extremely difficult, due to the heterogeneous progression of the neurodegenerative process and to the lack of established biomarkers.

Gene therapy arises as a powerful tool to treat neurodegenerative diseases, trying to restore missing functions or to improve neuronal homeostasis, or ultimately to enact the correction of the molecular pathogenesis. The potential of gene therapy in the treatment of neurodegenerative diseases was already highlighted throughout this book, for example, for spinal muscular atrophy (SMA). However, even if promising, there is still a long way to be traveled in order to increase the number of gene therapy solutions for the different neurodegenerative diseases.

Table 9.5 Advantages and disadvantages of different gene administration methods targeting cardiovascular diseases.

Gene administration method	Advantages	Disadvantages
Direct intramyocardial injection	Decreased risk of immune response	Expression of the gene limited to the local of the injection
	High density of gene transfer, limited to the cardiac tissue	Multiple injections are needed
	Simple and safe procedure	Potential damage from the needle
Anterograde arterial infusion	Simple and minimally invasive procedure with selectivity for the cardiac tissue	May be inefficient
	Can lead to a homogenous distribution of the gene	Can result in the systemic delivery of the transgene
Retrograde intravenous infusion	Can lead to high levels of transduction in the cardiac cells	Potential risk of ischemic events
	Can lead to a homogenous distribution of the gene	May be inefficient
Pericardial delivery	Can increase the transduction levels	Transgene expression limited to superficial epicardium
	Safe and minimally invasive	
Aortic cross-clamp left ventricular cavity infusion	Increased gene transfer efficiency	High risk of myocardial injury
		Not cardiac specific
		Requires open chest surgery

9.5.1 Administration Routes to the CNS

The choice of the administration route is an essential point for all gene therapy strategies, but it is especially critical in the context of neurodegenerative diseases due to the presence of the BBB. To overcome this barrier, the strategy most commonly used so far has been the local and direct delivery of the therapeutic gene into the brain parenchyma, through neurosurgical stereotaxic injections. However, an important limitation of the direct intracranial injection is the associated local trauma, as well as the stimulation of inflammation and toxicity-inducing events. Also, the procedure is associated with a poor spread of the vector, limited around the injection site. Of course, this limitation can be overcome with multiple injections, although this strongly increases the complexity of the surgery and its potential side effects. Additionally, the procedure is preferably performed only once due to the risks to the patient, thus precluding multiple administrations at different time points.

An alternative to this procedure can be the delivery of the vectors with the therapeutic gene into the cerebrospinal fluid (CSF), which can be accessed through a different route. This alternative will increase the spread of the vector, although it has an important degree of invasiveness and complexity in some of its routes. Maybe the one exception is the intrathecal injection, which is a common and relatively safe procedure in current medical practice. However, all these routes are relatively complex, and therefore non-invasive and peripherical routes of gene delivery were also developed and studied. For motor neuron diseases, an interesting route consists on the intramuscular injection of the vectors, which is a minimally invasive procedure that could reach the CNS by retrograde transport through the motor neurons.

However, the least invasive and probably most straightforward delivery route is the systemic administration of the vectors, especially through intravenous injection. This method has the advantage of being safe, allowing repeated administrations and the introduction of high amounts of vectors, if needed. However, the BBB limits the access of the systemic molecules to the CNS, and therefore the vector must have the ability to cross it. To achieve this, several studies have reported non-viral vectors coupled with molecules that facilitate or mediate the BBB crossing. On the

Table 9.6 Selected gene therapy clinical trials targeting cardiovascular diseases.

Condition	Therapeutic gene	Development stage	Vector	Delivery method	Number of patients	Findings
Coronary heart disease	*FGF2* (fibroblast growth factor 2)	Phase II	Naked plasmid DNA	Intracoronary infusion	337	No improvements in myocardial perfusion
Chronic stable angina	*FGF4* (fibroblast growth factor 4)	Phase I/II	Adenovirus	Intracoronary infusion	79	No major adverse events
Coronary heart disease	*VEGF-A* (vascular endothelial growth factor A)	Phase II	Adenovirus/ liposome	Intracoronary infusion	103	No major adverse events
Severe stable ischemic heart disease	*VEGF-A* (vascular endothelial growth factor A)	Phase II	Naked plasmid DNA	Direct intramyocardial injection	80	No improvements in myocardial perfusion
Severe and diffuse triple vessel coronary disease	*HGF* (Hepatocyte growth factor)	Phase I	Adenovirus	Intracoronary infusion	18	No adverse events
Hypoperfused area of viable ventricular muscle	*HIF1-alpha* (Hypoxia-inducible factor 1-alpha)	Phase I	Adenovirus	Intramyocardial injections	13	No safety concerns
Refractory coronary artery disease	VEGF+FGF	Phase II	Naked plasmid DNA	Percutaneous intramyocardial injection	52	Improved exercise tolerance
						No demonstrated improvement in cardiac perfusion

side of viral vectors, the discovery of the natural ability of AAV9 to bypass the BBB opened a new opportunity to treat neurodegenerative disease with viral vector-mediated gene therapy.

9.5.2 Candidate Conditions for Gene Therapy

Neurodegenerative disease is a broad term that includes several chronic conditions that cause a slow, progressive and irreversible loss of neurons. The brain regions and neuronal subtypes affected are different among these diseases, which translates into different phenotypes and symptoms. This group of diseases includes Alzheimer's disease (AD), Parkinson's disease (PD), amyotrophic lateral sclerosis (ALS) and other motor neuron diseases such as spinal muscular atrophy (SMA), and the polyglutamine diseases. This latter group constitutes the larger group of monogenic neurodegenerative diseases and comprises Huntington's diseases (HD), six forms of spinocerebellar ataxia (SCA1, 2, 3, 6, 7 and 17), dentatorubral-pallidoluysian atrophy (DRPLA), and spinal and bulbar muscular atrophy (SBMA). Aside from these chronic neurodegenerative diseases, other acute conditions affecting the CNS, such as stroke and spinal cord injury, can also be targeted by gene and cell therapy strategies.

9.5.3 Vectors for Delivering Genes into the CNS

Like for any other organ or tissue, the ideal vector used to deliver genes into the CNS should have specifically appropriate features, most of which have already been discussed throughout this book. Nonetheless, taking into consideration the important limitations pointed above, some of the conditions that are crucial for the success of gene therapy targeting the brain deserve a closer look. First, the gene introduced must be delivered efficiently to the brain. Second, since most of the neurodegenerative diseases result from a malfunction of specific regions and neuronal populations of the brain, the gene should be preferentially delivered to the affected neurons, avoiding the other cells of the CNS. The third important consideration concerns the levels and duration of gene expression. The access to the CNS is difficult; thus, ideally, gene delivery should be performed once, ensuring a therapeutic expression without cytotoxic side effects.

To meet all these conditions and overcome the particularities of the CNS, most of the studies use viral vectors to deliver genes. The herpes simplex virus type 1 (HSV-1) vector has a natural tropism to neurons, which makes it a good vector targeting the brain. However, its high cytotoxic profile makes it unsuitable for most of the neurodegenerative diseases. Currently, most of the gene therapy applications using HSV-1 vectors aim to take advantage of this feature to treat glioblastomas, which are brain tumors with a very poor prognosis. Even more limited is the use of adenoviral vectors for gene delivery to the brain, due to the strong immune response they elicit and their potential toxicity.

On the other hand, lentiviral vectors are attractive as gene delivery systems for the brain, allowing a good tropism to neurons or glial cells, depending on the envelope pseudotyping. One of the main advantages of lentiviral vectors is their ability to integrate the transgene into the host cell genome, facilitating the goal of one-time therapy. However, the possibility of insertional mutagenesis, along with their HIV-1 origin, constitute important safety limitations to their use in the treatment of neurodegenerative conditions.

Due to the limitations of both HSV-1 and lentiviral vectors, AAVs have emerged as good delivery vectors for the brain. AAVs can transduce nondividing cells and have the ability to confer long-term expression of the therapeutic gene without inflammation or toxicity. Moreover, the discovery of an AAV serotype that crosses the BBB provided an important boost to their use as vectors targeting the CNS. However, the use of AAVs is still limited, due to the presence of circulating pre-existing antibodies against several AAV serotypes. Moreover, their small cloning capacity constitutes another important disadvantage.

These considerations demonstrate that there is not a singularly perfect viral vector to deliver genes to the brain; however, current advances in lentiviral and AAV vectors make them the most probable and most common choice in both preclinical and clinical studies. In the next sections, several gene therapy strategies for different neurodegenerative diseases are described, although diverse examples have already been mentioned and discussed in other chapters throughout the book.

9.5.4 Alzheimer's Disease

Alzheimer's disease (AD) is the most common neurodegenerative disorder worldwide. The disease is mainly characterized by impairments in memory and behavior, and there are no therapeutic solutions available to stop the disease progression. Neuropathologically, it is characterized by a degeneration of cholinergic neurons, the formation of extracellular plaques of amyloid β peptide and the assembly of neurofibrillary tangles due to the accumulation of hyperphosphorylated tau protein. Despite being very common, less than 10% of AD cases are familial and, therefore, the intricate etiology of the disease complicates the development of therapeutic strategies.

Several gene therapy strategies have been tested in preclinical studies; however, the clinical studies focused mainly on the expression of neurotrophic factors, especially the nerve growth factor (NGF) as a way to improve AD-associated deficits. A pioneering study showed that an *ex vivo* gene therapy approach, based on the administration of fibroblasts expressing NGF into the brain, was able to increase the density of cholinergic neurons in aged monkeys [21]. This study was the basis for a phase I clinical trial developed in the USA in eight patients with early-stage AD. In this study, autologous fibroblasts modified by retroviral vectors to produce and secrete NGF [22] were transplanted into the nucleus basalis of Meynert of the patients. A 22-month follow-up of six of the patients showed no adverse effects, a reduction in the speed of the cognitive decline of around 50%, and a significant increase in cerebral metabolism, evaluated by positron emission tomography (PET) scan. Moreover, a brain autopsy of one of the patients showed robust expression of NGF and a dense concentration of cholinergic axons in the fibroblast graft. Although the results were promising, the high costs and complexity of the procedure prevented further development and the continuation of the studies in subsequent clinical trials.

More recently, a phase I clinical trial injected NGF-codifying AAV2 particles into the basal forebrain of 10 AD patients [23], upon stereotaxic surgery. A follow-up of 2 years showed that this *in vivo* gene therapy approach was well tolerated and was able to produce long-term expression of NGF. These promising results were the basis for a subsequent phase II clinical trial (NCT00876863). The first results of the study were somehow mixed, and 2 years after the procedure both groups showed a similar decline in cognitive function. Therefore, in 2015, Sangamo Therapeutics ended the development of this strategy, which was named CERE-110.

9.5.5 Parkinson's Disease

Parkinson's disease (PD) is the second most common neurodegenerative disease worldwide, arising as a heritable condition (in 5–10% of the cases) or, more commonly, as a sporadic disorder [24]. From a neuropathological point of view, it is characterized by neuronal loss in the substantia nigra, which causes striatal depletion of dopamine and the formation of intracellular inclusions of α-synuclein. Traditionally, motor symptoms associated with PD (e.g., rigidity, resting tremor) are mitigated by the reposition of dopamine levels with drugs such as levodopa. This treatment is able to counter some motor deficits; however, with disease progression, levodopa becomes less efficient. Therefore, there is an opportunity for gene therapy strategies and, in fact, several different approaches have been explored both in preclinical and clinical studies.

There are several gene therapy studies using nonhuman primates focusing on normalizing either the dopamine production or the abnormal firing in the basal ganglia, while other strategies aim to stop, or revert, neuronal loss [25]. In the same line, there are also several clinical trials performed in PD patients using gene therapy strategies (Table 9.7). Most of them were phase I clinical trials that passed the safety criteria and demonstrated some efficacy. However, in subsequent trials, almost all of them failed to show more efficiency than the placebo.

One exception was a phase II clinical trial based on the local injection of the *GAD* (glutamic acid decarboxylase) gene, delivered by AAV2 into the subthalamic nucleus in 22 PD patients [26]. There was an improvement in the AAV2-GAD-treated group compared with the control group; however, the effect was not better than the current standard of treatment, and therefore the development of the therapy was not continued. Another phase II clinical trial used AAV2 to deliver the neurturin (*NRTN*) gene into the putamen of 38 PD patients. Some encouraging results were observed, namely an increase of neurturin expression at the injection site, although its levels in the substantia nigra were not altered. Based on these results, a further phase IIb clinical trial was developed injecting the therapy both in the putamen and in the substantia nigra. However, there was no significant improvement in the clinical outcome of treated patients, and therefore the development of the therapy (named CERE-120) was not continued.

Another interesting strategy was developed by Oxford Biomedica, using a polycistronic lentiviral vector to express three enzymes essential for dopamine synthesis: tyrosine hydroxylase (*TH*), aromatic amino acid dopa decarboxylase (*AADC*), and GTP cyclohydrolase I (*GCH1*). The therapy (named ProSavin) was tested in a phase I/II clinical trial targeting the sensorimotor part of the striatum and the putamen. The therapy proved to be safe, and the majority of the treated patients displayed some improvements in motor behavior [27]. The company is now preparing a phase I/IIa clinical trial with an improved version of the therapy named OXB-102. The results of these and other clinical studies highlight the complexity of developing gene therapies for PD. However, the relevance of the disease and the continuous development of strategies in preclinical studies promise more clinical trials with gene therapies in the future.

9.5.6 Lysosomal Storage Diseases (LSDs)

Lysosomal storage diseases (LSDs) comprise a group of more than 50 different inherited metabolic diseases that result from a dysfunction of the lysosome, which leads to the accumulation of undigested or partially digested macromolecules. Despite being a metabolic diseases, around 70% of LSDs affect the CNS, leading to selective neurodegeneration in several brain regions and extensive neuroinflammation. LSDs are good candidates for gene therapy, as most of these diseases are autosomal recessive monogenic conditions. Therefore, there are several examples of preclinical and clinical studies using gene therapy for these diseases [28], including some strategies in the late stages of clinical development, which suggests that some gene therapies for LSDs may be approved in the near future.

Metachromatic leukodystrophy (MLD) is caused by a deficiency in arylsulfatase A (ARSA), leading to an accumulation of sulfatides and demyelination of the CNS and peripheral nervous system, which translates into severe motor and cognitive defects. In 2013, an *ex vivo* gene therapy clinical trial study was performed in three patients with presymptomatic MLD. Autologous hematopoietic stem cells (HSCs) were treated with a lentiviral vector containing the correct *ARSA* gene and then reintroduced into patients. In a more than 18-month follow-up, the disease did not manifest or progress in the three patients, which had a much better motor and cognitive function compared to their untreated siblings [29]. These results were the basis for a phase I/II clinical trial using the same strategy in 20 MLD patients, which were followed for 3 years post-intervention. The preliminary results of this study revealed that this *ex vivo* gene therapy strategy was safe and with a clear therapeutic effect [30]. An *in vivo* gene therapy strategy was also tested in a phase I/II clinical trial,

Table 9.7 Selected gene therapy clinical trials targeting Parkinson's disease.

Gene therapy				Clinical trial		
Name	Gene	Vector	Delivery route	Phase	Trial code	Current status
VY-AADC01	*AADC* (Aromatic L-amino acid decarboxylase)	AAV2	Injection in the striatum	I	NCT01973543	Active, not recruiting
AAV-GAD	*GAD* (Glutamic acid decarboxylase)	AAV2	Injection in the subthalamic nucleus	II	NCT00643890	Terminated
	GAD (Glutamic acid decarboxylase)	AAV2	Injection in the subthalamic nucleus	I	NCT00195143	Completed
AAV-hAADC-2	*AADC* (Aromatic L-amino acid decarboxylase)	AAV2	Injection in the putamen	I/II	NCT02418598	Recruiting
ProSavin	*TH* (Tyrosine hydroxylase), *AADC* (Aromatic amino acid dopa decarboxylase) and *GCH1* (GTP cyclohydrolase I)	Lentivirus	Injection in the striatum and the putamen	I/II	NCT00627588	Completed
AAV2-GDNF	*GDNF* (Glial cell line-derived neurotrophic factor)	AAV2	Injection in the putamen	I	NCT01621581	Active, not recruiting
OXB-102	*TH* (Tyrosine hydroxylase), *AADC* (Aromatic amino acid dopa decarboxylase), and *GCH1* (GTP cyclohydrolase I)	Lentivirus	Injection in the striatum	I/II	NCT03720418	Recruiting
CERE-120	*NRTN*, (Neurturin)	AAV2	Injection in the putamen	I	NCT00252850	Completed

based on the intracerebral AAV10-mediated delivery of the correct *ARSA* gene in 12 different sites in 5 patients with presymptomatic or early state confirmed MLD (NCT01801709). This study was based on extensive preclinical studies, including in nonhuman primates, which showed a safe and efficient gene transfer. However, until now, no results from the clinical trials were published.

Mucopolysaccharidosis type IIIA (MPSIIIA), also known as Sanfilippo syndrome type A, is caused by a recessive mutation in the *SGSH* (N-sulfoglucosamine sulfohydrolase) gene. It has a very early onset, around 2 years of age, with a progressive decline of cognitive and motor functions in the first decade of life, being fatal in the second decade. A first gene therapy study was launched by Lysogene, based on the intracerebral administration of AAV10 encoding both *SGSH* and *SUMF1* (sulfatase-modifying factor) genes. The therapy was performed in four children, and the vector was injected into six different brain areas. The results showed no adverse effects and revealed some improvements or the stabilization of selected clinical parameters in some patients [31]. The company is now recruiting patients for a phase II/III clinical trial (NCT03612869), based on the administration of AAV10 vectors containing the *SGSH* gene (LYS-SAF302). Another company, Abeona Therapeutics, initiated a phase I/II clinical trial (NCT02716246) based on the systemic administration of AAV9 encoding for the normal *SGSH* gene. No results were yet published, although the company states in its website that the therapy is well tolerated and that positive neurocognitive signals were detected.

9.5.7 Amyotrophic Lateral Sclerosis

Amyotrophic lateral sclerosis (ALS), also known as Lou Gehrig's disease, is an incurable and progressive neurodegenerative disease with adult onset and a relatively short course, causing death within 3 to 5 years after the diagnosis. Neuropathologically, it is characterized by the degeneration of motor neurons in the spinal cord, brain stem and motor cortex.

The disease includes a more common, sporadic form (affecting 90% of ALS patients), and a familial form, resulting from genetic mutations and affecting 10% of ALS patients. However, even for the familial cases, the genetic causes are complex, as ALS can arise from several mutations in at least nine different genes [32]. Maybe because of this, gene therapy studies are not abundant, with the few exceptions focused on using ASOs targeting the *SOD1* (superoxide dismutase 1) gene, which accounts for 12% of the familial cases, or the *C9ORF72* gene, that accounts for 40% of familial ALS cases. However, at least for the ASOs targeting the *SOD1* gene, the efficacy results were not as promising as expected.

For sporadic ALS forms, the development of gene therapy strategies is even more complex and difficult, as the causing factors are not clearly elucidated. Several preclinical studies in animal models showed important improvements in neuropathological and motor deficits upon viral delivery of trophic factors- coding genes, such as *GDNF* (glial cell line-derived neurotrophic factor), *IGF-1* (insulin-like growth factor 1) and *VEGF* (vascular endothelial growth factor). However, despite the positive results obtained, none of these strategies has yet advanced to clinical trials.

9.5.8 Polyglutamine Diseases

Polyglutamine diseases constitute the largest group of inherited monogenic neurodegenerative diseases. For this reason, and despite being rare, the nine polyglutamine diseases occupy a central stage in the development of gene therapy strategies. Several examples of preclinical and clinical studies were mentioned in Chaps. 6 and 7, especially for Huntington's disease and Machado-Joseph disease (also known as spinocerebellar ataxia type 3). Moreover, different approaches based both on cell and gene therapies were recently reviewed [33], highlighting the existence of extensive results from preclinical studies that support some ongoing clinical trials and certainly several others that will be developed in the future.

9.6 Future Prospects for Gene Therapy Studies

From this brief, and far from exhaustive, description of different gene therapy preclinical and clinical studies, it is clear that there is a huge interest in these approaches to the treatment of human disease. The technical advances of the last years and the improvements made in terms of delivery vectors' safety greatly contributed to this boost. Although several challenges are still ahead, including many ethical concerns and the question of the cost of gene-based treatment, gene therapy is now on a growing momentum. In fact, it is foreseen that, in the next few years, several gene therapy products will receive marketing authorization from regulatory agencies both in Europe and the USA. If that is the case, gene therapy will definitively establish itself as an effective therapeutic option for several diseases, providing the cure for many untreated diseases - especially for rare conditions -, and possibly even constitute an alternative to conventional therapies.

This Chapter in a Nutshell

- Preclinical studies normally refer to laboratory studies in cell or animal models that evaluate the safety and efficacy of a therapy.
- Clinical trials comprise a set of different studies that test the safety, efficacy and side effects of therapy in human subjects.
- Normally, before receiving marketing authorization, a therapy needs to undergo three different phases of a clinical trial.
- Gene therapy approaches targeting cancer are grouped into different categories: (1) introduction of suicide genes, which induce the generation of compounds that are toxic to tumor cells; (2) induction of cell lysis using modified viruses; (3) introduction of immunomodulatory genes, aiming to induce or increase the immune system response; and (4) introduction of tumor-suppressor genes, which will block cell division.
- Suicide gene therapy can be achieved using at least two strategies: an indirect approach, in which therapy is accomplished through an enzyme-activated prodrug that will be converted into a lethal drug in cells, or a direct approach entailing the delivery of proapoptotic genes.
- Oncolytic gene therapy strategies are based on replicative viruses that specifically target tumor cells.
- Immunomodulatory gene therapy is based on the recognition that growing tumors actively evade the immune system and that stimulating or enhancing antitumor immunity can be a therapeutic strategy to fight cancer.
- Tumor-suppressor gene therapy aims to functionally restore mutated/deleted genes and prevent uncontrolled cellular growth.
- Gene therapy targeting different conditions affecting the eye is among the most advanced strategies developed, mainly due to (1) the easy access to the eye, (2) the easy visualization of the administered therapy, (3) the fact that the eye is an immunologically privileged site, and (4) the fact that the existence of two eyes allows maintaining an untreated contralateral eye as a valuable control.
- Several gene therapy studies were developed targeting cardiovascular diseases; however, until now no therapy is currently approved.
- Neurodegenerative diseases are defined by a progressive dysfunction of the nervous system, which is normally translated into severe symptoms and a debilitating phenotype.
- Due to several important particularities of neurodegenerative diseases, gene therapy arises as a powerful tool to treat them, aiming to restore missing functions, to improve neuronal homeostasis, or ultimately block the molecular pathogenesis.
- There are multiple examples of gene therapy strategies, developed in both preclinical and clinical studies, that target different neurodegenerative conditions, including Alzheimer's disease, Parkinson's disease and amyotrophic lateral sclerosis, among others.

Review Questions

1. What is the main aim of a phase I clinical trial testing of a gene therapy strategy?
   (a) Access therapy safety
   (b) Access therapy efficacy
   (c) Compare the therapy with the gold standard therapy
   (d) Access the long-term effect of the therapy
   (e) None of the above
2. Which of the following is not a gene therapy strategy targeting cancer?
   (a) Suicide gene therapy
   (b) Oncolytic gene therapy
   (c) Tumor-suppressor gene therapy
   (d) Immunomodulatory gene therapy
   (e) Neurotrophic gene therapy
3. Which feature makes the eye a good target for gene therapy?
   (a) Difficult access
   (b) Immunological privilege
   (c) Is part of the nervous system
   (d) Non-viral vectors can be used to mediated delivery
   (e) Has only one type of cells
4. Which of the following viral vectors is more suitable for a direct administration of a gene therapy targeting the brain neurons?
   (a) Lentiviral vectors
   (b) Adenoviral vectors
   (c) Gamma retrovirus-based vectors
   (d) Adenovirus gutless vectors
   (e) Baculovirus-based vectors

References and Further Reading

1. Dale O, Salo M (1996) The Helsinki declaration, research guidelines and regulations: present and future editorial aspects. Acta Anaesthesiol Scand 40:771–772
2. Brandt AM (1978) Racism and research: the case of the Tuskegee Syphilis Study. Hast Cent Rep 8:21–29
3. Kane JR, Miska J, Young JS, Kanojia D, Kim JW, Lesniak MS (2015) Sui generis: gene therapy and delivery systems for the treatment of glioblastoma. Neuro Oncol *17(Suppl 2)*:ii24–ii36
4. Gangi A, Zager JS (2017) The safety of talimogene laherparepvec for the treatment of advanced melanoma. Expert Opin Drug Saf 16:265–269
5. Navarro SA, Carrillo E, Grinan-Lison C, Martin A, Peran M, Marchal JA, Boulaiz H (2016) Cancer suicide gene therapy: a patent review. Expert Opin Ther Pat 26:1095–1104
6. Fillat C, Carrio M, Cascante A, Sangro B (2003) Suicide gene therapy mediated by the Herpes Simplex virus thymidine kinase gene/Ganciclovir system: fifteen years of application. Curr Gene Ther 3:13–26
7. Ungerechts G, Bossow S, Leuchs B, Holm PS, Rommelaere J, Coffey M, Coffin R, Bell J, Nettelbeck DM (2016) Moving oncolytic viruses into the clinic: clinical-grade production, purification, and characterization of diverse oncolytic viruses. Mol Ther Methods Clin Dev 3:16018
8. Liang M (2018) Oncorine, the world first oncolytic virus medicine and its update in China. Curr Cancer Drug Targets 18:171–176
9. Matsuda T, Karube H, Aruga A (2018) A comparative safety profile assessment of oncolytic virus therapy based on clinical trials. Ther Innov Regul Sci 52:430–437
10. Karjoo Z, Chen X, Hatefi A (2016) Progress and problems with the use of suicide genes for targeted cancer therapy. Adv Drug Deliv Rev 99:113–128
11. Amer MH (2014) Gene therapy for cancer: present status and future perspective. Mol Cell Ther 2:27
12. Marshall HT, Djamgoz MBA (2018) Immuno-oncology: emerging targets and combination therapies. Front Oncol 8:315
13. Liu Z, Ravindranathan R, Kalinski P, Guo ZS, Bartlett DL (2017) Rational combination of oncolytic vaccinia virus and PD-L1 blockade works synergistically to enhance therapeutic efficacy. Nat Commun 8:14754
14. Zhou R, Caspi RR (2010) Ocular immune privilege. F1000 Biol Rep 2 p.3
15. Del Amo EM, Rimpela AK, Heikkinen E, Kari OK, Ramsay E, Lajunen T, Schmitt M, Pelkonen L, Bhattacharya M, Richardson D et al (2017) Pharmacokinetic aspects of retinal drug delivery. Prog Retin Eye Res 57:134–185
16. Planul A, Dalkara D (2017) Vectors and gene delivery to the retina. Annu Rev Vis Sci 3:121–140
17. Ku CA, Pennesi ME (2015) Retinal gene therapy: current progress and future prospects. Expert Rev Ophthalmol 10:281–299
18. Bennett J (2017) Taking stock of retinal gene therapy: looking back and moving forward. Mol Ther 25:1076–1094
19. Nabel EG, Plautz G, Nabel GJ (1990) Site-specific gene expression in vivo by direct gene transfer into the arterial wall. Science 249:1285–1288
20. Yla-Herttuala S, Baker AH (2017) Cardiovascular gene therapy: past, present, and future. Mol Ther 25:1095–1106
21. Conner JM, Darracq MA, Roberts J, Tuszynski MH (2001) Nontropic actions of neurotrophins: subcortical nerve growth factor gene delivery reverses age-related degeneration of primate cortical cholinergic innervation. Proc Natl Acad Sci U S A 98:1941–1946
22. Tuszynski MH, Thal L, Pay M, Salmon DP, U HS, Bakay R, Patel P, Blesch A, Vahlsing HL, Ho G et al

(2005) A phase 1 clinical trial of nerve growth factor gene therapy for Alzheimer disease. Nat Med 11:551–555

23. Rafii MS, Baumann TL, Bakay RA, Ostrove JM, Siffert J, Fleisher AS, Herzog CD, Barba D, Pay M, Salmon DP et al (2014) A phase1 study of stereotactic gene delivery of AAV2-NGF for Alzheimer's disease. Alzheimers Dement 10:571–581
24. Poewe W, Seppi K, Tanner CM, Halliday GM, Brundin P, Volkmann J, Schrag AE, Lang AE (2017) Parkinson disease. Nat Rev Dis Primers 3:17013
25. Axelsen TM, Woldbye DPD (2018) Gene therapy for Parkinson's disease, an update. J Parkinsons Dis 8:195–215
26. LeWitt PA, Rezai AR, Leehey MA, Ojemann SG, Flaherty AW, Eskandar EN, Kostyk SK, Thomas K, Sarkar A, Siddiqui MS et al (2011) AAV2-GAD gene therapy for advanced Parkinson's disease: a double-blind, sham-surgery controlled, randomised trial. Lancet Neurol 10:309–319
27. Palfi S, Gurruchaga JM, Lepetit H, Howard K, Ralph GS, Mason S, Gouello G, Domenech P, Buttery PC, Hantraye P et al (2018) Long-term follow-up of a phase I/II study of ProSavin, a lentiviral vector gene therapy for Parkinson's disease. Hum Gene Ther Clin Dev 29:148–155
28. Beck M (2018) Treatment strategies for lysosomal storage disorders. Dev Med Child Neurol 60:13–18
29. Biffi A, Montini E, Lorioli L, Cesani M, Fumagalli F, Plati T, Baldoli C, Martino S, Calabria A, Canale S et al (2013) Lentiviral hematopoietic stem cell gene therapy benefits metachromatic leukodystrophy. Science 341:1233158
30. Sessa M, Lorioli L, Fumagalli F, Acquati S, Redaelli D, Baldoli C, Canale S, Lopez ID, Morena F, Calabria A et al (2016) Lentiviral haemopoietic stem-cell gene therapy in early-onset metachromatic leukodystrophy: an ad-hoc analysis of a non-randomised, open-label, phase 1/2 trial. Lancet 388:476–487
31. Tardieu M, Zerah M, Husson B, de Bournonville S, Deiva K, Adamsbaum C, Vincent F, Hocquemiller M, Broissand C, Furlan V et al (2014) Intracerebral administration of adeno-associated viral vector serotype rh.10 carrying human SGSH and SUMF1 cDNAs in children with mucopolysaccharidosis type IIIA disease: results of a phase I/II trial. Hum Gene Ther 25:506–516
32. Renton AE, Chio A, Traynor BJ (2014) State of play in amyotrophic lateral sclerosis genetics. Nat Neurosci 17:17–23
33. Nóbrega C, Almeida LP (eds) (2018) Polyglutamine diseases. Springer: Gewerbestrasse, Cham, Switzerland
34. Stechschulte SU1, Joussen AM, von Recum HA, Poulaki V, Moromizato Y, Yuan J, D'Amato RJ, Kuo C, Adamis AP (2001) Rapid ocular angiogenic control via naked DNA delivery to cornea. Invest Ophthalmol Vis Sci 42(9):1975-1979
35. Touchard E1, Berdugo M, Bigey P, El Sanharawi M, Savoldelli M, Naud MC, Jeanny JC, Behar-Cohen F (2012) Suprachoroidal electrotransfer: a nonviral gene delivery method to transfect the choroid and the retina without detaching the retina. Mol Ther 20(8):1559–1570. doi: https://doi.org/10.1038/mt.2011.304. Epub 2012 Jan 17
36. Tanelian DL1, Barry MA, Johnston SA, Le T, Smith G (1997) Controlled gene gun delivery and expression of DNA within the cornea. Biotechniques 23(3):484–488
37. Sonoda S, Tachibana K, Yamashita T, Shirasawa M, Terasaki H, Uchino E, Suzuki R, Maruyama K, Sakamoto T (2012) J Ophthalmol. 012; 2012: 412752. Published online 012 Jan 31. doi: https://doi.org/10.1155/2012/412752 PMCID:PMC 3307015 PMID: 22518277.
38. Rajala A1, Wang Y, Zhu Y, Ranjo-Bishop M, Ma JX, Mao C, Rajala RV (2014) Nanoparticle-assisted targeted delivery of eye-specific genes to eyes significantly improves the vision of blind mice in vivo. Nano Lett 14(9):5257–5263. doi: 10.1021/nl502275s. Epub 2014 Aug 12.
39. Gomes dos Santos AL1, Bochot A, Doyle A, Tsapis N, Siepmann J, Siepmann F, Schmaler J, Besnard M, Behar-Cohen F, Fattal E (2006) Sustained release of nanosized complexes of polyethylenimine and anti-TGF-beta 2 oligonucleotide improves the outcome of glaucoma surgery. J Control Release. 112(3):369–381. Epub 2006 Mar 6

Solutions to the Review Questions of Each Chapter

Chapter 1

1. e
2. d
3. b
4. c
5. a, d

Chapter 2

1. b
2. a
3. e
4. e
5. b

Chapter 3

1. b
2. c
3. c
4. b
5. e
6. a
7. c
8. e

Chapter 4

1. b
2. c
3. b
4. b

Chapter 5

1. b
2. d
3. d
4. e
5. e

Chapter 6

1. (a) F
   (b) T
   (c) F
   (d) F
   (e) T
2. c
3. c, d, e
4. b

Chapter 7

1. e
2. d
3. e
4. b
5. b
6. d

C. Nóbrega et al., *A Handbook of Gene and Cell Therapy*,
https://doi.org/10.1007/978-3-030-41333-0

Chapter 8

1. e
2. c, e, f, h
3. b
4. a, b, c, d, e

Chapter 9

1. a
2. e
3. b
4. a

www.ingramcontent.com/pod-product-compliance
Lightning Source LLC
Chambersburg PA
CBHW081205241225
37172CB00006BA/346

* 9 7 8 3 0 3 0 4 1 3 3 2 3 *